Indigenous-Settler Relations in Australia and the World

Volume 10

The series, Indigenous-Settler Relations in Australia and the World, brings together scholars interested in examining contemporary Indigenous affairs through questions of relationality. This is a unique approach that represents a deliberate move away from both settler-colonial studies, which examines historical and present impacts of settler states upon Indigenous peoples, and from postcolonial and decolonial scholarship, which is predominantly interested in how Indigenous peoples speak back to the settler state. Closely connected to, but with meaningful contrast to these approaches, the Indigenous-Settler Relations series focuses sharply upon questions about what informs, shapes and gives social, legal and political life to relations between Indigenous peoples and non-Indigenous peoples, both in Australia and globally.

This is an important and timely endeavour. In Australia, relations between Aboriginal and Torres Strait Islander peoples and the state are at an impasse. In the wake of the government's rejection of the Uluru Statement in 2017 there is no shared view on how Indigenous-settler relationships might be 'reset', or even if this is possible. The contemporary Indigenous affairs policy domain is characterised by confusion, frustration and disappointment that, despite a seemingly endless succession of policy regimes, efforts to 'close the gap' between Aboriginal and Torres Strait Islander peoples and other Australians have not resulted in progress.

It is into this contested space that the Indigenous-Settler Relations series seeks to intervene with new, agenda-setting research. The series editors are based in a research unit in the Faculty of Arts at the University of Melbourne—the Indigenous Settler Relations Collaboration. The series will build on the work of the Collaboration in bringing together scholars and practitioners from around Australia, and around the world—particularly other Anglophone settler colonial societies such as Canada, the United States and New Zealand—whose work is concerned with Indigenous-settler relations across a range of disciplines. The multi-faceted approach to Indigenous-Settler Relations that defines the series seeks to capture how the question of relationality is already being asked by scholars across disciplines including political science, history, sociology, law, media, and cultural studies.

Readers of this series will look to it for fresh perspectives and new ideas about how to transform Indigenous-settler relations in Australia and elsewhere. They will learn from the leading lights in an emerging field who will connect their rich, multi-disciplinary scholarship to urgent social and political questions at the heart of Indigenous-Settler relations.

Melissa Kennedy • Erin O'Donnell
Editors

People, Place and Nature in Indigenous-Settler Relations

Recentring the More-Than-Human World

Featuring a Foreword by Heather Tanana
and Reflections by Maddison Miller

Springer

Editors
Melissa Kennedy
School of Social Science
Monash University
Clayton, VIC, Australia

Erin O'Donnell
Melbourne Law School
University of Melbourne
Parkville, VIC, Australia

ISSN 2524-5767 ISSN 2524-5775 (electronic)
Indigenous-Settler Relations in Australia and the World
ISBN 978-3-032-03599-8 ISBN 978-3-032-03600-1 (eBook)
https://doi.org/10.1007/978-3-032-03600-1

This Springer imprint is published by the registered company Springer Nature Switzerland AG
The registered company address is: Gewerbestrasse 11, 6330 Cham, Switzerland

Foreword: Recentring the More Than Human World

I remember when I first heard my Australian colleagues speak of "Country." As an American, the word initially evoked images of patriotism and political boundaries. Yet as they continued, their meaning revealed itself as something profoundly different: they were speaking of relationship—deep, reciprocal, and living connections with lands, waters, mountains, and all beings that share these spaces. As a Diné woman, I immediately understood this concept of Country not as territory to be owned, but as a web of relationships that defines who we are and how we belong. What my Australian colleagues described resonated with the teachings passed down to me through generations from my Diné ancestors—that we are fundamentally shaped by and responsible to the places that nurture us, and that these relationships form the core of our identity and purpose.

We stand at a crucial moment when predominant settler colonial approaches to environmental stewardship have proven disastrously inadequate, as evidenced by accelerating climate impacts and deteriorating ecosystem and community health. Despite significant progress made by Indigenous peoples in reclaiming traditional knowledges and their rightful roles in decision-making, powerful political forces continue to wage ideological warfare against Indigenous sovereignty. This attack comes precisely when our environmental crises require multiple knowledge systems and approaches to resolve.

The title, Recentring the More Than Human World, speaks directly to the transformative vision this book embodies. By shifting our perspective away from human-centered frameworks, we can recognize our place within an intricate network of kinship that includes land, water, plants, animals, and cultural dimensions of existence. This recentring is not merely theoretical—it constitutes a fundamental reimagining of systems that have failed to protect Mother Earth, harmed our children, and threaten our futures. It calls for transformation that begins with words and extends to governance, requiring us to confront colonial narratives while creating space for Indigenous knowledge systems to lead the way toward environmental justice and healing.

Through my extensive work on Tribal water rights in the United States, I have witnessed inequities firsthand—from severe water insecurity experienced in Tribal households to persistent patriarchal governance structures that exclude Indigenous voices from decision-making. These experiences deepen my appreciation for this groundbreaking anthology and how it centers Indigenous voices and knowledge systems that have sustained communities for millennia. It challenges the notion that Western frameworks represent the only or best approach, demonstrating instead how Indigenous knowledge systems—often dismissed as primitive or irrelevant—have maintained sustainable relationships with land and water for generations and continue to offer practical wisdom for addressing contemporary challenges.

This vision resonates with me personally. Under Diné worldviews, we recognize tó (water), ch'il (plants), níłch'i (air), and jóhonaa'éí ("sun") as nihiDyin dine'é (our Holy People). In Diné bi beehaz'áanii (Navajo Fundamental Law), we are taught that "Mother Earth and Father Sky is part of us as the Diné and the Diné is part of Mother Earth and Father Sky; The Diné must treat this sacred bond with love and respect without exerting dominance for we do not own our mother or father." Though our specific traditions differ, I recognized my own community's teachings reflected in the diverse voices represented in this book. Each contributor offers practical pathways toward transformation, ways of confronting colonial legacies and implementing Indigenous governance models. Reciprocity rather than dominance underlies many of the governance models explored—from cultural water allocations in the Murray- Darling Basin to forest management practices in Aotearoa that honor nature as relatives rather than purely resources for consumption and use. The reclaiming of voice and space is another powerful thread woven throughout this collection, demonstrating how Indigenous knowledge can inform more just and sustainable relationships with the more-than-human world. When my Diné elders taught me about our responsibilities to water, they were passing down practical wisdom for maintaining balanced relationships with the elements essential to survival. These teachings continue to offer crucial guidance in contexts where conventional resource management has prioritized short-term extraction over long-term relationship and care.

My personal work has shown me the devastating consequences of excluding Indigenous voices from decision-making. On the Navajo Nation where I was born, we face severe water insecurity—Navajo households are 67 times more likely to lack water access than white households across the United States, surviving on just 9 gallons per day while the average American uses 75–100 gallons daily. The cruel irony is that the Navajo Nation, like many tribes in the Colorado River Basin, holds significant legal rights to water from a river system now in crisis—overallocated, drying up, and long governed without meaningful tribal participation despite our ancestral connections and legal entitlements. Similar challenges face Indigenous peoples in Aotearoa and Australia, as documented throughout this collection.

These realities bring me back to a fundamental Navajo teaching: Diné Bilá Ashdlá'—as "five-fingered people," we are united as human beings and stand in relation to one another. As political climates increasingly threaten environmental protections and Indigenous rights, this book serves as both witness and catalyst.

May it inspire us all to question the dominant narratives that have separated humanity from the more-than-human world, to embrace the wisdom contained in these pages, and to work toward governance models that honor the agency and rights of all living beings. As Diné Bilá Ashdlá', we share responsibility for creating more just relationships with each other and with the living world that sustains us.

University of Denver Sturm College of Law, Denver, CO Heather Tanana

Acknowledgement

In creating this work, we acknowledge the First Peoples of Australia who have belonged to Country for over 65,000 years.

We pay respect to Ancestors past, who nourished their human and non-human relationships.

We pay deepest respects to Elders, present and emerging, who maintain unbroken connections to Country despite centuries of colonial violence.

We acknowledge all First Peoples globally whose lands were unlawfully stolen, whose waters and skies were claimed without consent, and whose peoples were forcibly displaced.

While the many injustices of colonisation remain unresolved, we commit to supporting First Nations in the pursuit of sustainable futures, truth-telling, and restoring rightful authority for Country.

Contents

Foreword: Recentring the More Than Human World v
Heather Tanana

1 **Introduction: Amplifying Connection, Care, and Decoloniality.** 1
Melissa Kennedy and Erin O'Donnell

Part I Words of the Land: Renaming, Resistance, and Renewal

2 **Unsettling the Settler: Lessons in Belonging and Relationality by Listening to Country** 13
Melissa Kennedy and Amy McCoy

3 **How Does It All Come Back Now? Re-organising Naturalised Histories** 29
Katrina Schlunke

4 **Waka Memory: Ancestral Connections** 47
Jade Hadfield

5 **Gender Responsibility Country: How Aboriginal People Define Their Role in Caring for the Universe.** 65
Nola Turner-Jensen

Part II Unmaking Extraction: Toward Restoration and Repair

6 **The Root of The Matter: Forests and Colonial Histories in Aotearoa New Zealand** 85
Anne Salmond and Manu Caddie

7 **Cultural Waters: Integrating Indigenous Knowledge and Governance for Justice and Sustainability in Central Queensland** .. 105
Clint J. Hansen and Matthew J. Currell

8 The River is an Island: (Re)imagining a More-than-human Future for the Waimatā Catchment 141
Megan Thomas, Billie Lythberg, Dan Hikuroa, and Gary Brierley

9 Whānau Care: A Model for Equitable Support for kaitiaki whānau 161
Jo Voice

Part III Restoring Country, Restoring Justice: Pathways to Transformation

10 First Nations Water Holder: The Future for Cultural Water in the Murray Darling Basin 177
Melissa Kennedy

11 Walking Together: Djaara Women's Journey to Restore Food Plants and Heal Country 195
Nina Roberts, Aunty Marilyne Nicholls, the Djaara Women's Knowledge Group, Julie Andrews, and Susan Lawrence

12 Securing Water Access: Economic Futures for Victoria's Traditional Owners 221
Cormac Mercer

13 *First Peoples, Living Waters* Cultural Exchange 237
Melissa Kennedy, Erin O'Donnell, Amy McCoy, Nicky Hudson, and Ben Muir

Reflections 257
Maddison Miller

Notes on Contributors

Julie Andrews OAM is a descendant of the Yorta Yorta people; her community and family connection is to the Cummeragunja Aboriginal Reserve. She is also a descendant of the Wiradjuri and Wurrundjeri Woiwurrung peoples. Her family has been instrumental in establishing Melbourne Aboriginal organisations and Aboriginal women's network across Australia. She is a Professor at La Trobe University, and her academic research has been focused on specific geographical areas because of their relevance to her life, cultural history, family, and community. Professor Andrews' research has primarily been anthropologically focused, but she is able to apply anthropological themes across a variety of research disciplines specific to Aboriginal Australia.

Gary Brierley is the Chair of Physical Geography in the School of Environment at the University of Auckland (Waipapa Taumata Rau), Aotearoa New Zealand. He is a co-developer of the River Styles Framework. He has published over 200 international journal articles and book chapters, co-edited books on landscapes and ecosystems of the Upper Yellow River and River Futures, co-wrote a textbook entitled Geomorphic Analysis of River Systems, and a reflective book entitled Finding the Voice of the River.

Manu Caddie (Ngāti Pūkenga, Ngāti Hauā, Ngāi Te Rangi) is a passionate entrepreneur, activist, and academic dedicated to Indigenous rights and sustainable development. Raised in Tauranga Moana, Manu has spent the last 25 years contributing significantly to community development and sustainable industry in Te Tairāwhiti.

Matthew J. Currell is an Adjunct Professor at RMIT University in Australia. He is an active researcher in the field of hydrogeology and geochemistry. He has published more than 70 peer-reviewed journal articles, was lead editor of a book and has served on the editorial board of the Hydrogeology Journal. He supervises PhD and Masters projects in hydrogeology, working with partners in the water industry and environmental regulators. He has conducted extensive work as an independent expert witness in the field of hydrogeology, including for government inquiries and committees in the field of groundwater.

Jade Hadfield (Ngāti Kahungunu ki Heretaunga, Ngāti Whātua ki Kaipara), a curator and conservator residing in Narrm (Melbourne), has over 20 years of invaluable experience gained from prominent cultural institutions, including Te Papa, ICCROM, The Koorie Heritage Trust, Museums Victoria, and her current position at State Library Victoria. Her extensive conservation training has played a crucial role in facilitating the return of her hapū Pou Tūpuna, coupled with a commitment to their ongoing care.

Clint J. Hansen (Yiman/Iman) is an Indigenous research fellow at Monash University in Australia, pursuing a PhD in Environmental Engineering. He holds a BEng (H1) and completed his honours in Sustainable Systems Engineering in 2020. His current research is guided by his Yiman/Iman heritage and Elders, focusing on protecting the cultural and ecological values of water in Coal Seam Gas (CSG)-affected areas of Central Queensland. By blending traditional Indigenous knowledge with scientific methods, Hansen advocates for Indigenous water rights and seeks to promote regenerative futures and justice for communities impacted by resource extraction.

Dan Hikuroa (Ngāti Maniapoto, Ngaati Whanaunga, Ngāti Mahuta, Pākehā) is a father, surfer, paddle-boarder, gardener, loves the taiao and is an Associate Professor in Māori Studies, Waipapa Taumata Rau-University of Auckland. Dan is an established world expert on weaving indigenous knowledge and science to realise the dreams and solve challenges of the communities he works with. Dan has been spearheading alternative ways of undertaking development and assessing sustainability, including braiding indigenous knowledge and epistemologies with science and into policies, assessment frameworks and decision-support tools. Dan's research includes how concepts of mauri contribute to ideas about "life", kinship as practical ontology, how Indigenous concepts can be integrated with techno-scientific rationales to imagine new futures for our rivers and waterways, and Māori concepts of time, space and the "universe". Dan is UNESCO New Zealand Commissioner for Culture, member of Pou Herenga, Māori Advisory to the Climate Change Commission.

Nicky Hudson is a Gunditjmara woman, and the Aboriginal Water Officer leading the work to care for and restore waterways at Gunditj Mirring Traditional Owner Aboriginal Corporation on the Budj Bim UNESCO World Heritage Site, which was inscribed in 2019 due to its Outstanding Universal Values and registered purely for its Aboriginal cultural significance. The Budj Bim Lava flows has enabled Gunditjmara to develop one of the largest and oldest freshwater aquaculture systems in the world. The system comprises of elaborate channels, weirs and dams which are used to contain floodwaters and create basins to trap, store and harvest aquatic species, including the culturally significant species Kooyang (Short-finned Eel). Nicky also assists with waterway management within the Gunditjmara Native Title Boundary across Southwest Victoria. She is the Co-chair of the Aboriginal Water Officers Network in Victoria.

Melissa Kennedy is Tati Tati from Murray River Country, a PhD candidate at Monash University, and a Senior Atlantic Fellow for Social Equity, a global community collaborating across borders and disciplines to address the root causes of inequity. Her research centres on better understanding and documenting the relationship between people, water, and wellbeing, to enable policy change and enduring transfers of power. Melissa is also a Director of Tati Tati Kaiejin, an Indigenous-owned and operated non-profit conservation organisation that applies place-based Indigenous ecological science to heal waterways and land.

Susan Lawrence is acting Co-Director of the Centre for the Study of the Inland and a Professor in the Department of Archaeology and History at La Trobe University in Australia. Her expertise is in the physical evidence of the recent human past, and she is internationally recognised for her research on mining, rivers, and British colonisation. Susan's 2019 book 'Sludge: Disaster on the Victorian Goldfields' (co-authored with Peter Davies) was shortlisted for the Prime Minister's Literary Awards. In her work Susan emphasises the importance of collaborative relationships and inter-disciplinary approaches, and she works with environmental scientists, ecologists, visual artists, heritage professionals, historians, Aboriginal Studies scholars, and Traditional Owners. Susan is also working with Djaara women on the co-designed 'Yam Paddock Project', a philanthropically funded program to use cultural fire and digging sticks to return yams and other traditional foods to Country in central Victoria.

Billie Lythberg is a Senior Lecturer at the University of Auckland (Waipapa Taumata Rau) where she is a member of the Organisation Studies disciplinary area. She is known for transdisciplinary work that advances Māori and Pacific kaupapa through heterogeneous research philosophies and methodologies. Billie has had roles on six Royal Society of NZ Marsden Projects and an MBIE National Science Challenge and is currently an investigator on the Marsden Projects 'Let the River Speak' and 'Artefacts of Relations'.

Amy McCoy is a woman of American, white, UK-European, settler/invader heritage living on the lands of the Ute Peoples in Colorado Springs in the USA, where she works in partnership with Tribes in the Colorado River Basin on water justice reform. Amy is a founding partner of Fluent Freshwater Insights, where she applies strategic research and creative communications to the confounding social, legal, and practical challenges of water scarcity and climate change. She also serves as an Adjunct Research Scientist with the University of Arizona Southwest Center and participates in a wide range of ecological, climate, and water-focused research collaborations.

Cormac Mercer is a Gadigal and Dharug man with strong community connections to the Bundjalung community on the north coast of New South Wales. Cormac is a Juris Doctor graduate of the University of Melbourne and an Associate at Pinsent Masons in Melbourne, Australia. He has a strong interest in the renewable energy transition in Australia and globally, in particular, all forms of renewable energy projects, including from the front end of the project lifecycle to contentious matters, having assisted on different projects to this end.

Maddison Miller is a Dharug woman living and working on unceded Wurundjeri Woi Wurrung Wilam Biik. Maddi is an archaeologist, artist, and a lecturer at the University of Melbourne, Australia. Her research explores storying in bringing together multiple ways of knowing to foster healthy Country outcomes. Maddi is a Counselor with the Biodiversity Council of Australia and a theme lead for Healthy Country at the Indigenous Knowledge Institute.

Ben Muir is a Wotjobaluk man and a Gatjin (Water) Officer responsible for caring for water at DJAARA (Dja Dja Wurrung Clans Aboriginal Corporation). He has previously worked for the Wimmera Catchment Management Authority and Barengi Gadjin Land Corporation.

Aunty Marilyne Nicholls is a painter and master weaver in Indigenous traditional techniques. Aunty Marilyne's work has been exhibited around Australia and the world and is found in public and private collection including the British Museum in London and the Koori Heritage Trust in Melbourne. Aunty Marilyne is a recognised elder amongst her people, sharing her knowledge, skills, and stories with grace. She has connections to fresh water and saltwater Country: Wadi Wadi, Barapa Barapa, Latji Latji, Jupagalk, Dja Dja Wurrung, Yorta Yorta, and Ngarrindjeri.

Erin O'Donnell is a woman of white, UK-European, settler/invader heritage living on Wurundjeri Woi Wurrung Country in Australia. She is a Senior Lecturer and ARC DECRA Fellow at the University of Melbourne Law School, where she researches water law and policy, including Indigenous water justice and the legal rights of rivers. She collaborates with Traditional Owners in Australia in her academic research on water justice. Since 2018, Erin has been a member of the Birrarung Council, the voice of the Birrarung (Yarra River) in Melbourne.

Nina Roberts is of coloniser/immigrant descent and a graduate researcher at La Trobe University. She was born and grew up on the northeastern fringes of Naarm (Melbourne) within a nature-loving family. She studied and worked in the fields of botany, ecology, and creative artistic practice intermittently for over two decades before commencing post graduate research on cultural food plants in collaboration with the Djaara Women's Knowledge Group. This ongoing doctoral research explores the potential for co-creating culturally informed contemporary ecological knowledge for healing Country and supporting Indigenous leadership.

Anne Salmond is a Distinguished Professor of Māori Studies and Anthropology at the University of Auckland (Waipapa Taumata Rau), and a leading social scientist in Aotearoa New Zealand. She is the first New Zealander to be elected a fellow of both the US National Academy of Sciences and the British Academy, and a member of the American Philosophical Society. In 2013 she was named 'Cultural Anthropologist of the Year' by Anthropology Works, US. Anne has won many international awards and written a series of prize-winning books about Māori life, European voyaging and cross-cultural encounters in the Pacific. She has a lifelong engagement with te ao Māori, working alongside kuia and kaumātua and presenting evidence in the Muriwhenua Land and Fisheries Treaty claims, the Ngāpuhi claim for Te Tiriti o Waitangi, and the first test case of the Treaty clause of the Resource Management Act.

Katrina Schlunke and writes and researches about the interconnections between cultural forms and Indigenous interventions. Her most recent essay is The Practices of Care: Extinction and De-Colonisation in the Natural History Museum (2024), her most recent poem was 'Burning Captain Cook' (2022). and she is a joint Chief Investigator on the ARC Discovery project 'Beyond Extinction: Reconstructing the Thylacine (Tasmanian Tiger) Archive'. She is an Adjunct Associate Professor at the University of Sydney (Department of Gender and Cultural Studies) and Adjunct Senior Researcher at the University of Tasmania (School of Creative Arts and Media).

Heather Tanana (Diné) is an Assistant Professor at the University of Denver Sturm College of Law. She is also an associate faculty member of with the Center for Indigenous Health at Johns Hopkins University School of Public Health. Heather's research interests revolve around the intersection of environmental and health policy. Much of her work focuses on tribal water issues, from climate change impacts to Colorado River management. She also leads the Universal Access to Clean Water initiative, which seeks to bring awareness to the lack of clean, safe, and reliable drinking water in Indian country and to make tangible progress on securing water access for all tribal communities.

the Djaara Women's Knowledge Group is a group of Djaara women of different ages and family groups who have been coming together to re-activate women's business on Djandak (Dja Dja Wurrung Country, Central Victoria). The group's activities – including ceremony, cultural burning and use of digging sticks – are focused on healing Country and building knowledge of plants and seasons among the women. The presence of Djaara women on Country strengthens connections with ancestorial knowledge and practices and empowers them to become knowledge keepers in caring for Country.

Megan Thomas is a PhD candidate at the University of Auckland in the School of Environment, while working part-time as an environmental scientist. Upon completion of her BSc (Honours) in Geography, she began a PhD project focusing on Anthropocene Riverscapes in New Zealand and the use of multiple knowledge forms to engender improved outcomes for living with rivers.

Nola Turner-Jensen (Gibson/Naden Family line), West Galari Wiradyuri speaking Ngurambang (clan estate). Nola is always learning respectfully from her Ancestors, Wiradjuri elders and knowledge holders regarding the repatriation of Wiradyuri kinship, ceremony, and Lore knowledge from around the world. Nola is Co-chair of the West Galari Wiradyuri Traditional Owners Corporation. She is also the founder, content writer and e-learning designer for over a decade for Crackerjack Education, Australia's first and leading Indigenous run education company for teachers and educators. An accomplished writer Nola has published four Aboriginal children's books. She was awarded a Fellowship of the First Nation Australian Writers Network in 2019 and, in 2021, was named a Community Changemaker within the Yanalangami: Strong Women, Strong Communities Changemakers Program.

Jo Voice is of European descent, and lives and works in Whanganui, New Zealand. Her career has involved working with children, young people and their families in various settings, mainly with children and their caregivers in foster care. Jo is currently employed as the Manager of Whanganui Women's Refuge, where she leads a team of advocates who provide support to women and children who are experiencing family violence. She is a qualified social worker who is passionate about creating equity in support for extended whānau who agree to care for tamariki so they do not need to enter state care. This passion led Jo to the Atlantic Fellowship, where she completed her Masters in Social Change Leadership to help inform her project to support whānau caregivers.

Chapter 1
Introduction: Amplifying Connection, Care, and Decoloniality

Melissa Kennedy and Erin O'Donnell

Abstract As we write this book in 2025, we face uncertain times. The rise of fascism and the far-right in many countries is amplifying the voices of white supremacy and empire. Significant and hard-fought gains made by Indigenous Peoples are being undermined, while unabated global heating is increasing threats to both the human and more-than-human communities (IPBES, 2019; IPCC, 2022). This book focuses on Indigenous-Settler relations with the more-than-human community in the settler colonial contexts of Australia, Aotearoa New Zealand, and the USA. In all three places, these relationships are increasingly under pressure.

1.1 Introduction

As we write this book in 2025, we face uncertain times. The rise of fascism and the far-right in many countries is amplifying the voices of white supremacy and empire. Significant and hard-fought gains made by Indigenous Peoples are being undermined, while unabated global heating is increasing threats to both the human and more-than-human communities (IPBES, 2019; IPCC, 2022). This book focuses on Indigenous-Settler relations with the more-than-human community in the settler colonial contexts of Australia, Aotearoa New Zealand, and the USA. In all three places, these relationships are increasingly under pressure.

M. Kennedy (✉)
School of Social Science, Monash University, Clayton, VIC, Australia
e-mail: melissa.kennedy1@monash.edu

E. O'Donnell
Melbourne Law School, University of Melbourne, Parkville, VIC, Australia
e-mail: erin.odonnell@unimelb.edu.au

M. Kennedy, E. O'Donnell (eds.), *People, Place and Nature in Indigenous-Settler Relations*, Indigenous-Settler Relations in Australia and the World 10, https://doi.org/10.1007/978-3-032-03600-1_1

In 2017, Indigenous Peoples gathered at Uluru, in the heart of Australia, and collectively called for Voice, Truth, and Treaty (Uluru Statement from the Heart, 2017). Although this call was comprehensively rejected by consecutive conservative governments, in 2022, the newly elected Labor Party committed to reforming Australia's constitution to enact an enduring 'Voice' for Aboriginal and Torres Strait Islander Peoples. The Voice requirement in the Uluru Statement was in recognition of the need to ensure that Aboriginal and Torres Strait Islander Peoples were effectively engaged and consulted about laws that could affect them, and that unlike earlier consultative models, the Voice could not be dismissed by an incoming government (Harris, 2024). The Labor Party made no formal commitment regarding Treaty, but did commit to establish a Makarrata Commission to progress Treaty and Truth-Telling (Erickson, 2022). In October 2022, 65% of Australians supported this constitutional reform, but following a withdrawal of bipartisan support and a 'No' campaign riddled with misinformation, only 39.9% of Australians voted 'Yes' in October 2023 (Carson et al., 2024). This was devastating for Indigenous communities across Australia. Aboriginal and Torres Strait Islander Social Justice Commissioner Katie Kiss stated 'the referendum result has given certain cohorts within the community a sense of freedom to perpetuate racism, racial hatred and racial vilification' (Knowles, 2024).

In 2023, Truth and Treaty processes were underway in Victoria, South Australia, Western Australia, Queensland, and the Northern Territory, but following the defeat of the Voice referendum, truth-telling commissions and pre-Treaty processes were formally abolished in Western Australia, Queensland, and the Northern Territory. South Australia's Voice was formally established in legislation prior to the referendum, but it is not clear whether this indicates an appetite for Treaty or Truth. However, in Victoria, the First Peoples' Assembly (established in legislation in 2018) has commenced formal negotiations with the Victorian state government on Treaty. The Yoorrook Justice Commission, established in 2021 with the full powers of a Royal Commission to investigate historical and ongoing injustices committed against Aboriginal Victorians since colonisation (Bracka, 2024), has already issued one report, with another on land and water justice expected in 2025. Following the referendum, the conservative opposition party in Victoria has withdrawn their support for the Treaty process, and pressure on the Treaty process to deliver outcomes and avoid becoming a political football is likely to mount as Victorians approach an election in November 2026. Opposition to Indigenous voices and Indigenous power is increasingly heard in the Australia of 2025, with calls for federal government funding for Welcome to Country ceremonies to be scrapped (Brennan, 2025), attacks in national newspapers on the inclusion of Indigenous content in law school curricula (Ross, 2025), and entrenched, systematic failure to address rising rates of Indigenous incarceration, children taken from families by welfare agencies, and provision of basic services to remote Indigenous communities (Productivity Commission, 2024).

In Aotearoa, the coalition conservative government elected in 2023 committed to preparing a bill for parliament that 'seeks to abandon a set of well-established principles that guide the relationship between Māori and ruling authorities' (Corlett,

2025). This is an extraordinary attack on the founding document of Aotearoa New Zealand, Te Tiriti o Waitangi/The Treaty of Waitangi. Although unlikely to be passed into law, the mere fact that this bill was brought before parliament is a travesty, undermining decades of progress towards the genuine participatory partnership envisaged in the Treaty (O'Sullivan, 2024). Tens of thousands of people have gathered in multiple protests across Aotearoa since this bill was announced. Over 300,000 submissions have been received by the select committee charged with reviewing the bill, with over 90% opposed. Protestors framed their opposition not just in terms of Māori rights, but for the collective New Zealand community: '[w]e are here to have a voice and back those who are collectively working to make things right, not just for us Māori but for the people of New Zealand' (Christina Taurua, quoted in Armstrong, 2024). While some are now hailing the bill as a surprisingly unifying move, building coalitions of support for the Treaty across the country (Armstrong, 2025), this has come at an enormous cost. Meanwhile, the conservative government has already acted to abolish the Māori Health Authority (Te Kāhui Tika Tangata Human Rights Commission, 2024), limit the use of te reo in public service (Tahana, 2024), and weaken the role of Māori in decision-making under the *Resource Management Act* (Environmental Defence Society, 2025).

In the United States of America, far-right Republican Donald Trump was elected for a second term in November 2024. Since forming government in January 2025, he has released a series of executive orders that, despite ongoing legal challenges, are overturning diversity, equity and inclusion policies across the nation, including federal funding for research that focuses on environmental justice, climate change, Native Americans, or women (Mervis, 2025). Massive funding cuts include a direction to the General Services Administration to 'terminate the leases of… 25 regional offices of the Bureau of Indian Affairs' (Walker, 2025). President Trump has also 'rescinded a Biden-era executive order, which aimed to strengthen tribal sovereignty for the 574 federally recognized tribes in the nation [and] federal agencies have flagged hundreds of words to limit or avoid, including, "indigenous community", "tribal", and "Native American"' (Walker, 2025). Tribal leaders have given evidence of 'real fear' in Tribal communities (Silversmith, 2025), and Tribal members are being advised to carry identification at all times, amid increasing removals of US citizens and permanent residents by U.S. Immigration and Customs Enforcement (Bardwell, 2025). Yet despite the shocking speed at which the Trump Administration is dismantling the fundamentals of US democracy (including a recent refusal to consider an unlawful third term, Nicholas et al., 2025), we can see this as an extension of many of the existing policies under former President Joe Biden. For instance, the extraordinary and violent response to student protests against what the UN Special Committee defined as 'genocide' in Palestine (Paragraph 69, Special Committee, 2024) set the stage by undermining both freedom of speech and academic freedom, two fundamental US freedoms that continue to be threatened in 2025.

In this context, this book offers an alternative vision, both for the future and the present. Rather than succumbing to fear and the growing sense of powerlessness, we come together to build community. In the following chapters, we share new and transformative ideas, meaningful insights rooted in ancient and Indigenous knowledges, and forge new connections as an act of resistance to the forces of unrestrained capitalism, white supremacy, and colonialism.

1.2 People, Place, and the More-Than-Human

This book features Indigenous and non-Indigenous scholars writing about Indigenous-Settler relations in Australia, Aotearoa, and the United States of America. We are immensely proud that the authorship is majority Indigenous, and that there are multiple chapters featuring the work of Indigenous-Settler collaborations. We have also chosen to amplify the voices of early career academics and practitioners, alongside those of established leaders in this field.

1.2.1 Part 1: Words of the Land: Renaming, Resistance, and Renewal

In Part 1, we bring together powerful voices that challenge colonial logics and reimagine our relationships with land, memory, and knowledge through Indigenous worldviews. At its core, this section explores how words—spoken, written, sung, or remembered—carry the potential to unsettle colonial narratives, rename spaces and histories, and renew responsibilities to Country. Across four chapters, authors engage with the enduring legacies of colonial institutions and worldviews while offering Indigenous-led methodologies and frameworks that centre relationality, accountability, and regeneration.

In Chapter 2, *Unsettling the Settler: Lessons in Belonging and Relationality by Listening to Country*, Melissa Kennedy and Amy McCoy invite readers into a reflective and relational dialogue on what it means to listen deeply to Country. They introduce unsettling not as a moment of disruption alone, but as an ongoing, relational process through which non-Indigenous people might learn to reckon with their complicity in settler colonial systems. Their conversation intertwines professional insight with personal experience, foregrounding the necessity of Indigenous perspectives in shaping sustainable and just futures. Through this approach, the authors challenge dominant frameworks and ask readers to shift from extractive relationships with land and knowledge to ones rooted in reciprocity, care, and humility.

Katrina Schlunke continues this critical interrogation in Chapter 3, *How does it all come back now? Re-organising naturalised histories*. Schlunke offers a provocative analysis of how 'nature' has been constructed in colonial institutions—especially museums—as a depoliticised, decontextualised object of study. She exposes how the classification of Indigenous Country as 'natural' space erases its dynamic histories and stories, transforming it into something to be observed, explained, and consumed. Schlunke turns attention to the creative and intellectual resistance of Indigenous curators, artists, and knowledge holders, who disrupt these dominant narratives by reanimating Country and restoring its rightful place as a living, storied, and sovereign entities. This chapter argues for a radical re-organisation of the museum space—one that relinquishes control and reclaims the relational essence of Country.

Chapter 4, *Waka Memory: Ancestral Connections*, by Jade Hadfield, extends the discussion into the domain of galleries, libraries, archives, and museums—institutions deeply complicit in colonial violence. Hadfield examines how these spaces continue to shape visibility and invisibility for Oceania peoples and knowledge systems. The chapter introduces waka memory, a collaborative exhibition framework grounded in Indigenous practices such as wayfinding, Talanoa, yarning, and wānanga. Waka memory is both methodology and movement—a way of reuniting Moana and Aboriginal peoples through shared ancestral ties and reciprocal practices. Hadfield positions waka memory as a tool for dismantling hierarchical structures of knowledge and restoring Indigenous presence within institutional frameworks that have historically excluded or misrepresented it. The chapter is a compelling call for transformation led by Indigenous methodologies of relationality and resurgence.

The final chapter in Part 1, *Gender Responsibility Country: How Aboriginal people define their role in caring for the universe*, Nola Joyce Turner-Jensen offers a deeply rooted perspective on the spiritual and gendered dimensions of Aboriginal knowledge systems. She outlines Gender Responsibility Country—an Indigenous framework that sees Country as a living relative, entwined with ancestral obligations passed down through song, ceremony, and story. Turner-Jensen reclaims the significance of matriarchal knowledge and the fluidity of ancestral gender roles, particularly highlighting the often-erased role of Old Mother, the gender-fluid Creation Ancestor. Her work critiques both the colonial dismissal of women's ceremonial authority and the academic silencing of Indigenous cosmologies. Through personal reflection and ancestral wisdom, this chapter demands the restoration of Aboriginal knowledge systems to the centre of cultural and institutional life.

Together, these chapters offer more than critique—they offer pathways. They illuminate how land, story, and Indigenous knowledge remain active forces of resistance and renewal through listening, renaming, remembering, and reimagining. Part I reminds us that words matter, and when grounded in Indigenous ways of being, they become vessels of transformation.

1.2.2 Part 2: Unmaking Extraction: Toward Restoration and Repair

In Part 2, we move into chapters that address the ongoing ecological, cultural, and political consequences of extractive practices, both past and present. This section foregrounds Indigenous-led responses to the intertwined harms of colonisation and environmental degradation, asking what it means to resist extraction and actively regenerate relationships with land, water, whānau, and future generations. Each chapter brings a distinct context and method, yet united in the call for a shift from extractivism toward Indigenous-led restoration grounded in relationality, justice, and care.

In Chapter 6, *The Root of the Matter: Forests and Colonial Histories in Aotearoa, New Zealand*, Dame Anne Salmond and Manu Caddie take us deep into the tangled roots of industrial forestry and colonial land management in Aotearoa. Their chapter explores how Māori knowledge systems offer alternative understandings of forests—not as commodities or resources, but as kin within complex, interdependent webs of life. The authors trace the environmental devastation caused by monocultural plantation forestry, especially in the erosion-prone landscapes of Tairāwhiti, culminating in the damage caused by Cyclone Gabrielle in 2023. They also spotlight the grassroots work of Mana Taiao Tairāwhiti, a Māori-led response to this ecological crisis, and the visionary programme Recloaking Papatūānuku, which aims to heal landscapes through native reforestation. This chapter is a compelling illustration of how decolonial futures are enacted through Indigenous ecological leadership.

Chapter 7, *Cultural Waters: Integrating Indigenous Knowledge and Governance for Justice and Sustainability in Central Queensland*, by Clint Hansen and Matthew J. Currell, explores the deep cultural, spiritual, and ecological significance of water for the Iman People. Focusing on the impacts of Coal Seam Gas extraction, this chapter highlights how extractive industries disrupt ecosystems and Indigenous cultural practices and rights. By centring the lived experiences of the Iman People, Hansen and Currell advocate for the incorporation of Indigenous cultural values into water governance frameworks. They argue that sustainable futures demand a fusion of cultural and scientific knowledge that respects Indigenous sovereignty and leads to more equitable and resilient management of water systems. This chapter underscores the urgent need for governance systems that recognise Indigenous Peoples as knowledge holders and rightful decision-makers in environmental futures.

In Chapter 8, *The River is an Island: (Re)imagining a More-than-human Future for the Waimatā Catchment*, Megan Thomas, Billie Lythberg, Dan Hikuroa, and Gary Brierley offer a radically different perspective on riverscapes that moves beyond anthropocentrism to embrace a more-than-human worldview. Focusing on the Waimatā River in Gisborne, Aotearoa, the authors blend geomorphology with Indigenous knowledge, art, and poetry to explore how rivers might be understood as active agents within cultural and ecological networks. They argue that academic and governance systems must radically shift to match the scale of transformation already occurring in landscapes. The chapter's "mountains to the sea" approach calls for a deepened awareness of interconnectedness, urging us to listen to rivers as storytellers and participants in their own futures.

Chapter 9, the last chapter in Part 2, *Whānau Care: A Model for Equitable Support for Kaitiaki Whānau*, by Jo Voice, turns our attention to the social consequences of colonisation, particularly within child protection systems. Despite policy reforms, Māori children remain disproportionately represented in state care, often removed from their cultural contexts and familial networks. Voice critiques these systemic failures and outlines the Whānau Care project, a co-designed initiative with Māori communities aimed at supporting kaitiaki whānau (Māori family caregivers). This model prioritises collective well-being, cultural continuity, and self-determination, challenging the deficit-focused logic of mainstream child welfare systems. It offers a restorative framework rooted in mana-enhancing care and Māori values of interconnectedness and responsibility.

Collectively, these chapters offer more than critique; they offer blueprints for renewal. By centring Indigenous worldviews and community-led responses, Part II charts a path toward restorative futures where the land, waters, and people are not merely protected but actively restored and re-empowered. These contributions remind us that true repair is not just ecological but relational, cultural, and political.

1.2.3 Part 3: Restoring Country, Restoring Justice: Pathways to Transformation

Finally, in Part III of this collection, we gather contributions that speak to the enduring strength and resilience of First Nations peoples in their efforts to reclaim sovereignty, restore cultural practice, and transform colonial institutions. Central to these chapters is the understanding that justice cannot be achieved without land and water justice. Transformation must be led by those who hold ancestral knowledge, responsibilities, and custodianship of Country. This section amplifies Indigenous-led governance models, stories of return and renewal, and powerful critiques of extractive policy structures that continue to exclude or marginalise Indigenous authority over land and water.

Chapter 10, *First Nations Water Holder: The Future for Cultural Water in the Murray-Darling Basin*, by Melissa Kennedy, sets the tone for this section by calling for the advancement of a First Nations Water Holder as a transformative model for Indigenous water governance. Kennedy critiques the dominant economic and environmental frameworks that have long structured Australia's water management systems, noting how these have systematically excluded Indigenous perspectives and spiritual relationships with water. The concept represents a reclamation of authority, enabling communities to manage water according to cultural needs and responsibilities—through ceremonial use, ecological restoration, or re-establishing water-dependent cultural economies. This chapter calls for a systemic rethinking of water governance, where Indigenous sovereignty and cultural knowledge are not added in as afterthoughts but recognised as central pillars of sustainability and justice.

In Chapter 11, *Djaara Women Returning Food Plants and Healing Country*, Nina Roberts, Aunty Marilyne Nicholls, the Djaara Women's Knowledge Group, Julie Andrews and Susan Lawrence tell a powerful story of renewal and reconnection. This deeply collaborative project centres on returning traditional food plants to Dja Dja Wurrung Country and revitalising women's cultural knowledge. The chapter offers a model for cross-cultural partnership rooted in trust, relationality, and embodied knowledge. It highlights how academic research can work in respectful dialogue with Indigenous priorities when cultural safety, reciprocity, and shared authority are foregrounded. The narrative reflects both the profound disruptions caused by colonisation and the strength of Djaara women's leadership in healing Country, communities, and themselves through land-based practices.

Chapter 12, *Securing Water Access: Economic Futures for Victoria's Traditional Owners*, by Cormac Mercer, provides a structural critique of water governance in Victoria. It explores how legal and policy systems—built on the false doctrine of aqua nullius—exclude Traditional Owner groups from equitable participation in water economies. Mercer interrogates how, while symbolically important, cultural flows often fail to deliver real opportunities for Indigenous groups to sustain livelihoods and economies that align with cultural values. The chapter calls for greater access to water as an essential resource for Indigenous economic self-determination, emphasising the need to remove systemic barriers and redistribute water entitlements in ways that acknowledge and redress colonial dispossession.

Finally, Chapter 13, *First Peoples, Living Waters: A Cultural Dialogue*, by Melissa Kennedy, Dr Erin O'Donnell, Dr Amy McCoy, Nicky Hudson, and Ben Muir, closes the section with a conversation that weaves together personal reflection, legal critique, and trans-Indigenous dialogue to explore how treaty-making and agreement processes can reshape water law and challenge settler state authority. It emphasises that in the face of climate change, restoring Indigenous water rights is a matter of cultural survival and ecological necessity. The authors model a decolonial method rooted in relationship, listening, and respect by presenting this dialogue in a conversational format.

Across these chapters, Part III offers a powerful vision of transformation. Restoration is not a metaphor; it is lived work. It is found in the return of seeds to soil, the assertion of ancestral authority over rivers, the weaving together of knowledge across continents, and the legal and political reimaginings of what governance could be. These chapters are a call to action for a future in which justice is deeply grounded in Country, and Country is once again held by those who know how to care for it.

1.3 Conclusion

> Our relationship with Country is one of obligation, reciprocation, habitation and affection. We do not 'use' Country, we live in it. We depend on it. We care for it. We love it. It is part of the very fabric of who we are. (Pascoe et al., 2023)

Across its three parts, this collection brings together a rich tapestry of Indigenous-led scholarship, storytelling, activism, and practice that collectively reclaims space from colonial institutions, reorienting us toward futures grounded in relationality, justice, and care. Whether through the renaming of landscapes, the unmaking of extractive systems, or the restoration of cultural authority over land and water, the chapters in this book offer a unified message: transformation begins with listening to Country and honouring Indigenous sovereignty in all its forms. This work resists settler colonial logics by exposing harm and enacting renewal through words, relationships, and the radical act of returning to knowledge systems that care for the earth, its waters, and its peoples. At its heart, this collection affirms that justice is not abstract nor distant—it is intimate, embodied, and already unfolding wherever Indigenous Peoples lead the way (Tanana, 2023).

This book is a vital intervention in a world teetering on the edge, weaving together Indigenous-settler relations and human and more-than-human kinships. With insights from the USA, Aotearoa, and Australia, this book centres voices that speak from survival and resurgence, from Indigenous authors who live in a post-post-apocalyptic reality while sustaining deep cultural and ecological relationships, and from non-Indigenous academics and allies pushing back on colonial narratives. It invites readers to confront short-term thinking and isolated individualism, offering instead a vision rooted in long-term memory, kinship, and solidarity—both human and more-than-human—as a path forward.

References

Armstrong, K. (2024). Thousands rally in NZ in support of Māori rights. *BBC*. https://www.bbc.com/news/articles/cv227wzjgy8o

Armstrong, K. (2025, April 10). New Zealand: Treaty Principles Bill voted down after widespread outrage. *BBC*. https://www.bbc.com/news/articles/c8je9013m0ro

Bardwell, N. (2025). Tribal Nations Urge Citizens to Carry ID as ICE Raids Increase. *Native News Online*. https://nativenewsonline.net/currents/tribal-nations-urge-citizens-to-carry-id-as-ice-raids-increase

Bracka, J. M. (2024). 'It Begins with Victoria'—The Yoorrook Justice Commission & Innovating transitional justice for first nations Australia. *Griffith Journal of Law & Human Dignity, 12*(1), 29–65. https://doi.org/10.69970/gjlhd.v12i1.1261

Brennan, D. (2025, January 31). Welcome to Country ceremony funding to be cut if Coalition elected. *National Indigenous Times*. https://nit.com.au/31-01-2025/16000/welcome-to-country-ceremony-funding-to-be-cut-if-coalition-elected

Carson, A., Strating, R., & Jackman, S. (2024). Why did the Voice referendum fail? We crunched the data and found 6 reasons. *The Conversation*. http://theconversation.com/why-did-the-voice-referendum-fail-we-crunched-the-data-and-found-6-reasons-228383

Corlett, E. (2025). New Zealand's controversial bill to reinterpret treaty with Māori should be scrapped, committee finds. *The Guardian*. https://www.theguardian.com/world/2025/apr/04/new-zealands-controversial-bill-to-reinterpret-treaty-with-maori-should-be-scrapped-committee-finds

Environmental Defence Society. (2025). *What, Why, How: Exploring the fundamental questions for resource management law reform in Aotearoa New Zealand (Paper 1, Context)*. Environment Defence Society. https://eds.org.nz/wp-content/uploads/2025/03/Context-of-resource-management-reform.pdf

Erickson, P. (2022). *Labor's commitment to First Nations peoples*. Parliamentary Information. https://parlinfo.aph.gov.au/parlInfo/search/display/display.w3p;query=Id%3A%22library%2Fpartypol%2F8638543%22

Harris, B. (2024). The voice to parliament. In B. Harris (Ed.), *Indigenous peoples and constitutional reform in Australia: Beyond mere recognition* (pp. 217–250). Springer Nature. https://doi.org/10.1007/978-981-99-7121-3_8

IPBES. (2019). *Global assessment report on biodiversity and ecosystem services of the Intergovernmental Science-Policy Platform on Biodiversity and Ecosystem Services*. IPBES Secretariat. https://doi.org/10.5281/zenodo.3831673

IPCC. (2022). *IPCC sixth assessment report: Impacts, adaptation and vulnerability*. Intergovernmental Panel on Climate Change. https://www.ipcc.ch/report/ar6/wg2/

Knowles, R. (2024). The Voice referendum failed and racism rose. Is misinformation responsible? *NITV*. https://www.sbs.com.au/nitv/article/the-referendum-failed-and-racism-rose-is-misinformation-responsible/occpe57xw

Mervis, J. (2025). Trump orders cause chaos at science agencies. *Science Magazine*. https://www.science.org/content/article/trump-orders-cause-chaos-science-agencies

Nicholas, P., Lee, C. E., Gomez, H. J., & Dixon, M. (2025). Trump's third-term talk freezes the potential 2028 Republican field. *NBC News*. https://www.nbcnews.com/politics/donald-trump/trump-third-term-talk-freezes-potential-2028-republican-field-rcna198878

O'Sullivan, D. (2024). Te Tiriti o Waitangi—Neither "one people" nor partners: Rangatiratanga and inclusive democratic kāwanatanga. *AlterNative, 20*(4), 609–617. https://doi.org/10.1177/11771801241279332

Pascoe, J., Shanks, M., Pascoe, B., Clarke, J., Goolmeer, T., Moggridge, B., Williamson, B., Miller, M., Costello, O., & Fletcher, M.-S. (2023). Lighting a pathway: Our obligation to culture and Country. *Ecological Management & Restoration, 24*(2–3), 153–155. https://doi.org/10.1111/emr.12592

Productivity Commission. (2024). *Review of the National Agreement on Closing the Gap*. Australian Government. https://www.pc.gov.au/inquiries/completed/closing-the-gap-review/report/closing-the-gap-review-report.pdf

Ross, J. (2025). Indigenous course requirements prompt 'indoctrination' claims. *Times Higher Education (THE)*. https://www.timeshighereducation.com/news/indigenous-course-requirements-prompt-indoctrination-claims

Silversmith, S. (2025). Tribal leaders sound alarm as Trump's federal cuts threaten Indigenous communities. *Arizona Mirror*. https://azmirror.com/2025/03/03/tribal-leaders-sound-alarm-as-trumps-federal-cuts-threaten-indigenous-communities/

Special Committee. (2024). *A/79/363: Report of the Special Committee to Investigate Israeli Practices Affecting the Human Rights of the Palestinian People and Other Arabs of the Occupied Territories*. United Nations General Assembly. https://documents.un.org/doc/undoc/gen/n24/271/19/pdf/n2427119.pdf

Tahana, J. (2024). Māori language 'at risk' as a result of government policies, commissioner says. *The Guardian*. https://www.theguardian.com/world/2024/sep/23/maori-language-at-risk-government-policies-new-zealand

Tanana, H. (2023). Voice of the river: The rise of indigenous women leaders in the Colorado River Basin. *Colorado Environmental Law Journal, 34*(2), 265–298.

Te Kāhui Tika Tangata Human Rights Commission. (2024). *Loss of Māori Health Authority out of step with te Tiriti and human rights*. https://tikatangata.org.nz//news/government-out-of-step-with-erasure-of-maori-health-authority

Uluru Statement from the Heart. (2017). https://ulurustatement.org/the-statement/view-the-statement/

Walker, A. R. (2025). 'A slap in the face': Activists reel as Trump administration removes crucial missing Indigenous peoples report. *The Guardian*. https://www.theguardian.com/us-news/2025/mar/20/trump-missing-murdered-indigenous-peoples-report-removed

Part I
Words of the Land: Renaming, Resistance, and Renewal

Chapter 2
Unsettling the Settler: Lessons in Belonging and Relationality by Listening to Country

Melissa Kennedy and Amy McCoy

Abstract This chapter investigates the transformative possibilities that emerge when relationality guides our interactions with land, water, and one another. It also considers the concept of "unsettling" as a critical tool for challenging the entrenched dynamics of settler colonialism and reshaping our relationships with land, water, and Indigenous peoples. Unsettling calls on non-Indigenous peoples to critically examine their role within colonial systems, confronting the discomfort of complicity and rethinking their relationships with the natural world. By foregrounding Indigenous perspectives as central to sustainable and just ways of living, we invite readers to reflect on the limitations of Western worldviews, frameworks and the possibilities for dismantling colonial legacies. Through a reflective conversation between Melissa and Amy, who are professional colleagues and personal friends, this chapter examines the critical role of relationality in fostering belonging and justice.

Keywords Indigenous · Relationality · Belonging · Unsettling

2.1 Introduction

Settler colonialism is a pervasive and insidious structural process that shapes the political, social, and environmental landscapes of many parts of the world (Coulthard, 2014). The legacies of dispossession, violence, and cultural erasure persist, often embedded within systems that continue to marginalise Indigenous

M. Kennedy
School of Social Science, Monash University, Clayton, VIC, Australia
e-mail: melissa.kennedy1@monash.edu

A. McCoy (✉)
Fluent Freshwater Insights, Bend, OR, USA
e-mail: amy@fluentfreshwaterinsights.com

M. Kennedy, E. O'Donnell (eds.), *People, Place and Nature in Indigenous-Settler Relations*, Indigenous-Settler Relations in Australia and the World 10, https://doi.org/10.1007/978-3-032-03600-1_2

peoples and their knowledge systems (Parsons et al., 2021; Wolfe, 2006). Scholars and activists have called for a more nuanced understanding of these dynamics in recent years, advocating for approaches beyond traditional decolonisation frameworks. One such approach is "unsettling," a concept that challenges non-Indigenous people to interrogate their position within colonial structures and to recognise their complicity in the ongoing violence of settler colonialism (Tuck & Yang, 2012).

Unlike decolonisation, often framed as the material return of land and power to Indigenous peoples (Tuck & Yang, 2012), unsettling centres on the internal, relational processes required to unlearn colonial assumptions and behaviours. It encourages people of settler heritage to examine their complicity, shifting from a mindset of ownership to one of responsibility (Tuck & McKenzie, 2014). Unsettling is not merely a theoretical exercise but an applied, ongoing process that demands active engagement with Indigenous epistemologies and an openness to rethinking fundamental aspects of identity, belonging, and governance (Simpson, 2017). Through this process, unsettling creates space for the resurgence of Indigenous autonomy and expression, which is the key theme of exploration in this chapter.

The authors come from diverse cultural, and life experience backgrounds, yet share many moral and ethical values and worldviews. Kennedy is a Tati Tati Traditional Owner from Murray River Country, Australia, with kinship relations to Wadi Wadi, Mutti Mutti, and Latji Latji, belonging to the waters of the Murrumbidgee and Milloo (Murray River). She holds an inheritance of belonging as the descendant of river people with enduring responsibilities of speaking for and with Country. McCoy is a White American—with lost Scottish ancestry—who grew up in the foothills of the Colorado Mountains. Working for two decades in the Colorado River Basin around Tribal water rights and water sustainability, her journey developed by being deeply troubled by the exclusionary nature of water laws in the US, particularly how they marginalise Indigenous peoples. This realisation motivated her work to understand, challenge, and transform the exclusive culture of water management in the United States and globally.

2.1.1 Centring Indigenous Ways of Knowing: In Practice

In this chapter, we share our efforts to honour Indigenous ways of knowing and being as a process in life, not just academia. In deference to Indigenous perspectives, we have intentionally documented our reflections and thoughts as we work to unsettle our own ways of being and knowing. These reflections are highlighted in the boxed text; however, we have chosen not to attribute sections to specific authors who come from diverse backgrounds. This invites readers to see themselves in our words and supports their own journey of unsettling. It is our intentional decision not to attribute ownership over our words. Indigenous ways of knowing and learning are not static (Kimmerer, 2013) and do not belong to any of us. Background information and contextual descriptions are provided in the other sections of the chapter.

We have also actively minimised the standard academic expectations and terminology, such as quoting authors' specific reflections, methodologies and ontologies. We are guided by the work of Indigenous authors such as Robin Kimmerer, who reminds us how the English language itself is structured around fixed nouns, reflecting and reinforcing a dominant, global worldview of possession and ownership: '[s]cience is language of distance which reduces a being to its working parts; it is a language of objects' (Kimmerer, 2013, p. 49). We are also influenced by Tuck and Yang's (2012) work exploring how refusal within research is critical to protecting Indigenous sovereignty, resisting colonial harm, and shifting power dynamics in institutions of education.

This experience of learning from and with Indigenous peoples—whether as a guest, host, or facilitator—is a potent reminder that unsettling is not a singular or isolated moment but an ongoing, transformative process of introspection and relationality. By actively engaging in unsettling practices, we are called to undo the colonial narratives that prioritise control, exploitation, and ownership, instead embracing the wisdom of relationality and interconnectedness that has long been embodied in Indigenous ways of knowing (Kimmerer, 2013; Poelina et al., 2019). In this context, unsettling becomes a transformative tool, inviting non-Indigenous peoples to reflect on their roles in perpetuating colonial systems, as well as their responsibility for dismantling colonial structures, both materially and ideologically (Lowman & Barker, 2015; Tuck & Yang, 2012).

2.2 Unsettling Settler Assumptions: Rethinking Power, Place, and Responsibility

In this next section, we delve deeper into how unsettling can be actualised in practice, emphasising the importance of transformative, relational approaches to environmental governance, land management, and the ongoing process of unlearning colonial frameworks that regard land and water as commodities (Kanngieser & Todd, 2020). By considering unsettling as an ongoing, introspective practice, people can begin to shift from ownership to responsibility, ultimately fostering more ethical relationships with the land, water, and Indigenous peoples. As Zoe Todd (2016, p. 5) states in her words discussing how ontology is just another word for colonialism:

> As an Indigenous feminist, I think it is time we take the Ontological Turn, and the European academy more broadly, head on.

2.2.1 The Concept of Unsettling

Unsettling challenges people of settler heritage (non-Indigenous people) to interrogate their positionality within colonial systems rather than focusing on the act of "decolonisation," which is often misunderstood or misused (Tuck & McKenzie, 2014). It demands an active engagement with discomfort and a willingness to shift one's assumptions about history, identity, and relationships with land and water. Unsettling as a framework seeks to dismantle the perceived normalcy of Western practices and values. This includes questioning the dominance of Western ways of knowing, which define "natural resources" as inert objects to be owned rather than living relatives deserving of respect and care (Moreton-Robinson, 2015; Poelina et al., 2019). Drawing on Tuck & McKenzie's (2014) work, unsettling requires non-Indigenous peoples to resist the instinct to "solve" Indigenous issues from within colonial systems, instead focusing on disrupting those systems and opening space for Indigenous resurgence and their own self-reflection as guests on Indigenous lands.

The following sections present a practical manifestation of the unsettling process, where we, as the authors, confront our roles within colonial systems and explore how we engage with Indigenous knowledge systems in a meaningful way.

2.2.2 A Personal Journey of Unsettling

In November 2023, Melissa Kennedy (Tati Tati and Wadi Wadi, Australian First Nations) and Amy McCoy (White Settler, U.S.A.) travelled from Naarm (Melbourne) to the floodplains and banks of Millu (the River Murray) with a group of Indigenous River advocates and a few non-Indigenous partners. During this trip, the First Nations Peoples were actively and deliberately recognised as the authoritative Knowledge Holders. We name this as too often non-Indigenous peoples and standpoints are accustomed to viewing Indigenous perspectives as interesting and novel, something different from the "normal" way of doing, being, and knowing. We are committed to flipping this script and positioning the Indigenous ways of relating to Country, culture, history, and life as the time-tested and proven sustainable way of being. In turn, the settler modes of acting were highlighted as different and resistant to change, embodying the key principles of unsettling.

Accepting this invitation as a visitor on Country, I approached the experience with humility and a readiness to learn from knowledge systems deeply rooted in these lands while reflecting on my cultural and professional lineage as a White Settler from the United States. As Puerto Rican writer and activist Aurora Levins Morales (2019) reminds us,

> [w]e can't alter the actions of our ancestors, but we can decide what to do with the social relations they left us.

These words guided me as I navigated the tension between inherited legacies of harm and the potential for meaningful transformation. I carry a commitment to challenging colonial systems and fostering more just, reciprocal relationships with the lands, waters, and peoples I encounter. This requires me to pursue two intertwined efforts: first, to interrogate the cultural assumptions and implicit power structures embedded in (white) settler American society that resist change and sustain inequity, and second, to listen deeply to Indigenous ways of knowing—not to appropriate, imitate, or extract, but to allow these teachings to reshape my understanding and guide my actions to dismantle violent legacies while embracing the possibilities for new, respectful, and transformative relationships. These efforts exemplify how unsettling is an applied and transformative tool for rethinking and restructuring relationships within settler-colonial contexts.

The time spent on First Nations lands during this journey was more than a physical passage; it was a celebration of Indigenous excellence and a testament to our deep love and commitment for land, water, animals, and place. It also starkly illuminated the divide between Western and Indigenous ontologies. Where Western frameworks often seek to control and exploit, Indigenous ways of knowing embrace interconnectedness and responsibility, emphasising a sustainable relationship with all living things. This journey reaffirmed the significance of relationality in our ways of being. By engaging with guests, I was able to showcase how our practices and values stem from a deep understanding of our interconnectedness with all life. These teachings offered more than knowledge; they provided an invitation to reimagine ways of being grounded in respect, reciprocity, and care.

2.2.3 A Critical Examination of Our Roles

My roles on this journey and throughout my life are as complex and multifaceted as the landscapes we traversed. I carried with me the lived experience of being an Indigenous person, navigating a world shaped by colonial control while steadfastly rooted in the strength and wisdom of my heritage. As a community leader and facilitator for social and environmental justice programs, I aimed to create spaces for dialogue that challenge entrenched colonial assumptions and amplify Indigenous perspectives. My role as a university researcher, State and Federal water policy advisor, and environmental advocate requires me to bridge the gap between academic discourse and the lived realities of Indigenous communities. At the same time, as a program organiser, participant, host, and guest, I engaged with others in ways that honoured the principles of reciprocity and respect.

As a host for this exchange, I humbly welcomed guests, recognising the importance of creating a space for shared learning. As a guest on Country that was not my own, I walked with respect, guided by the wisdom of the Knowledge Holders who generously shared their insights. These dual roles underscored the importance of balance, acknowledging the responsibilities of being both a learner and a teacher, a host and a guest.

This journey called on me to play multiple roles. As a logistical organiser, the first was central to my conviction that care is not just in attending to the details of a trip but in creating a space where participants felt safe, supported, and free to focus on the experience and building relationships. The second role required setting expectations for white, non-Indigenous participants, especially American men, whose cultural conditioning encourages a confident assertion of power, authority, and personal opinion while blinding them to the introspection this journey demanded. Witnessing and managing their struggles underscored how deeply ingrained the American worldview is—a patriarchal mindset that defines itself as "normal" and others as "lesser."

However, the most challenging role I took on for myself was to examine my own upbringing and the cultural rituals that shaped me. From patriotic celebrations to the belief that land exists only for economic use, I was taught that success meant achievement, control, and disconnection from the natural world. These lessons sit uneasily with me now. I am on a lifelong journey to name the harm these values perpetuate and to seek a way of living that enables me to embrace connection over control.

2.3 Place and Belonging

> To belong is to honour the interconnectedness of life, to know that the land, water, and sky are not separate from us but are integral to who we are. It is to see the spiritual and cultural dimensions of place and understand that they endure, even when ignored by those who treat the land as a mere resource.

Settler assumptions about land and power often operate through a lens of entitlement and extraction (Moreton-Robinson, 2015; Todd, 2022). Land becomes a symbol of individual success and power, while water is commodified to fuel industrial and agricultural growth. Scholars such as Wolfe (2006) have argued that settler colonialism is a structure that continuously reasserts itself, even as its more overt forms, such as violent dispossession, are often falsely perceived as having receded into history. Unsettling requires addressing this persistence by rethinking the foundational myths of settler societies, including the narratives of meritocracy, exceptionalism, and the moral superiority of colonial "progress."

This process involves questioning the frameworks that underpin settler governance and land ownership. As Coulthard (2014) notes, the settler state often seeks to co-opt Indigenous resistance through symbolic recognition without relinquishing power or resources. Unsettling challenges non-Indigenous peoples to move beyond these symbolic gestures, focusing instead on meaningful relationships that respect Indigenous sovereignty and epistemologies.

2.3.1 Listen to Country and Find Your Place

An exploration of belonging and connection to place directly aligns with the concept of unsettling, as it challenges the prevailing colonial mindset that often separates people from the land, treating it as a resource to be owned and exploited. This shift is not just theoretical; it requires people of settler descent to actively learn from and engage with Indigenous ways of knowing and being, emphasising belonging through a deep spiritual and cultural connection to place. The concept of unsettling becomes central in this journey, offering a pathway for personal transformation and collective healing as non-Indigenous peoples confront their complicity in colonial narratives and embrace a more sustainable, ethical relationship with the land.

Colonialism disrupted—is disrupting—Indigenous connections to place, imposing narratives that render First Nations people as nomadic and detached from fixed property, a falsehood used to justify dispossession and erasure under Western legal systems in Australia and the United States (Burdon et al., 2015). The act of severing people from place not only undermined Indigenous sovereignty but also actively strips place of its cultural and spiritual dimensions, reframing it as a commodity for exploitation.

Place holds profound significance within Indigenous knowledge systems, shaping cultural identity, customary law, and relationships with the environment (Todd, 2022). Indigenous understandings of place are deeply rooted in place, serving as an epistemological and cultural foundation that cannot be seamlessly integrated into Western frameworks (Tuck & McKenzie, 2014). This divergence highlights a critical challenge in aligning Indigenous and Western perspectives on place; Indigenous interpretations understand land as a living, interconnected entity, while colonial frameworks often reduce the place to a mere surface for human activity, devoid of more profound spiritual and relational meaning (Tomateo & Grabowski, 2024).

Amid the peaceful floodplains of the River Murray, in a place known as Tol Tol, dwells Paka Piyali, an ancient spirit of wind and wisdom. She is long-lived and ever-present, a living essence rich with history, kinship, and purpose. She is the guardian and the nurturer, bound in harmony with all she watches over. To the weary, she offers shelter; to the hungry, sustenance; to the vulnerable, protection; and to the lost, a welcome guide. From the fierce foraging ant to the proud white-bellied sea eagle to the Old Man Weed that rises and falls with the floodwaters, Paka Piyali endures steadfast and unwavering.

Paka Piyali observes, her heart full and her soul content. She is profoundly respected and revered as an elder of the Tati Tati people, the stewards of the river Country that Paka watches over. She has borne witness to countless Tati Tati children taking their first steps, catching their first fish. She has seen girls grow into women, fathers into grandfathers, and has watched as all return to the land that first cradled them, finding their eternal rest. Her lasting presence stands as a testament to her wisdom and strength. Her song resonates through tranquil nights, the oral and aural histories of the people rustling through her leaves.

For Paka Piyali is Grandmother Tree, the ancient Eucalyptus Camaldulensis, the Murray River Red Gum. Grandmother, still standing today, is 1000 years old. She has seen flood, drought, fire, hail and shine. She has seen life, and birth, and sorrow, and loss. Paka Piyali endures, steadfast and serene.

Grandmother endures because her resilience is tied to the strength of river Country. She has a living connection to the land and to her ancestors, a crossing between the seen and unseen spirit of all things. She, who stands firm against the destructive gales that blow from the West, carrying with it whispers of destruction, is not uprooted. It is her connection that grounds her; her deep, ancient roots stretch far and wide.

Grandmother watches the white man come.

White man claims that Paka Piyali is just a tree, they show no respect for the sacredness of her Country. They do not recognise her as a living ancestor; they do not recognise her people as sovereign. They proclaim that this land now belongs to the Crown and that the Crown answers only to itself. Yet still, Paka Piyali endures, steadfast and grave.

But then they infest Grandmother's lands with the greed of their ancestors, insatiable and relentless in their desire to take, conquer, and claim. Like a plague, they swarm, overrun, and feast. They consume Grandmother's River Country, crushed beneath the burdens of the white man's desire. The waters, once vibrant and life-giving, are poisoned, the land strangled by neglect. Country is in pain; Grandmother is in pain. With each wound the river suffers, Grandmother suffers. She endures, but she suffers. She endures, but she weeps.

She endures, but...

—Poem by Melissa Kennedy, 2023

As is common in America, my family stories only extend back four generations, with little to no knowledge of our European ancestors. Lost connections to European homelands nurture relationships and allegiances not to land but rather to sacrosanct principles of American democracy, Judeo-Christian doctrines, capitalism, and Manifest Destiny—values deeply ingrained in the community of my upbringing.

Growing up in the foothills of the Rocky Mountains, I developed an appreciation for the beauty of the peaks through recreational activities like camping, hiking, and fishing. Yet, I was also taught that land's value was not recognised as a relational connection with place but measured through ownership, private property, and its potential for resource extraction and economic profit.

While travelling through Country, Melissa described the land, noting slight changes in the slope of hills, the presence or absence of rocks, and the density of Eucalyptus stands. She also acknowledged which Mob lived in each area, explaining how their ways of living were directly attuned to the characteristics of the land itself so that how they lived would reveal where they and their ancestors lived. I felt as if I was learning a new language of relationship in hearing that Mobs know where they belong based on their relationship to, and knowledge of, land. From what I understood, for Mobs, relationality equals belonging.

As an American, relationality was never about land or belonging to place; instead, it was a map to show me how to engage with the primary structures of society—how to relate to my family, community, church, school, and national country.

This journey provided an opportunity to rejuvenate cultural connections and foster an understanding of how we belong to place. By guiding guests through our places, practices, and histories, I sought to share the richness of Indigenous knowledge systems and their rootedness in relationality. Every step on Country carried the potential for learning, not only for our guests but for myself, as I was reminded of the power and resilience of our cultures. Along the way, it was repeatedly demonstrated how the intricacies of our connection to place are understood: how the slope of a hill, the density of trees, or the course of a river reflect the lives and stories of the people who belong to that land.

Belonging is not just about where I am but how I am with the place. It is about listening to the land and feeling its heartbeat in my own. Tol Tol is more than a site; it is a living entity, a keeper of stories, memories, and lessons passed down through generations. It teaches me that place is reciprocal; it nurtures us as we nurture it. The connection between place and belonging is more than physical; it is spiritual, ancestral, and enduring. Standing beneath Paka Piyali, the Grandmother Tree, I feel her presence, a sentinel of time whose wisdom flows through the river's currents and rustles in the whispering leaves. Her steadfast resilience mirrors the strength of the Tati Tati people, my people, who draw purpose and identity from this land.

2.4 Relationship and Responsibility

Unsettling entails shifting from a framework of ownership to one of responsibility. This echoes Kimmerer (2013), who emphasises the importance of reciprocal relationships with land and water. In settler contexts, unsettling means recognising that true belonging cannot be achieved through ownership or domination but through humility, accountability, and an acknowledgment of harm.

For instance, Whyte (2018) explores how unsettling practices can reshape environmental governance, advocating for including Indigenous epistemologies that view water as a living relative. Such an approach challenges dominant settler paradigms, which separate land and water into discrete legal and administrative categories, reinforcing the commodification of natural resources (O'Donnell et al., 2023). Embracing relationality creates space for more ethical and sustainable practices that honour the interconnectedness of life.

Ancient waters speakSalt of the land and oceansDo you hear her sing?Wrath, unforgivingPrehistoric rock erodesSpirits cleansed anew.She flows ever onMy ancestor of creationHealing sacred lands.—Poem by Melissa Kennedy, 2023

2.4.1 Relationality: Essential to Unsettling

As described by Marshall (2019), Indigenous Australian ontologies recognise people as being in kinship relationships with Country, including its waterways. In this view, water is a being to be respected, nurtured, and protected, not merely a physical entity to be utilised. Many Indigenous belief systems regard water as a "more-than-human" relative, central to spiritual and communal life (Wilson & Inkster, 2018). These beliefs reflect a profound understanding of interconnectedness, where water, land, and people are inextricably linked.

Practising unsettling requires a commitment to lifelong learning and introspection. Non-Indigenous people must engage with Indigenous scholarship, listen to Indigenous voices, and resist the urge to centre themselves in conversations about justice and equity. Simpson (2017) highlights the importance of creating spaces for Indigenous resurgence while recognising that non-Indigenous people must relinquish control and accept their role as guests on Indigenous lands.

Unsettling is neither a linear nor smooth process. It requires people of settler heritage to sit with discomfort, confront their privileges, and actively work to dismantle systems that perpetuate harm. As Lowman and Barker (2015) argue, unsettling involves accepting that there may not be easy resolutions or clear pathways forward. Instead, it is an ongoing process of unlearning and reimagining relationships with place, people, and power.

Relationality is not just a worldview but a transformative framework for addressing modern water crises. Relationality is the understanding that all things are interconnected and can exist only in relationship to one another, a foundational principle across many Indigenous worldviews (Dudgeon & Bray, 2019). We know water as far more than a resource—it is a living relative, imbued with spirit and deeply tied to cultural and spiritual practices. This relational perspective establishes a sacred responsibility to care for water as kin.

The colonial settler perspective frames land and water as resources to be quantified, traded, and used to signal wealth or poverty (Moreton-Robinson, 2015). Language underpins this system: "rights" and "entitlements" suggest authority to control access, inclusion, and exclusion. Ownership is not merely possession but a relational map of power (Nakata, 2024). Owners make decisions about who is included and who is excluded; those without ownership are denied relationships, belonging, and voice. This logic is woven into the histories of the U.S. and Australia through policies like the Doctrine of Discovery (Miller et al., 2010), terra nullius and aqua nullius (Marshall, 2017), which conveniently declared land and water unoccupied, uncared for, and wholly available for settler claims.

As I reflected, a troubling truth rattled me: the American White settler relationship to land prioritises hierarchical control over relational belonging. As physical and legal constructs, fences symbolise this system, signalling ownership and imbuing those with the authority to manage, access, and utilise the land. Here, relationality is stripped of emotional resonance and reduced to legal, economic, and social constraints. The question is not how deeply you feel tied to the land or if it feels like home; what truly governs whether you are permitted to enter it by law, status, or wealth. Authority, not connection, belonging, or relational knowing, becomes the arbiter of access and enforcer of boundaries.

By redefining belonging to prioritise individual wealth over collective health, settler colonialism erased other relationships to land and water so thoroughly that for White Settler society, imagining alternative ways of being remains an abstract and disconnected exercise, entirely divorced from any lived experience. By adopting relationality as a guiding principle, societies can move beyond the commodification of water, embracing instead a model of stewardship that aligns with Indigenous philosophies of care and connection. This perspective challenges dominant narratives, offering pathways to more ethical and sustainable environmental management and governance approaches.

2.4.2 Unsettling Ownership

In Australia and abroad, political power and wealth have long been built on the colonial exploitation of land and water resources, read: ownership. To control these resources, past and present governments enforced policies that separated and dispossessed Indigenous peoples of their lands, waters, and cultures (Ardill, 2009). By restricting First Nations people's access to these resources, the government effectively controlled wealth, power, and socio-political outcomes that went hand in hand with land ownership (Cook, 2012). Today, control over resources has shifted to modern environmental management practices led by state authorities, which continue to benefit the economic elite (Hartwig et al., 2018).

The British occupation of Indigenous lands relied on the flawed legal concept of terra nullius—land belonging to no one. The lesser-known concept of aqua nullius, or water belonging to no one, mirrors the more widely recognised history of land appropriation; however, the difference is that the dispossession of water continues today (Moggridge et al., 2019). Wiradjuri scholar Dr Virginia Marshall argues that water rights, claimed unlawfully by the Crown, are a legal fiction with intergenerational impacts only recently acknowledged (Marshall, 2017). For many Basin First Nations, water is sacred and relational, a view fundamentally at odds with the Western commodification of water as a resource, creating what Arsenault et al. call a deep moral divide (2018).

In the western United States, the doctrine of prior appropriation governs water use under the principle of "first in time, first in right." This doctrine institutionalised a hierarchy of water rights, where White European Settlers who first diverted water secured senior claims, often at the expense of those who came later. Settler lawmakers designed this system to impose predictability on the region's inherently variable hydrology, ensuring water for agriculture, urban growth, and industry. Crucially, it allocates no water to the river or Indigenous communities who lived in the Basin long before Anglo settlers arrived. As such, Prior Appropriation might more accurately be understood as "third in time, first in right," privileging later settler claims while neglecting the earlier in time environmental health needs and the deep cultural practices and connections of Indigenous peoples to water.

The separation of land and water rights into distinct legal entities reflects a Western framework that contrasts sharply with Indigenous understandings of Country as a connected whole (O'Donnell & Kennedy, 2023). Lane et al. (2023) describe the rise of environmental management as the work of a technocratic elite—environmental policymakers who prioritise Western values of science and technology to benefit those in power while presenting an image of environmental stewardship. As a result, Indigenous Australian values, knowledge, and priorities are overlooked, and First Nations' access to resources and influence within these policies remains restricted. First Nations activists are now creating strategies to bring Indigenous perspectives and knowledge into water planning and management, seeking new and innovative approaches to these critical issues (Mooney & Cullen, 2019). Arsenault et al. (2018) highlight the potential of Indigenous epistemologies to reshape water governance, encouraging a shift from exploitative to reciprocal relationships with water. O'Donnell (2020) expands on this by advocating for relationality to guide political and legislative reforms, emphasising its capacity to inspire systemic change in how societies manage and protect water.

Unsettling compels us to reexamine the ethical and cultural assumptions underlying our relationships with land and water. By exposing the colonial underpinnings, unsettling reveals the need to replace systems rooted in domination and extraction with frameworks that honour relationality and reciprocity. This process redefines water not as a commodity but as a sacred, living presence interwoven with human and ecological systems. For non-Indigenous communities, unsettling is an urgent call to dismantle exploitative structures and embrace shared, respectful governance of resources as an ethical imperative.

2.5 Conclusion

The act of unsettling is not merely intellectual; it is a call to reshape the tangible structures of governance and resource management in ways that honour Indigenous knowledge, rights, and the interconnections between land, water, and people. The concept of unsettling offers a transformative lens through which to reimagine the relationships between non-Indigenous peoples, First Nations, and place. By challenging the deeply ingrained structures of settler colonialism, unsettling becomes a process of both deconstruction and creation. It dismantles systems of ownership and control while fostering frameworks rooted in relationality, responsibility, and reciprocity.

This shift is not only philosophical but also profoundly practical, demanding tangible changes to governance, environmental management, and the ways in which knowledge is produced and valued. This process requires humility and a willingness to engage with discomfort as non-Indigenous people confront the privileges and assumptions that have long shielded them from the impacts of colonial violence. At its core, unsettling emphasises the need to centre Indigenous knowledge systems, which offer deeply relational approaches to governance and environmental stewardship. By adopting these perspectives, societies can begin to address the ecological and social crises exacerbated by colonial worldviews.

Unsettling calls for a reimagining of belonging and identity. It invites people to move beyond narratives of entitlement and ownership, embracing instead a sense of place grounded in responsibility and respect. This transformation is not about appropriating Indigenous practices but about learning to exist as respectful guests on Indigenous lands. It requires non-Indigenous peoples to listen deeply, to unlearn colonial habits of thought, and to support the resurgence of Indigenous sovereignty without seeking to dominate or dictate its terms. Importantly, unsettling is not a one-time act but an ongoing journey. For both individuals and societies, unsettling offers a pathway toward more ethical and equitable relationships with the land and its original caretakers.

One question lingers as I reflect on this journey: "How did we get here?" What histories, ideologies, and systems allowed European settlers to claim and internalise moral and legal dominion over lands and waters, severing relationships in favour of ownership? These questions are far from abstract; they are the first steps toward dismantling colonial legacies that still shape how we live and govern. American exceptionalism—this insistence on our own supremacy—remains a toxic barrier to imagining a future where belonging replaces domination. As Patrick Wolfe wrote, "colonialism is a process, not an event" (Wolfe, 2006). Therefore, true transformation begins when I choose to question and invite other Americans into a conversation to unlearn our deeply engrained ways of being and see beyond the colonial mirror.

Ultimately, on this journey, I have been called to weave together the personal, cultural, and professional to embody the strength of our people while fostering opportunities for shared growth and transformation. It is an act of resistance, a celebration of resilience, and a reaffirmation of the power of Indigenous knowledge to inspire change. I am reminded that our teachings about belonging and place are not only for others to learn and adapt to but also for us to rejuvenate our spirit. It reaffirms our identity and strengthens our ties to the land. I am reminded that, like Grandmother, we endure. We suffer, we weep, but we stand, deeply rooted and connected to Country. Through these connections, we find strength, meaning, and the unyielding knowledge that we belong.

References

Ardill, A. (2009). Sociobiology, racism and Australian colonisation. *Griffith Law Review, 18*, 82–113.

Arsenault, R., Diver, S., McGregor, D., Witham, A., & Bourassa, C. (2018). Shifting the Framework of Canadian Water Governance through Indigenous Research Methods: Acknowledging the Past with an Eye on the Future. *Water, 10*(1), 49. https://doi.org/10.3390/w10010049

Burdon, P., Drew, G., Stubbs, M., Webster, A., & Barber, M. (2015). Decolonising Indigenous water 'rights' in Australia: Flow, difference, and the limits of law. *Settler Colonial Studies, 5*(4), 334–349. https://doi.org/10.1080/2201473X.2014.1000907

Cook, I. (2012). Critical theories. In R. Smith, A. Vromen, & I. Cook (Eds.), *Contemporary politics in Australia: Theories, practices and issues* (pp. 35–45). Cambridge University Press.

Coulthard, G. S. (2014). *Red skin, white masks: Rejecting the colonial politics of recognition*. University of Minnesota Press.

Dudgeon, P., & Bray, A. (2019). Indigenous relationality: Women, kinship and the law. *Genealogy, 3*(2), 23. https://doi.org/10.3390/genealogy3020023

Hartwig, L. D., Jackson, S., & Osborne, N. (2018). Recognition of Barkandji water rights in Australian settler-colonial water regimes. *Resources, 7*, 16.

Kanngieser, A., & Todd, Z. (2020). From environmental case study to environmental kin study. *History and Theory, 59*(3), 385–393.

Kimmerer, R. W. (2013). *Braiding sweetgrass: Indigenous wisdom, scientific knowledge, and the teachings of plants*. Milkweed Editions.

Lane, P. N., Benyon, R. G., Nolan, R. H., Keenan, R. J., & Zhang, L. (2023). Forests, fire and vegetation change impacts on Murray-Darling basin water resources. *Australasian Journal of Water Resources, 27*(1), 68–84.

Lowman, E. B., & Barker, A. J. (2015). *Settler: Identity and colonialism in 21st century Canada*. Fernwood Publishing.

Marshall, V. (2017). *Overturning aqua nullius*. Aboriginal Studies Press Canberra, ACT.

Marshall, V. (2019). Removing the Veil from the 'Rights of Nature': The Dichotomy between First Nations Customary Rights and Environmental Legal Personhood. *Australian Feminist Law Journal, 45*(2), 233–248. https://doi.org/10.1080/13200968.2019.1802154

Miller, R. J., Ruru, J., Behrendt, L., & Lindberg, T. (2010). *Discovering indigenous lands: The doctrine of discovery in the English colonies*. Oxford University Press.

Moggridge, B. J., Betterridge, L., & Thompson, R. M. (2019). Integrating Aboriginal cultural values into water planning: A case study from New South Wales, Australia. *Australasian Journal of Environmental Management, 26*(3), 273–286. https://doi.org/10.1080/14486563.2019.1650837

Mooney, W., & Cullen, A. (2019). Implementing the Aboriginal Waterways Assessment tool: Collaborations to engage and empower First Nations in waterway management. *Australasian Journal of Environmental Management, 26*(3), 197–215.

Moreton-Robinson, A. (2015). *The white possessive: Property, power, and Indigenous sovereignty*. U of Minnesota Press.

Nakata, N. M. (2024). Indigenous languages & education: Do we have the right agenda? *The Australian Educational Researcher, 51*(2), 719–732. https://doi.org/10.1007/s13384-023-00620-0

O'Donnell, E. (2020). Rivers as living beings: Rights in law, but no rights to water? *Griffith Law Review, 29*(4), 643–668. https://doi.org/10.1080/10383441.2020.1881304

O'Donnell, E., & Kennedy, M. (2023). The Legacy of Aqua Nullius is causing a sustainability disaster. *Pursuit*. University of Melbourne.

O'Donnell, E., Kennedy, M., Garrick, D., Horne, A., & Woods, R. (2023). Cultural water and Indigenous water science. *Science, 381*(6658), 619–621.

Parsons, M., Fisher, K., & Crease, R. P. (2021). *Decolonising Blue Spaces in the Anthropocene: Freshwater management in Aotearoa New Zealand*. Palgrave Macmillan. https://doi.org/10.1007/978-3-030-61071-5

Poelina, A., Taylor, K. S., & Perdrisat, I. (2019). Martuwarra Fitzroy River Council: An Indigenous cultural approach to collaborative water governance. *Australasian Journal of Environmental Management, 26*(3), 236–254. https://doi.org/10.1080/14486563.2019.1651226

Simpson, L. B. (2017). *As we have always done: Indigenous freedom through radical resistance*. University of Minnesota Press.

Todd, Z. (2016). An Indigenous feminist's take on the ontological turn: 'Ontology' is just another word for colonialism. *Journal of Historical Sociology, 29*(1), 4–22.

Todd, Z. (2022). Fossil Fuels and Fossil Kin: An environmental Kin Study of Weaponised Fossil Kin and Alberta's So-Called "Energy Resources Heritage". *Antipode, 1–25*. https://doi.org/10.1111/anti.12897

Tomateo, C., & Grabowski, Z. (2024). Indigenous justice frameworks for relational ethics in land-based design. *Ecosystems and People, 20*(1). https://doi.org/10.1080/26395916.2024.2409165

Tuck, E., & McKenzie, M. (2014). *Place in research: Theory, methodology, and methods*. Routledge.

Tuck, E., & Yang, K. W. (2012). Decolonisation is not a metaphor. *Decolonisation: Indigeneity, Education & Society, 1*(1), 1–40.

Whyte, K. P. (2018). The environmental injustice of the "eco-colonialism" of water governance: Indigenous water struggles and the settler state. *Environmental Sociology, 4*(3), 261–275.

Wilson, N. J., & Inkster, J. (2018). Respecting water: Indigenous water governance, ontologies, and the politics of kinship on the ground. *Environment and Planning E: Nature and Space, 1*(4), 516–538. https://doi.org/10.1177/2514848618789378

Wolfe, P. (2006). Settler colonialism and the elimination of the native. *Journal of Genocide Research, 8*(4), 387–409. https://doi.org/10.1080/14623520601056240

Chapter 3
How Does It All Come Back Now? Re-organising Naturalised Histories

Katrina Schlunke

Abstract There is nothing natural about 'natural' history. It is where nature becomes the object of study rather than something we might experience, live in or work with. When you consciously do the work of naming, explaining and studying nature, you become an exception to that nature—a naturalist. Indigenous Country, spaces filled with life, story, interconnections and histories, is fashioned as nature when encountered by non-Indigenous people. Once named 'nature', Indigenous Country becomes accessible to everyone at any time, explained only through the lens of the naturalist. This process of 'nature-ising' works alongside colonising, seeking to control Indigenous knowing and being. This chapter considers the interventions and creations of Indigenous curators, Elders, writers and artists that resist the colonial practice of isolating 'nature' from Country in museum contexts, especially natural history museums. These responses—creative installations, educational spaces—reconnect people and place through the interwoven worlds of Country. This re-centering of Country also re-centres the relationships between people, place and all that connects them. The chapter moves with the speculative fiction and poetry of Evelyn Araluen, Megan Cope's installation art, and Burra (eel), the Australian Museum's Country-driven learning space. It ultimately asks what a profound re-organisation of such museums and collections might produce.

Keywords Natural history · Indigenous Country · Museum · Nature · De-colonial

---

K. Schlunke (✉)
University of Sydney, Sydney, NSW, Australia

School of Creative Arts and Media, University of Tasmania, Hobart, TAS, Australia
e-mail: katrina.schlunke@sydney.edu.au

M. Kennedy, E. O'Donnell (eds.), *People, Place and Nature in Indigenous-Settler Relations*, Indigenous-Settler Relations in Australia and the World 10, https://doi.org/10.1007/978-3-032-03600-1_3

3.1 Introduction

This chapter title references the piece of speculative fiction by Evelyn Araluen titled Muyum, a transgression (2022), which in part features the return to Indigenous life of a natural history collection and ends with the telling and challenging line: 'This all come back now. Like,' (p. 33). That short story with its melange of reborn taxidermy mounts, bursting specimen jars and howling hollow turtles, depicts a potential transformation of 'natural history' when charged with the power of Indigenous forces that appear as a reassertion of Country.[1] The interventions, additions and transformations of museum natural history thinking and collections that do that transformative work, is the focus of this chapter.

As a non-Indigenous researcher, my work includes investigating histories of extinction from a postcolonial perspective, in part supported by an ARC Discovery Grant (DP2001018777). In doing so, I have developed de-colonial understandings of museum collections and 'nature' which have enabled a refiguration by me of both museum practices and ideas of ecological repair. The impacts of centuries of European imaginings of Indigenous Country as 'nature' and a site to be subject to the organisational taxonomies of 'natural history' can be clearly seen within a natural history museum context. Such museums often perpetuate and rely upon that colonial system of organisation to display and educate in what is often understood as a 'neutral' setting, as Murawski among many others have critiqued (Murawski, 2019). Producing alternatives to that assumed neutrality requires a multi-disciplinary, multi-faceted approach. As Sara Kian-Judge (2022), of the Australian Museum writes:

> A long history of Western separation between what is 'human' and what is 'nature' has left us with a sickness in our relationships with Country. All life depends on Country, so a sickness in those relationships is a sickness that reaches into our health and wellbeing too. This sickness makes us see animals and ecosystems as things instead of beings connected together in life-giving, life-creating, life-supporting patterns. We think this way without even noticing – ask yourself how many times you have referred to an animal as 'it', or in the context of the Museum, how often we haphazardly use words like 'specimens', 'collections', and 'objects' when talking about animal bodies. As a Walbanja Yuin woman, the idea of Country without personhood, aliveness and identity does not make sense. The animal bodies in the Museum are my Ancestors, family, teachers and friends.

In the midst of an extinction crisis and climate change induced by capitalism and empire, suggesting how the key element of that crisis—human relationships with the more-than-human world—could be re-figured through an idea of Country is a high stakes intervention with effects well beyond the museum world.

[1] Capitalisation of 'Country' follows Sara Kian-Judge (2022) as a means of showing respect for 'who' Country is and to differentiate Indigenous Country from national country.

3.2 Making Country into 'Nature'

In Australia, the particular effects of making a fundamental division between 'nature' and 'culture' attempted to divide Country, which always includes both. This recasting of Country created division across multiple systems and sites. Kian-Judge's (2022) powerful critique of the ongoing effects of disconnection that the organisational and pedagogical practices of 'natural history' has produced is borne out in Gumillya Baker's critical-creative-collective work 'relating to ideas of the Museum of Un-Natural History.' Gumillya Baker writes (2020, pp. 859–60):

> Un-natural because the recent history of the colonisation of this country is not and never will be a natural unfolding. Un-natural because the construction of race is not a natural reproductive process, these genealogical labels are but one view of humanity, and these ways of viewing people and our relationships to each other and our lands have done much to destroy the social, spiritual and cultural fabric of my people… Un-natural because there is no grand narrative of a civilised progression of scientific knowing. Despite and because of these un-natural ideas we continue to have institutions of 'natural history', institutions that have enormous power in the construction of knowledge for our children.

Gumillya Baker, while focussing on physical anthropology and the abuse of Indigenous peoples in museum practices, also reveals here the wide ranging and hard-hitting effects that the practices of 'natural history' has had. Not only is its thinking suffused with racist imaginings and practices but that thinking, and those practices have become 'naturalised' in many instances. Versions of natural history have invented lines of progression that marked Australian marsupial animals as primitive but also diverse Indigenous, enslaved, working class, sexually diverse and female human communities were also marked as primitive and so ordered into particular pernicious hierarchies (Schlunke & Schwarz, 2025; Ashby, 2021; Moyal, 2010).

Although much of the blatantly racist practices of natural history have been debunked and discarded, the material enactment of traditional taxonomies, when applied to the more-than-human world, continues in many museums. These museums are key educational sites for children, as Gumillya Baker states. Visits to museums, perhaps particularly natural history ones—or the natural history sections which are more usual in Australia—are seen as benign and educational. The place where kids will see stilled (as in taxidermy) versions of animals only otherwise met in books or zoos, and where children supposedly increase their perspective of the animals of the world by learning to place these animals in orders that connect them to other abstracted species, but rarely to specific Country or peoples. In these collections the key divisions between mammals and reptiles, or living and extinct, are displayed, as are the distinctness, even peculiarities of particular 'specimens'. Key to the educating power of such collections is the learning of names and differentiation. How to tell a parrot from a pigeon, how to know a monotreme is a particular class of marsupial—how to know through colonially imposed, classificatory names. As Araluen writes in her poem *Learning Bundjalung on Tharawal* (2021, p. 8):

and I know jalwahn and bilin bilin and ngarehr
but the rest are just nunganybil,
the rest are just: 'bird'

It is hard to unlearn a language:
to unspeak the empire,
to teach my voice to rise and fall like landscape,
a topographic intonation.

While this poem emerges from a scene of learning and remembering Indigenous languages, for non-Indigenous readers such as myself, it also shows how standard English will not follow or fall in with the land we live on. Meanwhile, the language of natural history does not simply name but repetitively inscribes a way of knowing the world that is deeply entwined with colonialism. Museum visitors will come to understand that to 'really' (truly?) know an animal such as a bird, it needs to have been recorded by a colonist, an example killed and examined inside and out and compared with similar animals across the world to place it within an historical system that emerged from a European cultural setting. Araluen's poem has the narrator knowing 'bilin bilin', which I discover means 'parrot' in Bundjalung language per the Macquarie, Aboriginal Words (1994). Parrot arrived via a Greek man, Ctesias, who was living in Persia and reported on a 'bittacos' that spoke Indian! This became psittacos (Greek) and psittacus (Latin) which is the name for the parrot family. But according to the Oxford English Dictionary (2024), 'parrot' is '[o]f uncertain origin. Perhaps a borrowing from French.'

When used in place or in a community, 'bilin bilin' may also describe the culturescape of the bird, its known places of habitation, or its best-known characteristics. But translated to the English word 'parrot' and put into the ordering system of natural history, its name does other work. For example, we can observe the production of the King Parrot (or King's Parrot), which exists all up the east coast of Australia (including Araluen's Bundjalung Country) as a colonial entity through its naming process. Colonial naturalist George Caley named "King's" to this bird, often described as friendly and relatively quiet, after Governor Philip Gidley King, who governed New South Wales from 1800–1806 (Gray & Fraser, 2013). King is recorded in his biographical entry by Shaw (1967) as saying of Indigenous Australians:

> 'These, he told Governor William Bligh, he 'ever considered the real Proprietors of the Soil'. He refused to allow them to be worked as slaves, tried to protect their persons and their property and to preserve a 'good understanding' with them; but he found them 'very capricious', often 'sanguinary and cruel to each other', and like his contemporaries failed to understand what he called their 'most ungrateful and treacherous conduct'.

But King is also infamous for having issued an order in 1801 that Aborigines could be shot on sight and who offered a reward for Aboriginal resistance leader and warrior Pemulwuy, dead or alive. King wrote of Pemulwuy after he was shot that, "[a]ltho' a terrible pest to the colony, he [Pemulwuy] was a brave and independent character" and "an active, daring leader" of his people (Kohen, 2005). Yet King took the head of Pemulwuy, that had been removed upon his death, and sent it to noted naturalist and collector Sir Joseph Banks. Banks (quoted in Vincent Smith, 2010) advised King in April 1803 that:

> the head of one of your subjects, which is said to have caused some Comical consequences when opened at the Customs house ... makes a figure in the Museum of the Late Mr Hunter.

Comical consequences? Seeing a fellow human head in a jar which did not frighten or affect Customs House workers would be far more terrible. Whatever these 'comical consequences' entailed, we can at least acknowledge that some humans, on viewing the head, reacted to this horror, and perhaps even made some embodied connection to the human suffering they witnessed, and the lack of care for the dead. In Bank's use of 'Comical', his superior knowing, his secured position per class and money, oozes from the text. In using 'comical' in a communication with King, he is providing the necessary inclusive anecdote that means this convened brotherhood of natural history will share a point of view via this brutal, insidious and frighteningly educative joke. This 'joke' will grant King (for a moment) inclusion in Banks's sophisticated elan. A world where 'getting the joke' builds connection and an indifference to the suffering their actions have caused. That Banks needed to report that not all the world shared their understanding of how to read what was in that jar may also suggest some order of insecurity, some tiny tear of doubt, but probably not. DOBBY (feat. Barkaa) reminds us as they sing/rap it in their song 'I can't breathe' (2020):

> *You gotta challenge the white settler narrative*
> *Got a lotta books that call us nomadic savages*
> *Maybe that's a connection to them attacking us.*

One of these men has organised the killing, and the other paid for the collection. Who kills and decapitates? Criminals? Terrorists? Who keeps, displays, and pays for human body parts? Fetishists? Perverts? Serial killers? But who does it in public? With all the world (as they knew it) watching? Right into the nineteenth century, Pemulwuy's head was known to be in the Hunterian Museum at the Royal College of Surgeons in London. Some suggest it was disposed of, and others say it was bombed when another war was being waged (Tovey, 2010). The head of this resistance fighter, this clever wise warrior of the Bidjigal peoples is still waiting for repatriation and return to place. In the Australian Museum collection there is a sacred artefact, Death Spear (2021) made by Raymond Timbery, Bidjigal Dharrawal and Joel Deaves, Gumea Dharrawal and evoking the ritual killing of the colony's gamekeeper by Pemulwuy. Timbery (McBride & Smith, 2021, p. 69) writes:

> This sacred artefact was created in representation of the strength and courage of our first resistance fighters, the bloodline of the Bidjigal people still exist today because of the strong warriors we had protecting our people.

Samuel Enderby wrote to Sir Joseph Banks from Paul's Wharf in London on 24 November 1802 (Kohen, 2005):

> Mesrs Enderbys present their most respectful compliments to Sir Joseph Banks and take the liberty of acquainting him that their Ship Speedy, Geo. Quested, Master is arrived from Port

> Jackson with 179 Tuns Sperma Ceti Oil – The Master says he has a Black Swan, a Native's Head in Spirits and a number of other things for Sir Joseph Banks.

If Pemulwuy's head was the example of colonial power on public display, then this list is a marker of the ubiquity of colonial terror, of the ways in which existing systems of commerce and trade seamlessly traded in stolen and violated Ancestral remains, captured Swans, and so very much more. Shipping orders, storage details, ship logs, customs and communications, and so many more lists carry the day-to-day action of the day-to-day violence of colonialism. The list neatly, shockingly, collapses the specificity of animal, human and matter. This matter-of-fact catalogue exemplifies the seemingly unlimited power of 'the list' to objectify knowledge that belongs not to people, place and listening but to the unspecified and unlimited space of the colonial imagining. As Ross Gibson, quoting Ong, writes in *South of the West: Postcolonialism and the Narrative Structure of Australia* (1992, p. 4):

> The greatest shift in the way of conceiving knowledge between the ancient and the modern world takes place in the movement from a pole where knowledge is conceived of in terms of discourse and hearing… to one where it is conceived of in terms of observation and sight and objects.

This list exemplifies the flow from the colonial periphery to the imperial centre. A flow that speaks to a new order of control through organised and targeted violence directed at individuals and a system of categorisation dependent upon measurement, examination and scientific description. This is a small and ordinary communiqué that is indicative of the efforts of Banks to create a natural history collection that will become a British institution—one way in which colonial 'curiosity' becomes an institutionalised seeing and naming. In this way, an imperial public came to look upon the stolen remains of Indigenous warriors as their own displayed property as they could look upon thieved animal and plant life as potential transplants and/or colonial oddities. So we see a King's Parrot. And so we have museum collections. Made by killer/collectors. And benefactors. As Araluen writes in her poem *Bad Taxidermy* (2021, p. 41):

> The brushtail dyed Tesco ice-chocolate brown and face stuffed forehead high like a squirrel.
> The tour that I left five minutes into the guide recounting all the middle names of its benefactors.

But giving the King's Parrot a name is not enough. The parrot must be scientifically described in relationship to other parrots. Other dead parrots. Other 'specimens'. A German, Hinrich Lichtenstein, the Professor of Zoology at the University of Berlin and director of the Berlin Zoological Museum, is credited with first describing the King's Parrot. Having never visited Australia, Lichtenstein's description must have been based upon specimens delivered to him. Näf-Gloor suggests 'that the objects were not only systematically ordered and labelled but also colour-coded according to the object's country of origin. It was the first time that this kind of geographical identification was used in a museum; previously, objects were labelled according to their donators or donors' (Näf-Gloor et al., 2019, p. 59). You can hear the echo of Kian-Judge's observation that within a museum setting, 'how often we haphazardly use words like 'specimens', 'collections', and 'objects' when talking about animal bodies' (Kian-Judge, 2022).

Transformed by language and 'collection', the life force of this blazing red and green feathered life becomes permanently marked by the global penetration of colonialism. Lichtenstein was lauded for at least providing some geographical specificity, but this is undercut by the fact that that specificity (Australia) was itself a silencing of the multi-nation Indigenous site from which the bird emerged. And how did Caley, who named the bird, get his specimen? Daniel Moowattin (c.1791–1816) was a local Indigenous man born in the Sydney region and adopted by a settler family at birth (Vincent Smith, 2005):

> Aged about 14, he became a helper for George Caley, the botanist employed by Sir Joseph Banks. Daniel was Caley's interpreter, bush guide, plant- and leaf-getter, bird-trapper, servant and companion on botanising expeditions around Sydney.

In the 'Sydney Language' as the Macquarie Aboriginal Words (1994) styles it, there is a word for 'king parrot': guma. It is the only translation for King Parrot listed in that dictionary. Is that name a trace of the presence, the physical and intellectual labour of Moowattin? An Indigenous naming written over by Caley currying favour with his Governor? The beautiful bird known for its beautiful head, all red or all green, now transformed from guma into "King's Parrot".

In such re-namings, the 'naturalising' of colonial history is learnt through its 'nature-ising' processes, repeated and passed on. Naturalise means making something appear inevitable and alternatives unthinkable. Naturalise in its suggestion that non-Indigenous people magically knew before Indigenous peoples the 'real' name of … everything; the interplay of this naming and recording and keeping and teaching within museums helping another kind of naturalising. 'Naturalise' in its biological meaning: 'establish (a plant or animal) so that it lives wild in a region where it is not indigenous' (2024). Another word for settlement? According to Gray and Fraser (2013) another name for 'King's Parrot, a name that immediately connects it to colonial or possible de-colonial activity, is—'Spud Parrot.' It is named for its actions of identifying grown potatoes, digging them up and damaging them. Unsettling the settler?

3.3 Can Art Do That?

Suppose I have been able to show, however briefly and inadequately through a single example, the ways in which European 'nature' was made from Indigenous Country through a violent process of division, removal, and the perpetuation and reiteration of the new regime through new names adhered to by the new order. In that case, there should be some sense of the complexity of the challenge to those histories and continuing systems. Thoughtful curators in natural history museums are already attempting to include, where possible, the work of Indigenous guides and other (sometimes, as in only sometimes paid) Black employees in the collecting of natural history specimens (Das & Lowe, 2018; Ashby & Machin, 2021, et al.). Such inclusions are very important but are not the radical life returning, the 'coming back' that is evoked by Araluen where the very idea of the museum is turned back to life, into Country.

In 2022, Indigenous artist Megan Cope created with her Quandamooka community a collaborative, decolonising work of restoring Country and reclaiming 'natureised' oysters. In an ABC TV segment (ABC Arts, 2022), she described her approach:

> Sometimes, presenting the facts doesn't lead to change. Yeah, I'd started to think a little bit radically about how do we make living sculptures that bring these things (oyster reefs) literally back to life? Can art do that?

Cope, on her webpage (Cope, 2024 describes what she did:

> Following from 6 years of research based sculptural practice investigating the impact of the early colonial lime burning industry and devastation of both Aboriginal middens and oyster reefs in Quandamooka Sea Country, Kinyingarra Guwinyanba (which means 'place of oyster rocks' in Jandai & Gowar language) is a hand built sculptural formation returning to Country to create living sculptures for the future.
>
> Resting on the intertidal zone near Myora, the project builds on the legacy of our ancestors, interrupted by colonisation, the legacy that we are now seeking to continue.
>
> Kinyingarra Guwinyanba is a living, generative land and sea artwork that demonstrates how art can physically heal country that has been colonised through the practice of ecologically restorative and ancestral processes.

Cope's sculpture takes the shape of locally sourced wooden poles that are strung with waste shells and then embedded into the sea, where they grow more and more oysters. The work is undertaken collectively and uses the knowledge of Country and appropriate places to attempt to recreate an oyster reef. But this work does more than that. Cope's process was one of moving from 'object' to life in a reversal of the processes of natural history that seek to create known 'objects' from diverse life forms. In her series Foundations 1, completed before Kinyingarra Guwinyanba, she used oyster shells and cement to make works that individualised particular shells in cast concrete. These were then arranged in lines and squares in eerie formations that seem to shift between ideas of industrial production and their ubiquitous manufacturing moulds and lines of tiny gravestones, each shell a lost shield in a deadly battle. Cope (2024) writes that:

> These works explore the foundations of cement production connected to the early lime burning industry that saw the mass burning of Aboriginal shell piles and middens throughout coastal Aboriginal communities.

The destruction of cultural historical markers such as shell middens in early colonial times was followed by the destruction of whole oyster reefs as the intensity of commodification shifted and demand for building grew. Then, as alternate sources of lime were found, oysters were wanted for eating by a vastly increased colonial population. By 1933, a paper read by Iredale and Roughley at the Linnean Society of NSW sought for this particular oyster, Saccostrea glomerata, (popularly known as the Sydney Rock oyster) to be re-named Ostrea commercialis. At the same meeting,

Roughley (1933), named an 'Economic Zoologist', reads a lengthy paper on the 'Life History of the Australian Oyster', beginning:

> The history of the oyster in every country of its occurrence shows that at some time or other the supply has fallen short of requirements demanded by commerce, and always scientific aid has been sought in an endeavour to account for the depletion, and if possible to provide means for ensuring that a similar shortage might be avoided in the future.

Every country? What of the Indigenous oyster gatherers of Quandamooka but also right across Australia, where oysters were eaten and used in tools and marked sites of gathering? Was there ever a record of lack of 'supply'? Wouldn't the vast middens suggest otherwise? The first colonists experienced the plenty of the oyster as managed by the Eora/Dharug people. Smyth (1788), a surgeon on the First Fleet, wrote of these plentiful oysters ('Badangi' in the 'Sydney language') as:

> All the rocks near the water are thick covered with oysters, which are very small but very finely flavoured. They also adhere to the branches of the mangrove trees. I frequently brought the branch of a tree, thus loaded with oysters, on board.

But in 1933, 'commercial' colonialism naturalises the lack that settler colonialism produced. In 1933, the 'life history' of the oyster is an account of the discovery of oyster reproduction and its efforts to increase it by knowing more about the temperatures and tides when the most spawning occurs. Who would know about the temperatures and times of oyster reproduction? Perhaps the peoples who had lived with and feasted without lack on those same oysters for thousands of years. And here is where the uncanny element of the invention of 'nature' emerges. To become a natural scientist, one must learn what Indigenous people already know. But without Indigenous guides the knowing of this new 'nature' is concentrated into a few heads and shared with those who will make money from the knowing. However, even in these commercial times, the effort to rename the Sydney rock oyster fails, and its Latin name remains Saccostrea glomerata. That name was given by Augustus Addison Gould, America's leading conchologist, in 1850 who first described the Sydney rock oyster. In this way the oysters known for millennia as Kinyingarra and Bandangi, are renamed by a white man living 16,000 kilometres away on the shores of an entirely different ocean.

Living inside this colonial and commercial past I begin to understand Cope's work with individual shells and cast concrete as a kind of reckoning with the isolating force of science. Thinking about America's 'leading conchologist' naming (and so removing from Indigenous histories and place, thousands of molluscs), I understand again the power of Cope's Foundations 1. For every colonial naming creates an archive. What Jeanine Leane describes in her poem *Cardboard Incarcerations* (2018, p. 3) as:

> This cardboard prison they call an archive is cold,
> airless and silent as death.

Unlike the prison of names within colonial archives, each of Cope's shell objects recalls a kinyingarra and re-members life. The memorialising of the destroyed middens and oyster reefs which mark the loss of community and the despoiling of

Country lends ever more force to the form of Kinyingarra Guwinyanba. Unlike colonial forms of knowing this form of making life is collective. Knowledge is not in single 'expert' heads but multiple bodies on Country. As Chen notes of those working with Cope, '[t]he contributing 'artists' were not selected because of their high-profile status within the art world, but through their connections, relationships, cultural knowledges and rights to tell the story' (Chen, 2024). The form of repair is not written in arcane scientific Latin but taught in groups.

In other works such as *A Monumental Public Artwork | Sydney Opera House 50th anniversary*, Cope uses that approach to include folk beyond the Indigenous community to make 'monumental' works (Cope, 2023

> *CALL TO ACTION:*
> Please come join us at our Oyster Workshop to give oyster shells a new life towards a monumental public artwork!
> You need to bring, water, comfortable clothes and your own lunch.
> If you have a drill please bring it. We have drill bits for you to use.

The premise of this work can sound like something a natural scientist might want to do (Cope, cited in Dow, 2023):

> Oysters fix carbon in their shells… and given the amount of water that they clean, if we can set ourselves a goal of restoring 40–50% of oyster reefs – not commercial oyster farms, but oysters that can stay in the water for 50 years without the threat of us taking them – then we might start turning a page, and also address other issues like ocean acidification and the ocean heatwave.

But Cope is not only wishing to do that. As the article continues (Dow, 2023):

> Quandamooka, the name of the Aboriginal nation to which Cope belongs and who are the traditional custodians of this area, is also the "mother spirit" of land and water. "She holds all life for us," says the artist. "She's the mode of transport, and she's also the producer of all the food our people have eaten for centuries. Our fundamental job is to look after her."

Oysters are centred here not as objects or money-making entities but collectively as kinyingarra, as part of Country not 'nature'. As Cope says elsewhere (Cope, cited in ABC Arts, 2022):

> Who's it for? It's for country. It's for the oysters.

3.3.1 *Burra (eel)*

At the Australian Museum there is a space and an experience that carries colonially invented 'nature' alongside known Country. It is called *Burra (eel)* and visitors are invited to: '[j]oin Burra (eel) on a fascinating learning journey through Country, at this exciting state-of-the-art education space that immerses you in Western science and First Nations knowledges' (Australian Museum, 2025a). The key First Nations Curator and Indigenous content producers on this project included Courtney Marsh and Sara Kian-Judge but the project also included First Nations Elders, members of the Aboriginal and Pacific Islander community, Indigenous and non-Indigenous designers, artists, makers, educators and others. This list already suggests that this educational space is a manifestation of multiple thinking and connecting. The experience themes within the spaces are: GadiSalt Water, Sweet Water (Fresh water), Bitter Water (Brackish water) in the first space and then Stone, Fire, Water, and Resilience in the second space.

This is one of the few museum-based 'natural history' experiences that is also Country-based. It is aimed at adults and children, but the majority of the images are of children in the space and when I visit in the morning it is populated by mothers/female carers and usually one child and in the afternoon a sea of children have arrived in different school holiday groups. There is a warning that 'Children under 16 must be accompanied by an adult and actively supervised at all times' (Australian Museum, 2025b) but it is as if this place also cares for its transient kin. In the corner a young boy sits high on a seat looking over the joyful cacophony before re-joining the colouring of eel cut-outs using real casts of different snake skins and in the room next door three girls about eight, look into the magical boxes of coral, jelly fish, shells et al and duck in-between to look from the side and underneath. Small groups are negotiating turns rowing the half boat and one boy calls out 'lady, lady' to me to offer a magnetic fish when he sees me making shapes with them on the wall. Around the corner there is a projection onto the floor of stylised eels and other sea animals moving along the currents and we fall about ourselves trying to follow Burra (or perhaps Gurangatch the creation eel) moving faster, then stopping, then jumping sideways when we see her/them coming into view again. This wanting to be with, or become eel as our own bodies turn and shift is infectious and little waves of kids keep rippling in and out as does the odd uncoordinated adult.

For adults, the space may also suggest a multi-dimensional way of learning that includes the senses and educational text that combines ethics and science. Or as it is better described by the producers 'a "many-ways" education' (Explore, 2021, p. 11). One of those ways is the use of 'scientific artefacts' to make their connection to Country. One example is a displayed drill core of Hawkesbury sandstone extracted from the north-east corner of the museum. The panel next to it says in part: 'Stone holds the stories of these lands from the first sunrise. It holds First Nations Lores and Ancestral Spirit in its layers…This core comes from Gadigal Country, five metres below the surface of the ground' (Gadigal to the Core, 2022). This is a very educative but also elegant way of making sure that everyone who reads that panel will know they are standing on Gadigal Country and that Country is not merely a surface or landscape but includes the earth below and exists across time. This idea is even more tellingly made for the situating of the drill core next to a high box filled with a topographic map made of a sand like substance that we are invited to move around to make valleys and islands and watch how the water, in the form of blue light flows after our fingers. At the same time the banks, hinterlands and islands we make become green and yellow. It is a sensual experience and most of us stop at some stage and look at our hands lit in rainbow light, feeling the power of shaping a world. As the accompanying panel suggests: 'Imagine your finger is Gurangatch the creation eel moving through the land as you draw in the sand of this topographic map. See how the water follows your movement. The story of Gurangatch is one way of understanding how water shapes land. Topography and hydro-geology are another way of understanding that same idea' (Many Ways of Knowing, 2022).

Thinking of Gumillya Baker's earlier comments about the importance of making museums safe places for Indigenous peoples and the education of Indigenous children, I also recognise the density of the undoing of natural history thinking for non-Indigenous peoples to attain an understanding of Country that is neither a romanticised version of European nature nor an easily understood field of potential commodities (see Calma, 2008; Bignall et al., 2016; Goldtooth, 2015).

The whole of the Burra (eel) space follows the shape of an eel, curved and with a patterned, sculptured ceiling suggesting other eels. A particularly large realistic eel sculpture with an enormous amount of rounded teeth can be touched and followed amid tables and other furniture curved as eels can. And then on the walls and floors and multiple screens are stylised eels drawn in the manner of a children's adventure book and artworks by Jasmine Miikika Craciun whose 'art-making process is informed by her diverse familial background (Barkindji, Malyangapa, Romanian, Austrian)' (Craciun, 2025). The visitor becomes a participant and is drawn to touch, feel, and even dance, following along with videoed Pasifika instructors. We are also invited to use magnifying glasses to examine life more closely or from another angle and to consider how particular Indigenous groups have learned to know and be with Country through understanding sea currents and other changing more-than-human phenomenon. The design team of Burra (eel) have described the space as (Designer Institute of New Zealand Best Design Awards, 2024):

> There is a heady mix of cultural and natural history interactives and physical activities: many of these are designed with a "moral" - for example, a tern nest highlights the bravery of terns in defending their nests from "bullies" like off-leash dogs, and a city section celebrates animals which some Sydneysiders look down on, like the ibis and the fruit bat, as street smart survivors that many First Nations people admire for the way they have adapted to the concreted-over incarnation of country which now surrounds them.

The difference between the Burra (eel) learning and play space, and the traditions of natural history museums that include single specimens in single display cases with some pedagogical text is immense. Immersion in non-linear storytelling and embodied responses to stories, sculptures, and diverse images is profoundly challenging to the taxonomical and colonial naming processes. The radical (in museum terms) of bringing together cultural stories or 'ethnography' and 'natural history' in a sustained experience showcases connections that are often hidden by natural history museum exhibitions. With its collaborative base of museum Indigenous advisers and their connections to Sydney Indigenous communities and thoughtful designers and artists, a Country centred space within a colonially created and sustained museum has emerged. In a northern American context, Jamara Wakefield, a writer and producer who creates stories about Black life, posed the question (Wakefield, 2019):

> What would a world look like where our art and culture institutions are our greatest liberation movement allies because they model de-colonial practices? Fuck with me on this one – think about it.

Wakefield's question is vast and yet acute. It reminds us of the power of all cultural institutions and how some challenge and some perpetuate colonial practices. In my positioning of Burra (eel) as one example of a de-colonial practice specifically in relation to natural history and Country I recognise that some expected arguments might be:

- it's aimed at kids
- it's a 'learning and play area'
- it's about Country, not natural history traditions.

To which I would say: Yes, yes and yes. And then… And so? Baker has already raised the ubiquity of Indigenous children's exposure to colonially informed museums that reproduce racist hierarchies and so one de-colonial act is to refuse to use the taxonomies that emerge from that tradition and re-centre Country but also the human community so key to Country. Burra (eel) also includes a big engraved possum cloak. McBride (cited in Farley-Fitzgerald, 2022) described its' creation as:

> One of the highlights of the project was a workshop we held with Aboriginal Elders and youth, who created a Possum Skin Cloak featuring stories and knowledge of Country and culture that they wanted to teach to visitors at the Museum

In this way both Elders and younger members of the community have ownership of what is shared and shown. I would also argue that the focus on 'children', the centring of 'play' and the embrace of Country can also be understood as critical interventions for all. In terms of a white, middle-class adult, more-than-middle-aged, queer, cis-identified, childfree woman (me), the space promises the unknowing involved in 'play'. That is, the opportunity to recognise that I am not necessarily in charge of what happens next or to know the rules in advance. It suggests having to at least acknowledge and possibly interact with carers and children, an often-unpredictable combination, via an appreciation of a more-than-human entity: Burra. And it promises pleasure. The possibility of learning—learning to dance, learning to touch and learning to be with diverse beings—in a shared space centred in Indigenous knowledge, not colonial histories. As McBride stated upon the opening of the space, 'the new space follows Burra on their epic migration journey with three levels of learning woven in: [W]estern science, Indigenous knowledge and a First Nations value—a moral or ethical lesson written in Dharawal, such as "give as much as you receive" and "Listen to all the voices of Country—especially those different to your own' (Pitt, 2022). The intentions of this space/learning area would seem to echo Poelina et al.'s (2024, p. 26) hope that:

> We trust the world will wake up to First Law, the Law of the Land, which shows us how to perform an ethics of care, love and response-ability with more-than-human beings.

Within *Burra (eel)* I can hear 'the voices of Country' that are very different to my own experiences of eels as frightening and/or delicious entities. Here, their mythological, practical, day-to-day and law-making powers are (at least partially) returned to them. I have a space to think and learn, and I feel awkward at my adult knowingness, which is not working as it usually does in a museum space. Here, colonially invented 'nature' is not isolated from Country in a museum that holds one of Australia's oldest 'natural history' collections. Who knows where *Burra (eel)* will take thinking about natural history collections? Who knows what ideas and experience of Country each of us who visits takes 'home'? Perhaps this is a space where '…Australian people would be able to move from 'sorry' to 'thankyou', and after a while, to 'please' as Tyson Yunkyaporta suggests may happen when Indigenous knowledges are respected as the 'most sustainable basis' for living in this country (Yunkaporta, 2019, p. 267).

What is clear is that *Burra (eel)* has shown how at least one museum with a natural history collection can also become a place where all visitors might come to appreciate the 'signals from Country' (Australian Museum, 2025a). These natural/cultural phenomena let Indigenous peoples know what is happening and remind non-Indigenous people of the complex interconnections between the human and more-than-human, that rumble through and about 'natural history' museums and collections.

3.4 Conclusion

> "The renaming of the world has never stopped." Tuhiwai Smith, L. (2021, *p. 94)*

> "See, we do not live anywhere that looks anything like this. We live on Country." (emphasis in original). (Brierty, M. & Muecke, 2023, *p. 6)*

In the novel *James* by Percival Everett (2024), which retells *The Adventures of Huckleberry Finn*

> by Mark Twain, the enslaved narrator Jim is teaching slave children how to do 'situational translations' to avoid annoying white folk and so avoid violence of all orders
> (p. 22):You're walking down the street and you see that Mrs Holiday's kitchen is on fire. She's standing in her yard, her back to the house, unaware. How do you tell her?
> "Fire, fire," January said.
> "Direct. And that's almost correct," I said.
> The youngest of them, lean and tall five-year-old Rachel, said, "Lawdy, missum! Looky dere."
> "Perfect," I said. "Why is that correct?"
> Lizzie raised her hand. "Because we must let the whites be the ones who name the trouble."
> "And why is that?" I asked.
> February said, "Because they need to know everything before us. Because they need to name everything."

In bringing together the colonial critique of Tuhiwai Smith and Brierty's response to colonially invented maps with the novelistic imaginings of Everett, I am trying to gesture to the continuing work that the naming and tradition of exhibitions of natural history collections do. When every bird, bee, tree, wind and cloud was renamed—into English, into Latin, into a system of knowing and then taught in schools, universities and museums, it 'magically' appeared as if non-Indigenous peoples knew not only the 'correct' name but knew it before Indigenous peoples (Brierty & Muecke 2023; Haraway, 1988). Indigenous peoples were not held apart from this system of naming but forced into the systems and complexes that such naming of all life, including the mapping of all natural formations, produced. However, as Everett suggests and Brierty shows, naming does not remove the consciousness of community or Country. 'Nature' continues to be naturalised even when there is an awareness that using the namings and traditions of 'natural history' can silence Indigenous histories, knowledges and solutions to ecological problems. And as Marcia Langton in her Boyer Lectures shows, beyond the museum setting, there is a problematic association of Indigenous knowledges not with Country but the idea of 'wilderness'. In her introduction, she writes: 'I also aimed to correct or raise some doubts about the veracity of the 'wilderness' campaigners' stance towards Aboriginal people, which relegates us to the status of the 'new noble savage' and enemies of their grand struggle to save nature' (Langton, 2012, p. 137). That 'nature' emerges from particular colonial practices, including establishing natural history collections.

In 2015, Jodi Dean (and other North American academics and activists) began a project called The Natural History Museum conceived as an 'ongoing performance' that adopts 'the legitimating aesthetics, pedagogical models, and presentation forms of natural history museums in support of a divisive perspective on science, nature and capitalism' (Dean, 2016, p. 76). They challenge 'climate denialism, obfuscation, evasion and greenwashing' and have helped remove fossil fuel money from museums (Dean, 2016, p. 92). Dean's (2016) stated perspective of looking at climate from 'the side' (p. 75) may be what understanding 'nature' as a stolen and renamed Country might do. However, their approach does not seem to have First Nations input and seems to accept the European idea of 'nature' with a firm focus on climate change. This means that some of their 'performances', such as teaching participants to 'take the view of museum anthropologists who are attuned to the social and political forces shaping nature' appear to be an uneasy continuity with colonial thinking (Dean, 2016, p. 76). They are not promising a re-enlivening but interventions into how climate change can be imagined within a 'natural history' frame.

This is unlike Araluen in her fiction, which I have understood to be a radical re-imagining of what a natural history collection would become when brought back to life through a re-vivification of every 'specimen's' ongoing connection with Country (Schlunke, 2024). Meanwhile, Cope has imagined and facilitated an artwork that has literally 'brought back life' and enriched Country, including the repair of ocean ecology. And at the Australian Museum Marsh and Kian-Judge et al. have created a 'learning and play' space where Country is centred in a museum where it is only intermittently acknowledged. All three examples work to undo the disconnecting power of naming that produces items that denies the complex whole of Country. That these three examples come from different 'disciplines'; writing, art and museum education show the breadth of responses to how Country can be re-centred. The work of understanding how Country became and continues to be 'nature-ised' requires a multi-disciplinary approach. It speaks to what a profound re-organisation of white institutional thinking about 'nature' would produce across the Country(s) called Australia. It has already created in these three examples an imagining and understanding of 'nature' and of museum 'specimens' as always within Indigenous worlds and a force to be reckoned with. Together, they offer a suite of ways to imagine how non-Indigenous Australia could re-imagine colonial 'nature' to create potentially very different futures.

Acknowledgements I would like to acknowledge the many people who have made this article possible. Julie Gough (Curator, Indigenous Cultures), artist and writer first drew my attention to the work of Megan Cope and added invaluable context to my thinking about 'nature, Anja Schwarz (Prof. Dr) alerted me to the existence of Burra (eel) when it first opened and saw its importance immediately. Very warm thanks to Sara Kian-Judge (Assistant Content Producer, First Nations) and Courtney Marsh (Manager of Strategic Projects and Curatorial, First Nations) who contextualised the making of Burra (eel) for me and were willing to share some of their experiences as producers which assisted my thinking enormously. Stephen Muecke was a valuable big picture reader. I would also like to thank the key insights of my two anonymous reviewers and the intellectual and editorial input of the two patient and elegant editors of this collection Melissa Kennedy and Erin O'Donnell. The chapter is far better as a result.

Thanks to the Australian Research Council and the ARC Discovery scheme for assisting this research DP200101877.

References

ABC Arts. (2022). *Megan Cope's regenerate artworks in Minjerribah.* Retrieved December 9, 2024, from https://www.youtube.com/watch?v=MDLi3CIUJII (0:48 of transcript—art do that)

Araluen, E. (2021). 'Learning Bundjalung on Tharawal' in Drop Bear (p. 8). University of Queensland Press, St Lucia.

Araluen, E. (2022). Muyum: This all come back now. In M. Saunders (Ed.), *This all come back now: An anthology of First Nations speculative fiction* (pp. 22–29). University of Queensland Press.

Ashby, J. (2021). The political platypus and colonial koala: Decolonising the way we talk about Australian animals. *Journal of Natural Science Collections, 9*, 35–45. http://www.natsca.org/article/2684

Ashby, J., & Machin, R. (2021). Legacies of colonial violence in natural history collections. *Journal of Natural Science Collections, 8*, 44–54. https://www.natsca.org/article/2631

Australian Museum. (2025a). *Burra.* Retrieved December 5, 2024, from https://Australian.Museum/Exhibition/Burra/

Australian Museum. (2025b). *Listening to country.* Retrieved December 5, 2024, from https://australian.museum/learn/first-nations/burra/listening-to-country/

Bignall, S., Hemming, S., & Rigney, D. (2016). Three ecosophies for the Anthropocene: Environmental governance, continental posthumanism and indigenous expressivism. *Deleuze Studies, 10*(4), 455–478. https://doi.org/10.3366/dls.2016.0239

Brierty, M., & Muecke, S. (2023, July 7). Whitefella Mischief: A tour of the museum of the magicians of reason. *Overland*, 6. https://overland.org.au/2023/07/whitefella-mischief-a-tour-of-the-museum-of-the-magicians-of-reason/

Calma, T. (2008, July 7). *Essentials for social justice: Land and culture—Economic development.* Archived Speech. https://humanrights.gov.au/about/news/speeches/essentials-social-justice-land-and-culture-economic-development

Chen, D. (2024). Commissioned by country. *Art Monthly Australia, 337*, 58–70. https://sydney.primo.exlibrisgroup.com/permalink/61USYD_INST/2rsddf/cdi_proquest_journals_2898101344

Cope, M. (2023). *A monumental public art work/Sydney Opera House 50th anniversary*. Retrieved December 12, 2024, from https://www.megancope.com.au/public-art/amonumentalpublicartwork

Cope, M. (2024). Kinyingarra Guwinyanba. Retrieved January 2, 2025, from https://www.megancope.com.au/works/kinyingarra-guwinyanba

Craciun, J. (2025). *About Jasmine Miikika Craciun*. Retrieved January 2, 2025, from https://www.jasminecraciun.com/about

Das, S., & Lowe, M. (2018). Nature read in black and white: De-colonial approaches to interpreting natural history collections. *Journal of Natural Science Collections, 6*, 4–14. https://natsca.org/article/2509

Dean, J. (2016). A view from the side: The natural history museum. *Cultural Critique, 94*(Fall), 74–101. https://doi.org/10.5749/culturalcritique.94.2016.0074

Designer Institute of New Zealand Best Design Awards. (2024). *Burra learning space*. Retrieved January 1, 2025, from https://bestawards.co.nz/spatial/exhibition-temporary-structures/story-inc-studio-cassells-ltd-australian/burra-learning-space/

DOBBY {feat. Barkaa}. (2020). *I can't breathe*. Single, written by Quayle and Clapham. Lyrics. Retrieved December 2, 2024, from https://genius.com/Dobby-i-cant-breathe-lyrics

Dow, S. (2023, July 22). Indigenous art unites Australians in a common cause: Abuse of the ocean. *Guardian*. https://link.gale.com/apps/doc/A758129209/AONE?u=usyd&sid=bookmark-AONE&xid=699710d4

Everett, P. (2024). *James* (p. 22). Doubleday.

Explore. (2021, Winter). *News*, Explore, *Australian Museum Magazine*. p. 11.

Farley-Fitzgerald, C. (2022, July 1). Australian Museum opens Burra: A ground-breaking, interactive learning and play space. *Australian Museum*. https://australian.museum/about/organisation/media-centre/burra-opening/#:~:text=

Gadigal to the Core. (2022, February). Information Panel, Burra Educational Space. *Australian Museum*, Sydney. Observed February 2025.

Gibson, R. (1992). *South of the West: Postcolonialism and the narrative construction of Australia*. Indiana University Press.

Goldtooth, T. (2015). Indigenous peoples cosmovision, conflicts of conquest and need for humanity to come back to mother earth. In S. Biggs, O. O. Lake, & T. B. K. Goldtooth (Eds.), *Rights of nature & mother earth: Rights-based law for systemic change* (p. 15). https://www.ienearth.org/wp-content/uploads/2017/11/RONME-RightsBasedLaw-final-1.pdf

Gray, J., & Fraser, I. (2013). *Australian bird names: A complete guide* (1st ed.). CSIRO Publishing.

Gumillya Baker, A., Ulalka Tur, S., Rosas Blanch, F., & Harkin, N. (2020). Repatriating love to our ancestors. In C. Fforde, C. T. McKeown, & H. Keeler (Eds.), *The Routledge companion to indigenous repatriation: Return, reconcile, renew* (pp. 854–873). Taylor and Francis.

Haraway, D. (1988). Situated knowledges: The Science question in feminism and the privilege of partial perspective. *Feminist Studies, 14*, 575–599.

Kian-Judge, S. (2022). *Who is country? Why I capitalise animals, rivers, coolamons and storms*. Retrieved January 2, 2025, from https://australian.museum/learn/first-nations/burra/who-is-country/

Kohen, J. L. (2005). Pemulwuy (1750–1802). *Australian dictionary of biography*. National Centre of Biography: Australian National University. Retrieved December 16, 2024, from https://adb.anu.edu.au/biography/pemulwuy-13147/text23797 (published first in hardcopy 2005).

Langton, M. (2012). *The Quiet Revolution: Indigenous people and the resources boom*. Boyer Lectures 2012. ABC/Harper and Collins, Kindle Version Introduction, Loc 137.

Leane, J. (2018). *Cardboard incarcerations* (p. 3). Walk Back Over. Cordite.

Many Ways of Knowing. (2022). Information panel, Burra educational space, *Australian Museum*, Sydney. Observed February 2025.

McBride, L., & Smith, M. (2021). *The Sydney Wars* (pp. 66–69). Unsettled. Australian Museum Trust.

Moyal, A. (2010). *Platypus* (New Edition). Allen & Unwin.

Murawski, M. (2019). 'Museums could be powerful, liberatory spaces': Interrupting white dominant culture in museums. *Singer Island*, Newstex. https://www.proquest.com/blogs-podcasts-websites/museums-could-be-powerful-liberatory-spaces/docview/2248400334/se-2

Näf-Gloor, S., Lengwiler, M., Penn, N., & Harries, P. (2019). A naturalist's career: Hinrich Lichtenstein (1780–1857). In M. Lengwiler, N. Penn, & P. Harries (Eds.), *Science, Africa and Europe: Processing information and creating knowledge* (pp. 47–65). Routledge.

"Naturalize". (2024, December). *Oxford English Dictionary*. Oxford University Press. https://doi.org/10.1093/OED/9850995989

"Parrot". (2024, December). *Oxford English Dictionary*, s.v. "parrot (*n.1*). Etymology. https://doi.org/10.1093/OED/5475432636

Pitt, H. (2022, July 1). The mysterious migration of eels unmasked by the Australian Museum. *Sydney Morning Herald*. https://www.smh.com.au/culture/art-and-design/the-mysterious-migration-of-eels-unmasked-by-the-australian-museum-20220701-p5ayb9.html

Poelina, A., Webb, B., Wooltorton, S., & Godden, N. J. (2024). Waking up the snake: Ancient wisdom for regeneration. In M. Lobo, E. Mayes, & L. Bedford (Eds.), *Planetary justice: Stories and studies of action, resistance and solidarity*. Bristol University Press. https://www.jstor.org/stable/jj.10677905.8

Roughley, T. C. (1933). The life history of the Australian oyster (*Ostrea commercialis*). *Proceedings of the Linnean Society of New South Wales, 58*, 279–333. https://www.biodiversitylibrary.org/part/47809

Schlunke, K. (2024). The practices of care: Extinction and de-colonisation in the natural history museum. *Continuum: Journal of Media & Cultural Studies*. https://doi.org/10.1080/10304312.2024.2384951

Schlunke, K., & Schwarz, A. (2025). The First Beutelwolf: How Berliners were taught to see the thylacine. *Journal of Natural Science Collections*, (13), 3–15. (forthcoming).

Shaw, A. G. L. (1967). King, Philip Gidley (1758–1808). *Australian dictionary of biography*. National Centre of Biography, Australian National University. Retrieved December 16, 2024, from https://adb.anu.edu.au/biography/king-philip-gidley-2309/text2991 (Published first in hardcopy 1967).

Smyth, A. B. (1788, January 28). *A Journal of a Voyage from Portsmouth to New South Wales and China*. E-Book was produced by Colin Choat. https://gutenberg.net.au/ebooks20/2000051h.html

Tovey, J. (2010, April 1). Off with his head: Prince pledges to return warrior's skull. *Sydney Morning Herald*. https://www.smh.com.au/national/off-with-his-head-prince-pledges-to-return-warriors-skull-20100331-rewp.html

Tuhiwai Smith, L. (2021). *Decolonizing methodologies: Research and indigenous peoples* (pp. 91–107). Zed Books. https://doi.org/10.5040/9781350225282.0009

Vincent Smith, K. (2005). Moowattin, Daniel (1791–1816). *Australian dictionary of biography*. National Centre of Biography, Australian National University. Retrieved January 2, 2025, from https://adb.anu.edu.au/biography/moowattin-daniel-13107/text23713 (First published in hardcopy 2005).

Vincent Smith, K. (2010). Pemulwuy. *Dictionary of Sydney*. Retrieved January 2, 2025, from https://dictionaryofsydney.org/entry/pemulwuy

Wakefield, J. (2019, May 14). Museums could be powerful, liberatory spaces if they let go of their colonial practices *Racebaitr*. Retrieved January 2, 2025, from https://racebaitr.com/2019/05/14/museums-could-be-powerful-liberatory-spaces-if-they-let-go-of-their-colonial-practices/

Yunkaporta, T. (2019). *Sand talk: How indigenous thinking can save the world*. Text Publishing.

Chapter 4
Waka Memory: Ancestral Connections

Jade Hadfield

Abstract In this chapter, I explore the contrast in visibility and experiences of Oceania peoples to the colonial narratives, perceptions, possessions, and the consequent formation of racial hierarchies in the arts sector—galleries, libraries, museums and archives. These institutions were historically created to uphold white supremacy and remain complicit in ongoing colonialism in Aotearoa, Australia, and Oceanic Nations. This chapter examines efforts to decolonise the arts sector, offering a brief history of Aotearoa and Australian Indigenous involvement in these institutions, and highlights the failure to value Indigenous-to-Indigenous relationships. Museums function as tools of power in the colonial project, reinforcing dominance over social and political power. In my work, I seek to break down colonial gatekeeping of knowledge and art, and decentre the presence of European invasion through waka memory. Waka memory is a collaborative approach to exhibition development using wayfinding methodologies and practices of talanoa, yarning or wānanga as critical engagement, centring Indigenous ways of being, knowing and doing. I proposed an exhibition entitled Waka Memory to create unification between Moana and Aboriginal peoples, recentring relationships and place, and shifting museum culture. This chapter explores the benefits of waka memory in reawakening ancestral connections and challenging knowledge hierarchies in art institutions.

Keywords Museums · Colonialism · Wayfinding methodology · Indigenous Knowledge · Ancestral connections

---

J. Hadfield (✉)
Melbourne Museum, Carlton, VIC, Australia

M. Kennedy, E. O'Donnell (eds.), *People, Place and Nature in Indigenous-Settler Relations*, Indigenous-Settler Relations in Australia and the World 10, https://doi.org/10.1007/978-3-032-03600-1_4

4.1 Grounding

I write these words as a wahine toa from the hapū Ngāti Paarau, situated on the lands of the Bunurong and Wurundjeri Woi-wurrung. It is a complex position many Māori and Moana people find themselves in, as visitors on unceded Aboriginal lands. On this whenua (land), I often prioritise my diasporic Moana identity over my Māori identity, sometimes out of ease, sometimes out of honouring the wider Moana (Enari & Haua, 2021). In my work, I seek to reawaken kinship links with the people of this land and the people of the Moana who have, and are still, navigating the complexities of colonialism, and how to unite against colonial structures. As Māori and Moana visitors here, we must understand our place on unceded Aboriginal land.

Like many Indigenous people, I walk two worlds and wear many hats. In one world, I descend from Rangatira. I am a wahine toa with a husband, two children and a very large, connected whanau. I am actively involved with my hapū, despite living in Australia. In the other world I am a cultural sector worker, a conservator and curator, with Eurocentric training in a predominantly and historically white sector. Collecting institutions were never made for Indigenous people, it is impossible to truly Indigenise an institution when their foundations were intentionally created on Indigenous erasure (Lee, 2022; see also Fig. 4.1). However, as long as these institutions hold our taonga, ancestral items, we need to be there to care for them to make decisions for and about them.

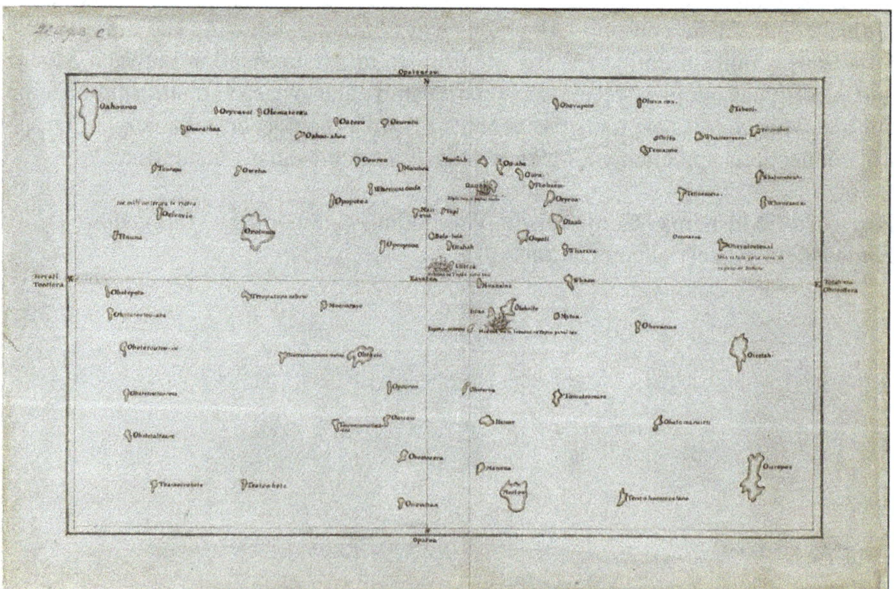

Fig. 4.1 Chart of the Society Islands, with Otaheite [Tahiti] in the centre, 1769, © The Trustees of the British Museum

For several reasons, including the challenges of working within colonial spaces, I have been unable to bring my proposed exhibition into being. I offer this chapter as insight into the possibilities that such an exhibit would create, and the power of waka memory as a way of connecting the art of Moana and Aboriginal peoples.

4.1.1 Waka Memory

Art is integral to Indigenous culture, intrinsically linked to our knowing and being and, arguably, has a different emphasis to the Western definition of art, which tends to emphasise the aesthetic. Across Oceania, art has been described as a 'central thread of culture' (Chitham et al., 2019, p. 23) or an intergenerational transmission of culture. In te ao Māori, Tā Hirini Moko Mead (1996) describes art as 'a part of the heritage of the Māori people that was handed to them from generations of ancestors. Māori art brings together all those forms of artistic expression that follow the styles, traditions, and values of Māori culture.' Aboriginal people have a similar concept of art, with AIATSIS (2020) stating that 'art is an expression of people's identity, culture, spirituality and relationships to Country. It tells stories of ceremony and creation and connects people to ancestors and kin.' Art is a powerful vehicle of transformation, as a vehicle for expression and culture with many benefits equating to 'pride, self-esteem, maintenance of culture, transmission of culture and inter-generational learning' (Commonwealth of Australia, 2007, p. 15). Art is a fundamental aspect of cultural expression, it is more than an aesthetic pursuit, but rather intricately intertwined with our understanding of self. Through the embedded nature and practice of art, it can be a transformative tool to communicate our knowledge and decolonise cultural institutions.

Alice Te Punga Somerville's (2012) book, *Once were Pacific: Māori connections to Oceania*, serves as inspiration for my proposed exhibition *Waka memory: ancestral connections*. Te Punga Somerville powerfully demonstrates the connection between Māori and their whanaunga across the ocean using an illustration by Tupaia, depicting a 'Māori chief' and Joseph Banks trading goods (Fig. 4.2). At first glance, many assume the item that Banks is trading is a white cloth; and from our western learnings in art history, the general assumption is that this was white cotton from Britain. However, Te Punga Somerville argues that the item is not in fact cotton, but rather it is ahu, a barkcloth from the Society Islands, acquired by Cook and his crew during their time in the Society Islands. The British perspective, always at the centre, had been the determining standpoint when viewing this work. However, with Te Punga Somerville's perceptions, the western lens has been shifted to the periphery.

In this moment, we have Tupaia the observer, the 'Māori chief' and the ahu, representing the ancestral Moana connections that Te Punga Somerville draws on through Robert Sullivan's concept of 'waka memory' (Sullivan, 2012 as cited in Te Punga Somerville, 2012). Te Punga Somerville uses the illustration to recentre the relation of Māori to Oceania. The book serves as inspiration for an exhibition which

Fig. 4.2 Māori trading a crayfish with Joseph Banks, Tupaia, 1769, © The Trustees of the British Museum

Fig. 4.3 Australian Aboriginal People in Bark Canoes Tupaia, 1769, © The Trustees of the British Museum

reconnects Indigenous people across land and waters (see e.g. Fig. 4.3). It is an opportunity to recentre Indigenous as self-determined people within the colonial project, and simultaneously decentre the settler colonial narrative.

4.1.2 The White Possessive

The settler's contemporary existence on colonised lands is fundamentally predicated upon the structures and mechanisms of the colonial system (Fanon, 2001). These colonised lands are rife with what Aileen Moreton-Robertson (2015) describes as the white possessive, 'by which whiteness operates possessively to define and construct itself as the pinnacle of its own racial hierarchy' (Moreton-Robinson, 2015). This can be seen with the building of the colony through statues of founding fathers, grand neoclassical buildings, and great exhibitions all which were used to create a hierarchy of whiteness, while collecting institutions like museums and libraries were built as a central mechanism to control knowledge (Lee, 2022). Colonisers in Oceania established institutions that advanced a possessive investment in whiteness. Collecting institutions were central to the propaganda responsible for upholding institutions founded on looting, theft and war

(Shringarpure, 2022 as cited in Lee, 2022) to elevate the status of colonialism and house and maintain the ideas of white supremacy.

Western colonisation and diaspora has disconnected Indigenous peoples of the Ocean from their greater Moana roots. Imperialism's technique of divide and rule has been successful in creating the barriers between us, creating a new siloed culture within the current western borders. The people within each border assert their Indigenous sovereignty and identity but are often separated through division tactics, separatist policies, setting Indigenous people against each other through competition of work and space (Te Punga Somerville, 2012). Intertwined with collecting institutions is identity and culture making; Nicholas (2022) states 'art and arts narratives had been vital to the making of colonialism … art thus helped fashion the very identities of colonists' (p. 14). These institutions were founded on the narratives that uphold colonial superiority as seen in the many paintings of tamed pastoral landscapes or scenes that depicting settler colony progress through built structures. All of these scenes are depicted with obvious absences of Indigenous people.

Museums can also be seen as a tool of the state to maintain their dominion over Indigenous bodies, and collect Indigenous cultural heritage under the imperialist idea of salvage anthropology conceived by Franz Boas (Lee, 2022). Salvage anthropology was premised on the notion that Indigenous cultures were doomed to disappear (Lee, 2022, p. 41). After the impacts of Indigenous genocide there was a drive to collect ancestral items, with the purpose of preserving dying cultures. Displaying colonial plunder and racial relics through major exhibitions was popular in the nineteenth century. This was during a time when Indigenous people were often live exhibits, curated by settlers as commodities and examples of exoticism (Scott & Laurie, 2007). My proposed exhibition was driven by the need to right the wrongs of these histories, and to reconnect people and looted taonga.

4.1.3 Casting the Colonial Lens to the Periphery

Institutions are complicit in upholding the harmful white narratives. Lopesi (2018) encourages us to look beyond the 'False Divides' that are served to separate the Moana Peoples, asking 'how do we get to know each other again?' (p. 21). I see this prompt from Lopesi as a call to waka memory as Lopesi looks at the Moana as a connector and ancestor, connecting the Pacific and extending to Taiwan. Christian Lugnan (2023) points out that we all connect through Country, 'Although we are separate people(s), Māori and Goori people (Aboriginal and Torres Strait Islander people), sea Country connects us all and keeps us grounded and together. It is our way of understanding each other.' Through our relationship to sea, land and sky, we connect. It is through this relationship we can reawaken our ancestral connection to global Indigenous unity. *Waka Memory* incorporates coming kanohi ki te kanohi, face to face with one another to remove the arbitrator deliberately and destructively enforcing this false divide.

Chang (2016) argues how the concepts of equality and diversity have been exploited and rendered meaningless, used as a corporate marketing tool as well as an evasive manoeuvre against more radical efforts at mitigating inequality (Chang, 2016). In this regard decolonisation is a preservation of whiteness (The Racial Imaginary Institute, 2022). The violence that fortified the power of museums was not simply attempted erasure of cultural memory and heritage. It is a specific settler colonial violence that is ongoing and underway (Shringarpure, 2022 as cited in Lee, 2022, p. 5). Enduring colonialism permeates throughout institutions, which lack the cognizance to address necessary structural reforms.

The pressing need for decolonisation within the institution therefore intersects with the need for structural reform and cultural representation. In Australian institutions, there is recognition from advocates in the arts space that reform needs to happen at all levels: "[i]t is not just hiring a bunch of non-white, non-Euro staff. It is interrogating your board, the way you run the marketing, the hierarchy itself" (Zhang as cited in Fairley, 2020). A change is needed at all levels of cultural institutions, addressing the inequity within the hierarchy of the institution, interrogating who the decision makers are and who are telling the stories. At a national level, the development of *Revive: A Place for Every Story, A Story for Every Place* (Department of Infrastructure, 2023) provides a new national arts policy that prioritises First Nations and diverse voices. *Revive* will be a critical tool in addressing inequities and resourcing decolonial initiatives.

In Australia and Aotearoa there is a move towards the idea of Indigenising the institution, or Indigenous Museum practices. However, with the uptake of Indigenisation as a concept along with a rising number of literature by non-Indigenous people, there is a risk of Indigenisation becoming a metaphor, much like how decolonisation has been subsumed by colonial frameworks (Tuck & Yang, 2012). In Aotearoa, the Director Audience Insights at Te Papa, Puawai Cairns (Ngāti Pūkenga, Ngāti Ranginui, Ngāi Te Rangi) is a vocal advocate for what she calls reMāorification, using decolonisation and Indigenisation as tools towards mana Motuhake. Cairns' blue-sky thinking is where we can all aim to be, utilising decolonisation and Indigenisation for mana motuhake or self-determination.

Within the State Library Victoria (SLV) there are some processes in place that serve Aboriginal people, established by the SLV's Victorian Indigenous Research Centre (VIRC). SLV follows the Aboriginal and Torres Strait Islander Library, Information and Resource Network protocols, protocols created to guide safe collection practices for Aboriginal people and collections. SLV also has the VIRC cultural permission program (National and State Libraries Australasia, n.d.), which aims to get permission from the people of or descendants of any Aboriginal people that are in requested photographs.

Over the past few decades, the slow pace of decolonisation has led to varied levels of Indigenous involvement in cultural institutions, from collaboration to telling our own stories to Indigenous decision making. In Aotearoa, Conal McCarthy (2011) traces the recent history of Māori in museums from decolonising politics in Aotearoa in the 1970's, to Te Māori exhibition in 1984, to the road to biculturalism. The significant transformative step from collaboration to Māori led and embedded, was the Te Māori exhibition in 1984. Mina McKenzie in McCarthy (McCarthy, 2011) thought the process of Te Māori was more important than the result, 'while the Americans and Douglas [Newton] were looking at Te Māori for the excellence of its sculptural form…Māori people were being reunited with their ancestors' (p. 69). Te Māori was one of the best-known examples internationally of community collaboration (McCarthy, 2011), from the top down it was led by Māori and included nationwide consultation with iwi.

Similarly in Australia, community engagement within museums has been driven by Aboriginal groups, with the impetus being the 1978 Adelaide UNESCO seminar on the role of museums in preserving Indigenous cultures, where Indigenous people had a say in the role of museums particularly for the care and repatriation of their cultural items (Ambrose, 1978). In museums in the 1990's (Australian Museum, 2020) there was a push for Aboriginal staff within institutions for the establishment of the Aboriginal Heritage Unit at the Australian Museum. Much like Aotearoa, government policy and cultural revival was the instigator for museum collaboration, with funding from the Aboriginal and Torres Strait Islander Commission (ATSIC) to enable museum outreach and collection access for Aboriginal communities (Australian Museum, 2022). Australian Government policy followed suit with the release of Creative Australia in 2013 (Cordes, 2015) identifying community engagement one of their major goals. The 2018, *First Peoples: a roadmap for enhancing Indigenous engagement in museums and galleries Indigenous roadmap* (Janke & Janke, 2018), is a ten-year road map committed to improving Indigenous engagement and employment in the Australian museum sector, resulting in an increase in identified Aboriginal positions within the cultural sector (Eshrāghi, 2021).

Globally, however, there is still an apparent absence of exhibitions that promote cross Indigenous unification, with European galleries often holding Indigenous shows. The British Museum, like many European institutions, place their looted collections thematically by region, consistent with their practices of displaying as an encyclopaedic museum. Rather than an effort to create meaningful global Indigenous exhibitions, their motivation for placing geographically and culturally diverse collections together also reflects a practice where Indigenous collections are the lowest priority with the least attention and resources (Hudson, 2017). Instead of a transformative exhibition, the presentations reflect tokenism rather than a considered curatorial process, also evident in the lack of Indigenous input. *Waka Memory* is an opportunity to enrich and connect community voice to collections that are held in Europe, not only Australia and Aotearoa.

4.1.4 Reclaiming the Narrative

Recentering Indigenous narratives represents an opportunity to reclaim cultural identity to counter the destructive influence of imperialism. By shifting the focus towards our place in the world and our relationships to each other, we challenge the dominant colonial narrative. In developing the proposal for the *Waka Memory* exhibition, I intended to use this exhibition as a way to recenter Indigeneity, as an opportunity for us to unlearn colonial teachings, remember and relearn our practices, and educate the population of our place here – who we are, where we are, values and what we create (see e.g. Fig. 4.4). I intended to create an exhibition that would change cultural institutional structures.

As a proposed exhibit, *Waka Memory* does not adhere to irrelevant and outdated Eurocentric modes of thought and assessment that uphold racist Eurocentric hierarchies and views of Indigenous people and practices (Rameka, 2021). Makere Stewart-Harawira (2005, p. xiv) explains the relevance of Indigenous knowledge, arguing that 'traditional Indigenous social, political and cosmological ontologies are profoundly important to the development of transformative alternative frameworks for global order and new ways of being'. Therefore, through Indigenous cosmological relationships and arts practice, I will centre Indigeneity, making Indigenous connections and knowledge relevant (Fig. 4.5).

Fig. 4.4 Dancer and costume of the Chief Mourner, Tupaia, 1769, © The Trustees of the British Museum

Fig. 4.5 A group of musicians from the Society Islands, 1769, © The Trustees of the British Museum

4.2 Wayfinding Methodology

He waka eke noa
We are all in this together
- Whakataukī (Mead & Grove, 2001)

Wayfinding is the process of orienting and traveling from place to place; wayfinding methodologies utilise the skills, behaviours, and ontologies of traditional of waka navigators or wayfinders to advance Indigenous leadership, decolonisation, and Indigenisation (Spiller et al., 2015). The wayfinder leads the waka into the unknown, drawing on multiple sources of knowledge to ensure a safe journey across the Moana until landfall. They have deep understanding of themselves, the crew, the waka and the environment. Wayfinding is intuitive and adaptable to work with the needs of many. Nainoa Thompson (n.d.) explains: 'Hawaiians call it na'au – your instincts, your feelings, rather than your mind, your intellect.' The wayfinder draws on all their senses and intuition to feel their way through the ocean, trusting the knowing in their puku, tapping into spiritual guidance by drawing upon ancestral wisdom, sensory perception, and logic for decision making. In addition to this, wayfinders are actively self-aware and can continually refresh their thinking. Spiller et al. (2015) describe this state as constant shedding, by challenging their own mindset, and leaving themselves open to innovative decision making.

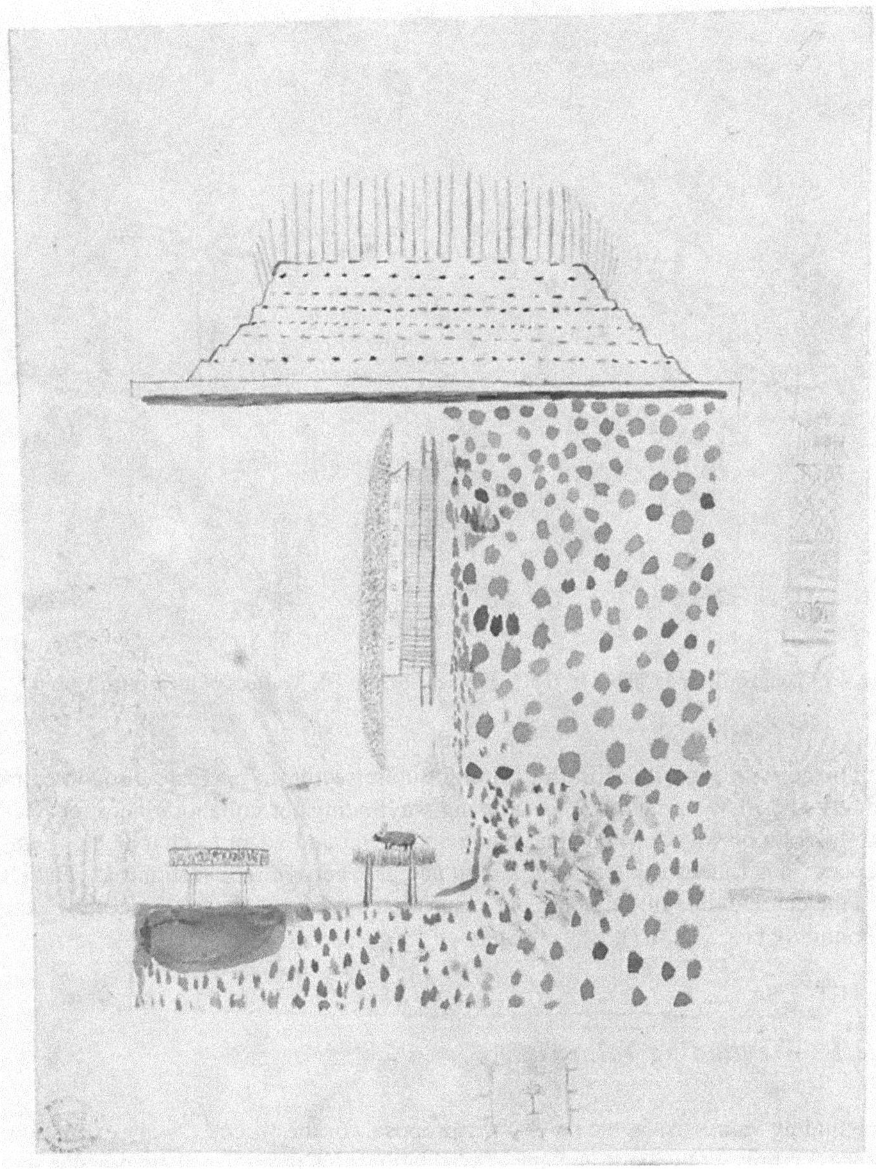

Fig. 4.6 Diagrammatic drawing of a marae: side elevation, 1769, © The Trustees of the British Museum

When working with communities as diverse as Australian and Moana peoples within a colonial institution, different knowledges must be drawn upon to navigate the unknown. The wayfinder draws on different knowledges to read the environment yet stands firm in their value system. I draw on my lived experience as a child of the marae (see Fig. 4.6) with observations of the paepae and community politics,

Fig. 4.7 Longhouse and Canoes in Tahiti, Tupaia, 1769, © The Trustees of the British Museum

combined with experience in the cultural public sector, to navigate through community and SLV bureaucracy. In utilising wayfinding for collaboration, it enables one to draw on ways of knowing, being and doing and adapt to changing circumstances. This framework creates a way to navigate between the institution, yourself and the community, guiding both but ensuring the safe journey for your 'crew', the community (see Fig. 4.7).

4.2.1 Wayfinding Values

Wayfinding values are at the core of my proposal for the *Waka Memory* exhibition, by wayfinding connections and ancestral relationships through artistic practice and storytelling. *Waka Memory* focuses on shared experiences, emphasising the significance of our connection to whenua, Country and cultural practices, reinforcing the interwoven fabric of our shared connections. They are humārietanga, rangatiratanga, manaakitanga, whanaungatanga, kaitiakitanga, wairuatanga and aroha:

- Humārietanga is humility, a highly regarded personal characteristic for Rangatira. By allowing your actions to speak for you and placing your ego aside enhances your mana. In Barclay Kerr's (Spiller et al., 2015, p. 59) words, 'let the deeds speak. And if what you have done has made a difference it will speak louder than anything you have to say'.

- Rangatiratanga pertains to inspirational leadership, weaving people together to work cohesively (Spiller et al., 2015).
- Manaakitanga is nurturing relationships, looking after people and being very careful about how others are treated. Mead (2003, p. 36) states that all tikanga are underpinned by the high value placed upon manaakitanga.
- Whanaungatanga is a fundamental principle that embraces whakapapa and focuses upon relationships (Mead, 2003).
- Kaitiakitanga is a popular term used in environmental care or within collections as guardian. Broadly speaking, it is our role of stewardship. It is interconnectedness, our kinship and maintenance with people and the environment (Spiller et al., 2015).
- Wairuatanga is the awareness and navigating the spiritual and physical realms.
- Aroha, much like and linked to manaakitanga, is often interpreted as love, however in many examples of translation, it means much more than that. Tai as cited in Spiller (2015) defines aroha as aro – the life principle, thought; ro – introspection, within; ha – life force, breath. A mindful concept connected to our lifeforce, the breath.

Although these principles are drawn from Māori culture, I offer them as a bridge to connect across Oceania. In doing so, the *Waka Memory* exhibition would create an opportunity to displace the colonial gaze and braid Indigenous knowledges.

4.2.2 Retaking the platform

Museums have been used as tools of power in the colonial project, in that they are effective at maintaining their dominant cultural power (Lee, 2022). In this project, I aim to use this tool of power to elevate Indigenous art. SLV has over 2 million visitors a year. It is a well-resourced institution, being one of the most visited libraries in the world and the most visited library in Australia (Temple, 2018). It is a significant platform that allows for impact exhibitions. This provides a great opportunity to educate new audiences, broadening the supporter base and generating advocacy for the issues being explored in the exhibition.

Exhibitions are powerful tools of education; they can be life changing and immersive. By using Tupaia (Fig. 4.7) as the entry point, instead of Cook's 'discovery', it is a way of departing from the imperialist narrative and focusing on the people of Oceania, our practices, our ontologies and cultural connections. By being clear about our message and the intended actions, being brave, hopeful, enacting collective thinking, positivity and being authentic, the exhibition can be a tool for transformation (Morris Hargreaves McIntyre, 2021). We can use the institution as a platform to build capacity for our people, elevate Indigenous art and challenge governance processes and create an equitable cultural sector.

When I developed the *Waka Memory* proposal, I was an employee of SLV. Through *Waka Memory*, I intended to explore how we could do things differently in the library, challenging the narrow focus of collecting institutions, pushing the boundaries of its entrenched systems. I hoped to influence other SLV teams to work in an inclusive way that prioritises Indigenous ontologies and epistemologies.

As a cultural sector professional that has in-depth knowledge of how institutions work, my role was the buffer between the community, project team and the institution, much like the Mohawk Council of Akwesasne, with an 'appearance of compliance' (Brimley et al., 2007). Decision-making for *Waka Memory* would have been undertaken through formal and informal talanoa or wānanga, with the project being informed in a collective nature. Internal policy would be developed that enabled appropriate governance processes for community engagement, cultural safety and manaakitanga [care and hospitality]. Indigenous methodology and methods would be practiced by the project team within the institution, and we would be led by wayfinding values that are relevant to cross-cultural Indigenous Peoples. We would have utilised holistic Indigenous theories and frameworks such as kaupapa Māori, whakapapa, talanoa or yarning, which can bring our stories to the forefront, and demonstrate how our cultures are dynamic, vibrant, living and evolving.

4.3 Conclusion

Through the Indigenous methods and recentering Indigeneity, *Waka Memory* holds the potential to bring about a profound shift in perspective challenging western centric historical narratives within cultural institutions. Although I have been unable to hold the exhibition in partnership with SLV, I hope to see *Waka Memory*, and the underpinning values and methodologies, be brought to life in the future.

The increasing involvement of Indigenous peoples within institutions reflects development in Indigenous decision making, co-design and telling our own stories. Initiatives like *Revive: A Place for Every Story, A Story for Every Place* underline the importance of Indigenous-led practices and policies, advocating for an overhaul of institutional structures. My proposed exhibition identifies a gap in Indigenous-to-Indigenous relationships in exhibitions and reveals the tokenistic approaches by global institutions. Using wayfinding methodology and values, I propose to curate an exhibition that creates *Waka Memory: Oceania connections*, where Indigenous can decentre the white gaze and narrative and reawaken ancestral relationships with each other through connection to Country and cultural practices.

Waka Memory aims to create a transformative experience that not only celebrates Indigenous culture but also serves as a catalyst for broader societal change. I seek to create Indigenous to Indigenous relationships, working with the Wurundjeri Woiwurrung on their land. *Waka Memory*, as a Moana-themed exhibition, could be used as a mass educator within cultural institutions, specifically the State Library Victoria to centre Indigeneity, reawaken ancestral connections and change the hierarchy of knowledge within cultural institutions. The journey through *Waka Memory: Oceania connections*, reveals a complex narrative of connectivity and ongoing colonisation. Revealing the impact of western colonisation and the tactics of division and separation, this calls for a reawakening of waka memory, a reconnection of the Indigenous people of Australia and the Moana.

Waka Memory holds the potential to bring about a shift in perspective within cultural institutions, and the broader public, aspiring to be a catalyst for societal change. Through unity, self-determination, and the power of storytelling, the project has the potential to reawaken ancestral connections and recentre Indigeneity.

Acknowledgement This chapter describes a conceptual exhibition that, while not yet realised, reflects a deeply powerful and important vision. It was conceived with care and intention, and although circumstances have prevented its materialisation to date, its spirit remains vibrant and necessary. The idea deserves to be shared not only as a blueprint for a future exhibition but also as a contribution to the ongoing discourse around Indigenous methodologies and the decolonisation of cultural institutions.

The strength of this chapter lies not in recounting a completed exhibition, but in presenting a compelling and grounded proposal that emerges from Indigenous-led thinking, grounded in relationships, and driven by a commitment to transforming the arts sector from within. While it may sit outside conventional expectations of academic chapters based on realised projects, we believe it holds valuable space as a provocation, an offering, and a possible pathway forward.

Every effort has been made to respectfully attribute and seek permission for the inclusion of all images and artworks referenced in this chapter. These visual elements are not simply illustrative; they are integral to the conceptual and cultural fabric of Waka Memory.

References

Ambrose, W. R. (1978). UNESCO seminar on the role of museums in preserving Indigenous cultures (Adelaide 10–15 September). *ICCM Bulletin, 4*(4), 22–24. https://doi.org/10.1179/ICCM.1978.4.4.003

AIATSIS (2020). *AIATSIS code of ethics for Aboriginal and Torres Strait Islander research.* Australian Institute of Aboriginal and Torres Strait Islander Studies.

Australian Museum. (2020). *Aboriginal & Torres Strait Islander collection timeline*. Australian Museum. https://australian.museum/learn/cultures/atsi-collection/timeline/

Australian Museum. (2022). *Reflections from Phil Gordon*. Australian Museum. https://australian.museum/about/history/people/oral-history-phil-gordon/

Brimley, S., Cornell, S., Flies-Away, J. T., Jorgensen, M., & Starks, R. R. (2007). *Resurgent justice: Rebuilding the Mohawk Justice System. A case study for the Mohawk Council of Akwesasne.* Native Nations Institute.

Chang, J. (2016). *We gon' be alright: Notes on race and resegregation*. Picador.

Chitham, K., Mahina-Tuai, K., Skinner, D., & Sorzano, R. (2019). *Crafting Aotearoa: A cultural history of making in New Zealand and the wider Moana Oceania*. Te Papa Press. https://www.tepapa.govt.nz/about/te-papa-press/history-books/crafting-aotearoa-cultural-history-making-new-zealand-and-wider

Commonwealth of Australia. (2007). Indigenous art—Securing the future Australia's Indigenous visual arts and craft sector. Parliament of Australia. https://www.aph.gov.

au/Parliamentary_Business/Committees/Senate/Environment_and_Communications/Completed_inquiries/2004-07/Indigenousarts/report/c03

Cordes, S. (2015). *A short history of community engagement in Australian museums*. https://www.academia.edu/35666491/A_Short_History_of_Community_Engagement_in_Australian_Museums#:~:text=The%20history%20of%20community%20engagement,vehemently%20pursued%20by%20Aboriginal%20elders.

Department of Infrastructure. (2023). *Revive a place for every story, a story for every place: Australia's cultural policy for the next five years*. Australian Government—Department of Infrastructure, Transport, Regional Development and Communications.

Enari, D., & Haua, I. (2021). A Māori and Pasifika Label—An old history, new context. *Genealogy, 5*(3), 70. https://doi.org/10.3390/GENEALOGY5030070

Eshrāghi, L. (2021). 'The colour chart': Ruminations on the white power status quo. *Art Monthly*, 108–112. https://doi.org/10.3316/informit.168212796955347

Fairley, G. (2020, August 25). *Racism in the arts—Reform or revolution?* ArtsHub Australia. https://www.artshub.com.au/news/news/racism-in-the-arts-reform-or-revolution-260963-2368270/

Fanon, F. (2001). *The wretched of the earth* (Issue 78). Penguin Modern Classics.

Hudson, D. J. (2017). The whiteness of practicality. In G. Schlesselman-Tarango (Ed.), *Topographies of whiteness: Mapping whiteness in library and information studies* (pp. 203–234). Library Juice Press. http://www.ustream.tv/recorded/92596956

Janke, T., & Janke, C. (2018). *First Peoples: A roadmap for enhancing Indigenous engagement in museums and galleries*. Australian Museums and Galleries Association.

Lee, S. (2022). *Decolonize museums*. OR Books.

Lopesi, L. (2018). *False divides*. Bridget Williams Books. https://doi.org/10.7810/9781988533858

Lugnan, C. (2023, November 10). *AFSE Fellow Christian Lugnan recognises how important on-Country learning was to his AFSE Fellowship experience*. Facebook. https://www.facebook.com/AFSocialEquity/posts/pfbid0dwqz5Vdybdp1sFeceGzPHnFbRJFwPzPgJYBdUHrXM-w8R3yFhEWiiDye9pQ64HLGdl

McCarthy, C. (2011). *Museums and Māori: Heritage professionals, Indigenous collections, current practice*. Te Papa Press.

Mead, H. (1996). Māori art restructured, reorganised, re-examined and reclaimed. *He Pukenga Korero, 2*(1), 1–7.

Mead, H. M., & Grove, N. (2001). *Ngā pēpeha a ngā tīpuna: The sayings of the ancestors* (8th ed.). Victoria University Press.

Mead, S. (2003). *Tikanga Māori: living by Māori values*. Huia Publishers.

Moreton-Robinson, A. (2015). *The white possessive: Property, power and Indigenous sovereignty*. University of Minnesota press.

Morris Hargreaves McIntyre. (2021). *Creating museum exhibitions to provoke meaningful behavioural change—CultureHive*. Culture Hive. https://www.culturehive.co.uk/resources/creating-museum-exhibitions-to-provoke-meaningful-behavioural-change/

National and State Libraries Australasia. (n.d.). *Creating a process for cultural permissions—National and State Libraries Australasia*. National and State Libraries Australasia. Retrieved November 11, 2023, from https://www.nsla.org.au/resources/cslp-collections/case-studies/slv-icip/

Nicholas, T. (2022). *Possessions: Indigenous art / colonial culture / decolonization* (2nd ed.). Thames and Hudson Ltd..

Rameka, L. K. (2021). Kaupapa Māori assessment: Reclaiming, reframing and realising Māori ways of knowing and being within early childhood education assessment theory and practice. *Frontiers in Education, 6*. https://doi.org/10.3389/feduc.2021.687601

Scott, J., & Laurie, R. (2007). Colonialism on display: Indigenous people and artefacts at an Australian agricultural show. *Aboriginal History, 31*, 45–62. https://doi.org/10.3316/ielapa.382174594499971

Spiller, C., Barclay-Kerr, H., & Panoho, J. (2015). *Wayfinding leadership: Ground-breaking wisdom for developing leaders*. Huia Publishers.

Stewart-Harawira, M. (2005). *The new imperial order: Indigenous responses to globalization*. Bloomsbury Publishing.

Te Punga Somerville, A. (2012). *Once were Pacific: Māori connections to Oceania*. University of Minnesota Press.

Temple, E. (2018, May 10). *The 12 Most Popular Libraries in the World*. Literary Hub. https://lithub.com/the-12-most-popular-libraries-in-the-world/

The Racial Imaginary Institute. (2022). *On whiteness*. Self Publish, Be Happy.

Thompson, N. (n.d.). *On Wayfinding*. Retrieved July 31, 2023, from https://archive.hokulea.com/ike/hookele/on_wayfinding.html

Tuck, E., & Yang, K. W. (2012). Decolonization is not a metaphor. *Decolonization: Indigeneity, Education & Society, 1*(1), 1–40.

Chapter 5
Gender Responsibility Country: How Aboriginal People Define Their Role in Caring for the Universe

Nola Turner-Jensen

Abstract This chapter presents the results of my research on playing a role in restoring the narrative relating to the reciprocal and integral role of the Three Creation phases of certain Ancestors in Aboriginal Oral Systems. It is to ensure voices of Aboriginal peoples Creation Ancestors voice is living again in Wiradyuri universal places. Places where, due to the banning of our handing down or telling of our Ceremonial knowledge for many generations by the white men, much of these teachings and wisdom have been erased. The Western System's way of mapping since 1788, is flawed, according to Aboriginal people, because it obscures the purpose and bloodlines of place. Aboriginal people use songs and dance to know where their responsibility to Country is located, revealing unique sound codes via ceremonial areas, memory language, sign posting, and inherited bloodline connection that continues today in 2025. In my people's Aboriginal transformative worldview, emphasis is placed on changing individuals' lives and reducing the impact of issues affecting them.

Keywords Power symbols · Gender responsibility · Aboriginal place · Creation ancestors · Universal Life Cycle

5.1 Gender Responsibility Country

> Honour the power essence every living thing holds inside, and you will feel its greatness grow within *your* soul.
>
> Sing again its ancient ancestry, it will always be with you, and it will come along with me—and I too will be whole.

N. Turner-Jensen (✉)
Faculty of Science, University of Melbourne, Melbourne, VIC, Australia
e-mail: nola.turnerjensen@unimelb.edu.au

M. Kennedy, E. O'Donnell (eds.), *People, Place and Nature in Indigenous-Settler Relations*, Indigenous-Settler Relations in Australia and the World 10, https://doi.org/10.1007/978-3-032-03600-1_5

This chapter presents the results of my research on playing a role in restoring the narrative relating to the reciprocal and integral role of the Three Creation phases of certain Ancestors in Aboriginal Oral Systems. It is to ensure voices of Aboriginal peoples Creation Ancestors voice is living again in Wiradyuri universal places. Places where, due to the banning of our handing down or telling of our Ceremonial knowledge for many generations by the white men, much of these teachings and wisdom have been erased. This is my attempt to honour the most powerful and important ancestors in our ancient oral system—the Power Symbols Creation Spirits of the universe. Using a mixed method methodology incorporating Aboriginal Spoken learning and communication styles underpinned by a pragmatic approach and an action-oriented transformative worldview that will allow me to gather a comprehensive understanding of both written and spoken research to provide a more holistic perspective. In my people's Aboriginal transformative worldview, emphasis is placed on changing individuals' lives and reducing the impact of issues affecting them. The theoretical framework underpinning this research is postcolonial theory—how the history of colonisation has and continues to shape people's lives, especially Aboriginal people and how their returned knowledge has been whitewashed by early ethnographers and colonial government employees.

5.1.1 Introduction

I am Wiradyuri, the second largest language group in this Country, now called Australia, located in the colonial named state of NSW. As an Aboriginal being I am a verbal universal communicator and learner. When I say "I", I actually mean "we". This is due to my belief that all that I am (including all that I have achieved) is due to the thousands of generations of Ancestors who created me. It is these many Ancestors who deserve the credit, not me. I acknowledge all amazing Aboriginal people across Australia, and you will find knowledge from both Wiradyuri and Warlpiri language groups in this chapter. I am a Senior Aboriginal Researcher who specialises in reconstructing pre-Colonised place using Ancestral Sound Coding Navigation mapping. I have been translating an Aboriginal navigation system of my people's Mother Country for over 15 years. My relational Aboriginal brain is designed for verbal knowledge hunting.

As a Wiradyuri Aboriginal being, I was born from the world's most ancient wombs: the sacred womb of a First Nations being that the Western System calls a woman. In what I have named the Aboriginal Universal Life Cycle System, Aboriginal people call living beings by titles relating to their relationship to others—sisters, daughters or mothers. There is no name such as women in our world. The blood pumping through my veins is the life force of a long line of First Nations mothers (life holders) who emerged and evolved from the Creator of this Universe, a being known as Old Mother in many language groups across Australia and called Mudyigali in my Wiradyuri peoples speak. It is a privilege to be raised in a First People system that honours the power of the First Nations Matriarchy: a Universal kinship that is framed upon values that uphold the role of life holders as equal to all other living things, a matriarchy whose power comes from an enduring love for the universe first, where individuality is discarded as irrelevant.

The Western System's way of mapping since 1788, is flawed, according to Aboriginal people, because it obscures the purpose and bloodlines of place. Aboriginal people use songs and dance to know where their responsibility to Country is located, revealing unique sound codes via ceremonial areas, memory language, sign posting, and inherited bloodline connection that continues today in 2025.

When Aboriginal people 'follow' the 'footprints' of Old Mother Earth, it is like how the Aboriginal hunters 'follow' animals' tracks. Except these Senior people are hunting for knowledge, not food. This entire responsibility to place stems from the Creation Ancestors' Universal Life Cycle. Each family awoke each day to maintain an ancestral resource and its habitat on their inherited piece of Mother Country. They did not benefit from its bounty, that was for others to do. This is their eternal contribution to the universe.

People often ask me why Aboriginal people did not use agriculture and become more "progressive"? I always answer that the universe, and all its majesty was designed and created by Aboriginal people's wisest and most powerful spirit ancestors. These Ancestors gifted to all living things everything they needed to maintain, honour and refresh the universe. To change their ancient design would most certainly dishonour and anger these Creation beings. Beings who sit in places unseen, observing the actions and behaviours of all.

As described by Browne et al., (6) the 'post' of postcolonial can be attributed to the movement of increasing Aboriginal control over research and research practices over the past few decades that continues today. I have been repatriating documents or knowledge relating to Wiradyuri Ceremony, Lore and Kinship from every institution and community I can access across the world for over 20 years. In this time, I have noticed multiple sound codes/patterns relating to our language when it is written down. I have named these systemic patterns "Sound Codes". Why the term Sound Code? I am doing this to separate Aboriginal expression rules from that of Western Systems linguistics rules. Western discourse created a system that prioritises the written word as the dominant form of record keeping and until recently, Westerners have generally considered oral societies to be peoples without history. Mere myths and legends instead of historical recorded facts. This could not be further from the truth. Oral societies record and document their histories in complex and sophisticated ways, including performative practices such as dancing and drumming. Aboriginal expression of their Creation Epistemology is spoken, sung, carved, signed and emulated; it was never designed to be written. Another difference between Aboriginal expression and English written language is the fact that our words cannot change. The ceremonial, kinship and Lore language in our Wiradyuri dictionary has been proven to be thousands of years old by western scientists, Aboriginal oral knowledge and linguists. This is showcased when the correct Ancestral translation of these words reveal eternal information, location and relationship knowledge.

Aboriginal Oral Knowledge Holding is fundamentally universal first teachings and knowledge. It is one of the few thinking systems that encompasses timeless ways of knowing, from the natural, physical, and social sciences, linguistics, astronomy to the humanities. Aboriginal expression is framed on Relational Processing Memory (RPM), that utilises the environment as visual cues to teach a cause-and-effect education stemming from a specific Ancestry within a calendar of ceremonial events. It is not human-centric, rather it includes and celebrates all living things contribution to the wellbeing of the entire universe. When we say all living things, we mean everything even rocks. We see rocks as living because they birth other rocks and hold inner purpose things such as stone axe heads that benefit the universe.

Since colonisation, many prominent writers (Western male dominant), including researchers from the 1800s to the 2020s who research traditional Aboriginal life have emphasised only the sacredness and secrecy of male ritual activities. These researchers witnessed a rigid contradiction between the sexes, particularly in the spiritual sphere, and have consequently categorised men as sacred and women as irrelevant regarding their roles of keeping Country healthy. These male writers ignored the reciprocal relationship between both sexes and downplayed and just plain ignored the foundational role of the original holder of the Universal Power Symbols. In Wiradyuri, The Three Power Symbols represent the Three transformations of things in the universe, Conception, Fertility and Rebirth. The Creation spirit that created the universe as it is today, known to many as Old Mother, this spiritual creator is a gender-fluid being who lived in the time before all.

This is despite the obvious dominant theme of fertility within all ceremonial events and activities, the role of fertility (rebirth), and the integral spiritual being that created all living things. according to my Wiradyuri people's ancient Lore knowledge the ring at ritual ceremonies represents the womb of the Old Mother, which honours the power symbol transformation phase of conception.

It is true that males of a certain ritual age now hold (are the custodians of) the Universal Power Symbols and play a lead role in most ceremonial practices. Therefore, it is the men that hold the associated teachings and songs that accompany this custodianship. However, this was not always the case. In the time before all, known as the beginning of the living in Wiradyuri teachings, the Creator Spirit of the Universe (Mudyigali) held responsibility for the Power Symbols and after her it was handed down to her daughters who held the symbols and fertilised the universe in the second transformation phase. In the current, third phase of Creation that we are in now, it is the men's turn to hold the Power Symbols that keep the universe in harmony and balance.

In the western desert of central Australia, and multiple other language groups, Elders talk openly that the women once possessed all sacred ceremony, which was later taken from them by the men. As Yolngu men say in reenacting the myth of the second Creation time of the Fertility Sisters,

> "But really, we have been stealing what belongs to them (the women), for it is mostly all women's business; and since it concerns them it belongs to them. Men have nothing to do really, except copulate, it belongs to the women. All that belonging to those Sisters, the baby, the blood, the yelling, their dancing, all that concerns the women; but every time we

> have to trick them. Women can't see what men are doing, although it really is their own business, but we can see their side. This is because all the Dreaming business came out of women—everything; only men take "picture" for that Julunggul [i.e., men make an artificial reproduction of the Snake]. In the beginning we had nothing, because men had been doing nothing; we took these things from women." (Berndt & Burndt, 1964)

This story of how men became the holders of sacred symbols is repeated across the land. It is accepted by the male elder knowledge holders that the reason they hold some parts of ceremony in secret is because the women know all the knowledge (songs and dances to make the power symbols living) and could take them back at any time.

The men had to hide the re-enactment of Old Mother, her two daughters and their daughter (who are known as the three sisters), so they created new language names that are only known to them and spoken in sacred ceremony where the women cannot hear. These language names replicate or mimic the events of Old Mother and her daughters. The men believe that if they spoke the same language names as the women the women would take back the Power Symbols to again hold the sacred ceremonies.

These ceremonies serve as an emotional outlet for Country and people. They establish good will between participants, and cement the bonds of friendship, bringing groups closer together. Ill will and jealousy cannot flourish at such bonding events. Moreover, it draws the women closer into the sacred responsibility within a Universal Life Cycle. All sacred ceremony is deeply rooted in the concept of procreation, brought about through the contact of male and female. All Ceremony substantiates Old Mother's Universal Life Cycle by eternally refreshing the Country and ancestors.

5.1.2 Why Aboriginal Gender Responsibility Country?

Sixteen years ago, a Wakka Wakka /Goreng Goreng woman asked me two questions:

"Where is your Mother Country?" was the first.

I replied it was on the Galari (Lachlan River) at Yuabalong in Western NSW.

"What are you doing for your Mother Country?" she asked next.

I told her I am developing the first Aboriginal online education company in the world, how I was trying to get Aboriginal knowledge in schools, how I am on numerous committees, how I go home when I can and even though I live off Country now, I am doing my best.

"So, nothing." She replied, with the look only a holder of secrets and wisdom can give.

I looked at her and, try as hard as I did to argue the point; in the end, I had to agree.

I asked her: "How can I help my Mother Country if I don't live there anymore?"

She told me to go and take some time to reflect on this question. I sat thinking for several days. I went back and I told her of many ideas I had come up with, like running programs out there, writing books for kids based on my Mother Country area and other things.

"You misunderstood the question," was her response, looking at me blandly in a matter-of-fact voice.

"I did not ask what you're doing for the *people* on your Mother Country," she told me impatiently.

"I asked you what you are doing for *your* Mother Country?" she said, raising her eyebrows and waiting for me to respond.

I asked her for some advice, and she told me to think hard about where the Ancestors had directed me to look, which made me uncomfortable about being there or what was revealed.

I started to write a very long list.

However, pondering where I felt uncomfortable or disturbed made me remember when I attended an event at a library in NSW. At this event, about celebrating language, they had hundreds of books, papers and artifacts about Wiradyuri people laid out on table after table.

I mentioned to a librarian how surprised I was at the large amount of research and artifacts about my Wiradyuri people that I had never seen before. She proudly replied to me, "That is just the tip of my little finger of how much we have about Wiradyuri here" and wandered off, content in her explanation.

I went back to this beautiful being and said proudly with purpose; I am going to spend three hours every day of my life finding, gathering and taking back from the world's Western institutions as much Wiradyuri knowledge and research as I can about my mother country. My goal in doing this is so this knowledge can be returned to the rightful people who hold responsibility for it to try and restore the handing down across generations to refresh, maintain and honour my mother country's important place within the universe.

She nodded and returned to watching her stories (soap operas) on TV.

Her nod is still one of the happiest moments of my life.

Twenty-five years later, I am still keeping my promise, and I am forever grateful for that lesson.

Extraordinary damage has been done to our beautiful Universal Life Cycle thinking. My Aboriginal family today has received so much learning and modelling of another culture's societal systemic standards and rules (colonial heritage) that it is changing or masking our Aboriginality. This masking is leading to many of us self-destructing. No Aboriginal person was ever meant to be individual, on our own or self-indulgent. No Aboriginal brain was originally wired to think about themselves first before their kin or worry about their outside more than their inside spirit. In my education from my Wiradyuri Elders, I have learnt that the Creation Ancestors developed an education system using cause-and-effect visual teaching and action learning. For example, when my Wiradyuri people see the Welcome Swallow (known as Yiraybunbinya in Wiradyuri expression) returns at the end of August each year, Wiradyuri women would prepare for the sun bringer ceremony and know that this is a sign that warm weather is returning. Yiray means sun and bunbinya means tired, come and rest. Aboriginal people's incredible observation knowledge and skills crafted and maintained since universal creation have the capacity to contribute greatly to the scientific and history education about this land in all schools today.

5.1.3 A Brief Explanation of Universal Creationism Belief Systems

The question of when God created the world is a matter of faith and interpretation, with different perspectives within religious and secular contexts, including the idea of a young Earth (around 6000–10,000 years old). The Genesis creation narrative in the Bible describes God creating the heavens and the Earth in six days, followed by a day of rest.

The National Academy of Sciences (1999) states that the origins of life cannot be dated as precisely. Still, there is evidence that bacteria-like organisms lived on Earth 3.5 billion years ago, and they may have existed even earlier, when the first solid crust formed almost 4 billion years ago. Today, all living organisms store and transmit hereditary information using two kinds of molecules: DNA and RNA.

In our Indigenous Creation belief system, we view life or the time of the universe's birth very differently. In fact, our understanding of time is nothing like the Western System's definition. Our Indigenous Creationism teaches that the world was originally covered completely in salt water and eventually the land rose up and the saltiness of the oceans was pushed down, and fresh water was created. This is known as Dry time in my Wiradyuri teachings and is a period when the mountains, valleys, deserts, forests and all earth (including earth's language) was created.

We believe all things within the universe are equal and loved equally. We honour and maintain the original Eternal Universal Life Cycle as it was established by our Creation Ancestor in the First Phase of Creation (Burumyinganha (boo rum yin ga nar)), 'Dry time' as it is known in my Wiradyuri speaking area of the universe. We have three Power Symbol phases of Creation in our universal oral system.

The Wiradyuri old people teach us that in the beginning, Mudyigali (maw ji gar lee) held all the living things of the universe within her watery womb. This was an incredibly painful and tiring burden for her to carry. Mudyigali moaned and cried out in pain. Her relief came from her husband, the Rainbow Serpent, who speared Mudyigali, creating a way for all these wondrous beings to leave her womb and become living upon the land, sea and air. This is how all life began, according to my Wiradyuri Aboriginal people. It is a historical recount of how all things came to be from the time before all.

We honour and respect all belief systems (religions) if they are acting in the best interest of all living things in the universe.

In the time before all—Wiradyuri's Dugaaybul Time, the Universe (known as Girr in my Wiradyuri people's teachings) Created a powerful Creator Spirit in the First Power Symbol Phase of Universal Creation. This is the time when the first holder of all life known as the Creator spirit Mudyigali was responsible for the three power symbols that were given to them by Girr and are key to making the world living. Whoever has responsibility for the three power symbols is responsible for maintaining the Universal Life Cycle of all things.

Mudyigali is our Old Mother or big rainbow. Mudyigali is the "why" of how everything concerning Wiradyuri oral philosophy, standards and rules exists today that have been handed down uninterrupted in song, dance, symbolically, ritually and through body talk prior to colonisation's violent and sudden disruption.

Girr taught Mudyigali a set of eternal rules and standards of the Universal Life Cycle that every living thing they created must live by. These include—

- An education system with rules and standards of how to live modestly in Kinship with all living things.
- A Lore system with rules and standards of how the Universal Life Cycle should be expressed honourably.
- A Ceremonial System with rules and standards of how to maintain and refresh the Universal Life Cycle fearlessly.

Mudyigali also taught each region of the earth to speak a unique dialect of her universal first language in every place by singing it into the Country. This way each region spoke a different dialect and maintained her universal philosophy at a local, manageable level. This is why there are so many languages (dialects) across the Universe. These oral expressions of our Universal Life Cycle identify where each family have an eternal responsibility to a place.

The Universe (Girr) gave the spirit three power symbols to hold and use to establish and maintain the Universal Life Cycle balance of all things:

1. The first is the inner essence of all living things (purpose/place);
2. The second holds the eternal life of all living beings (representing the shape of the uterus/rebirth); and
3. The third pierced the earth and made the place (homes) of all living things (representing such things as the male penis, waterways, mountains and all places underground).

The three living teachings are connected by an invisible string that connects all things. We name it Muraymin (moo ray min) in Wiradyuri country. It is Jukurrpa in Warlpiri and has often been translated in English as Dreaming, Dreamtime, or Ancestral Times (Mulvaney et al., 1997), but these translations obscure rather than explain the richness of a universal spoken cosmology.

When the British came without invitation and were gazing on what they saw as the new land, they saw the civilised story they wanted to make, the colonial fiction they would create: a story about property and value of the resources here for their King, not land and Ancestry or connection, about the "nation" to come, not the "Country" that is and always had been. The colonial story is humancentric and individualistic. The "new" story has been told in foreign languages to the ear of our Country. It is framed on foreign standards and rules of how to live and read it.

Following on from the foreigners coming, explorers and surveyors were the forerunners of colonisation for the next one hundred years, mapping out the new territory suitable for resource-hungry settlers, pastoralists, miners, and speculators (Mitchell, 1839; Oxley, 1820). The need to find water became an imminent goal, and the mapping of the rivers is a prime example of this. As the explorers and colonists pushed outward from Sydney, their English language moved with them, and as Country was "discovered," it was "named" by them accordingly to appease and impress their benefactors and governors.

Aboriginal people have lost more land to western mapping than through anything else.

The renaming of Old Mother's Country disguised its nature, obscured its meaning, stole its voice and took away its gender responsibility. It also took away our tongue because they banned our language. New maps showed nothing of the meaning of Country, or the ancient stories embedded in it or to it. The Western System maps are two-dimensional and scrollable views of place. This system tends to flatten places into lines, distances, and measurements instead of the journeys and activities of the Ancestral Creation Beings directly connected to that place. Instead of being *with* Country, Western maps only see the land from a bird's perspective.

Ethnographic literature from across the Central Australian region documents a Warlpiri intercultural history going back to the beginnings of the region's pastoral industry in the 1880s and hints at the rich history of ceremonial trade and exchange across a broad area of Central Australia (Berndt & Burndt, 1964; Meggitt, 1967). Many elders today also reflect on the past trading of ceremonies. They can give extensive oral histories of the places and Aboriginal groups involved in these social networks to time depths well beyond their memories.

Aboriginal peoples and Gender Responsible Country are enclosed within the fictional boundaries and borders of Western cartography: a fictional place with lots of names copied from other places and people, neither imaginative nor original and mainly in English or conveniently or perhaps as a novelty taking Aboriginal language names to name their farms (though they shortened or changed them according to what their English ear heard).

5.2 Gender Comparison: Aboriginal Universal Versus Western Individual System

In her book *Gender(s)*, Kathryn Bond Stockton (2021) points out that in the Western System, a child's sex, pronounced at birth, is a gender word: "It's a girl!" or "It's a boy!" The idiom is not "it's a male!" or "it's a female!" An entire world of *cultural* assumptions is rolled up into the word rolled onto the baby and its genitals.

Bond Stockton says the Western Individual System of Gender has been changing since it was invented. Gender's changeability isn't new—but in *perceiving* it, people are advancing it, sometimes fighting it, and undeniably accelerating it. Gender, let's remember, was invented in response to certain recognitions. Biology and medicine, after all, came to realise in their way that the binary of sex was imposed upon a range of human bodies and their lived desires. Fascinating, fraught, and intimate to everyone, gender increasingly has our attention as something newly morphing.

In the Aboriginal world, Gender is never singular. A Bigender perspective extends across the land and water and up into the SkyWorld. It is everywhere and within everything. Aboriginal gender is about the purpose a thing holds inside and where that purpose sits or belongs within Ceremony, Lore and Kinship. It ignores the outside shell and never separates one living thing from all others.

It is why many beings or living things in Aboriginal sky and earth worlds, for example, have both male and female parts that tell different stories and are expressed in different names. To Aboriginal people, it is not correct to have blanket names where many things are grouped as one. For example, most Aboriginal language groups do not have an individual language name that means tree, shrub, women, man, lake, river etc. The names of a particular living thing honour its purpose or the gifts it holds for others. It's inner essence or spirit/s. How can you honour something and know their inner essence if you group them all under one name?

The living thing's name is framed on the Ceremonial, Lore or Kinship relationship and the knowledge level the person viewing it has about what this living thing holds. So, is this (tree) living thing a holder of string, canoes or perhaps medicine? Its inner essence cannot be honoured by calling it a tree.

It is the same for humans. Humans hold many spiritual gifts and should be honoured for their inner contributions, just as living things with bark, leaves and roots (a tree) should.

As Bond Stockton (2021) points out, Western Science, and more specifically many restoration projects, have historically emphasised the 'educate' parts of the behavioural model (e.g. sharing knowledge of medicinal benefits of plants, plant form, classification, economic value, carbon offsets and natural capital benefits) as a means to encouraging action for nature. However, the pursuit of facts and education without a spiritual connection may create psychological distance for many individuals and organisations and create a barrier to action people cannot see the emotional relationship of trees to their local community or their Kinship.

5.2.1 Aboriginal Inner Purpose Understanding of Life

Aboriginal people's objective in life is to know and understand the inner purpose (called Malunma in Wiradyuri speak) of all things and how they can honour its purpose.

Let's take the English and Wiradyuri names for the River Red Gum. The Western system has Latin kinship names for this being. Aboriginal people say its kinship is from the Goanna Creation guardian. It is why goannas crawl up and all over it. Its common name in Wiradyuri language is Yarra.

It is prized and loved by Aboriginal women and is called Yarrawalang (yarra wa lung) when it plays its ancestral role in ancient ceremony of birthing. Thick layers of leaves are heaped on steam pits to smoke babies and new mothers, so they grow as strong and old as the Yarra itself. Walang means strong in Wiradyuri language. Each baby is made a Coolamon (original baby cradle) from the skin (bark) of the Yarrawalang Ancestor that the baby was born near to in Wiradyuri country.

This Ancestral being is honoured in men's language as one of the best holders of long, lasting shallow canoes, which are cut from the belly of the special beings that hold the right shape. a Canoe Holder or as we say in Wiradyuri—Gaambilanha Muriin (garm bill la nah moo reen), that is the perfect "tree" as most people call this ancestor now in the English language, to make a traditional bark canoe that our old people used to travel around waterways. My old people told me that Girawaa ancestor, who is known in English as a tree goanna, was the inventor of the first canoe in the universe, it is why you see him crawling all over the trees that Wiradyuri people call living things

Another of our names for River Red Gum Ancestor is Biyaala. Biyaala means forearm that way. To know why we call the River Red Gum forearm that way requires the speaker of that word to know much about where this being fits within our Universal Kinship System. This name is honouring its purpose as a direction tree in my Country. Direction or ring trees in Wiradyuri lands are a certain type of tree who are manipulated when they are a child with its forearm to be a circle or window for you to look through to navigate to important ceremony locations across the land.

The female within this ancestor is called Yarrawulay (yarra wall lay). This is when this Ancestor blossoms at certain times of the year. The translation means "calling out" for her husband who is a certain warm wind to come and kiss her. Yarrawulay announces her lover's arrival by bursting into flower to tell the world he has come to her.

River Red Gum living thing is honoured for so many purposes. Twigs were used by healers to beat away poison out of patients, and leaves were cooked with meat to add flavour. Manna oozing from the trunk was collected and eaten, and the dense wood was used to make boomerangs, clubs, coolamons and throwing targets.

Knowing the inner universal purposes (their essence) of all things is key to connecting and reading the Gender of Country.

If I think of my one English name—I think of how many names I should have. I am an oral Systems historian and researcher, writer, mother, sister, Aunty, daughter and friend. How does Nola encompass my inner purpose/essence?

If I was living prior to colonisation, I would have had many names based on the relationship I held with the other person I was speaking to.

These are just a few of the names I would have been known by in camp and at ceremony depending on who I was talking to—

Wiradyuri name	Universal Life Cycle relationship (for general camp & at ceremony)	Western relationship equivalent
Gunhi (gone knee)	Mother (Mother's eldest sisters in ceremony)	Mother
Munga	Daughter (sons & daughters in ceremony)	Son, daughter
Migay	Wife (Fathers eldest sisters' kids in ceremony)	Wife, cousin

There are other relationship terms I would have been known by once. Our Universal Life Cycle system is all about Ancestral relationships between all things; a single name does not cover or honour these multiple relationships (Box 5.1).

Box 5.1: Can a caterpillar fly?
If you think about the caterpillar. First it is a caterpillar, then it turns and becomes a chrysalis and then it turns into a butterfly. That butterfly is still the caterpillar according to our Universal Life Cycle standards and rules.

Aboriginal people believe that the outside appearance can go through transformations, however, the inner Universal relationship/purpose can never change.

5.3 The Responsibility Elements within the Three Creation Phases of the Universal Life Cycle

Each of the Creation Phases of the Universal Life Cycle relates to who hold the responsibility for maintaining the Three Power Symbols that keep the Universe well and in balance. The first phase (Dry time) was when Old Mother was responsible. That responsibility was passed on to her children (Fertility time), and finally now we are in the phase where the men have taken responsibility (Ritual Time).

The ideological goal for all living things is to learn the Three Inner Essences of the Universe that stem from Old Mother's teachings that gift you an understanding of what is right for the universe that they created.

This Ancient Philosophy teaches humans how to ensure their inner essence glitters as brightly from within as possible to ensure they share their inner purpose for being, which is to make the universe healthy.

Muraymin represents the three key elements of Aboriginal people's ancient universal system that Old Mother created: Lore, Ceremony, and Skin (also called Kinship). Wuurraganha (woo rah gar na) in my Wiradyuri language is a concept that highlights the primary relationships between these elements while also creating an awareness of their deeper complexities. Understanding Muraymin has many benefits for Aboriginal people and those who work with them.

5.3.1 Ngiyawaygunhanha—Old Mother Power Symbol Responsibility Time

This Creation Phase 1 describes the time when Old Mother held responsibility for the Universal Power Symbols of life. Old Mother created, taught and prepared the Universe for fertilisation.

All womb-holding beings have inherited Mudyigali's ability to hold life within their bodies. Not all women of course, choose to or can fulfil this purpose. Some through choice and some through medical reasons. Muraymin teaches this is just one of the spiritual purposes of women or any living thing and this story is focused on honouring all the inner purposes of all living things. Living your life to the Muraymin standard and rules of Mudyigali is the goal. Having children is not the end goal of womanhood.

One of our all-knowing sky fathers' wives, Binhalbang had no children. Her inner universal purpose was to gift the womb-holding beings of the world the tools, morals and knowledge to contribute to the wellbeing and health of the entire universe so they could come and live with her in Barranggalinya (bar rang ga lin ya—translates to white go up—white haired old people) Sky world.

The language, ceremonies and songs were strategically created for each family to be able to orally express Muraymin within a small area of land, thus allowing them to play their part within their universal responsibility to know and honour the inner purpose of all living things, and ensuring that every living thing contributes to the maintenance of the ancient universal oral system created so long ago.

Even in the sky world there is an allocation of responsibility for all resources (things)—men and women have separate and shared ceremonial areas to honour the purpose of all things.

A human body is a fleeting shell that expresses its place within the universal Muraymin for a short period of time. It does not hold Muraymin. Muraymin enters their body in the conception place (country) of its beginning (known as Birrel in Wiradyuri speaking country).

5.3.2 Dabu-yarra-murrun (Plant, Sing, Life)—Fertilisation Creation Time

This Creation Phase 2 describes the Fertilisation Creation Time and the Sisters Power Symbol responsibility time. Old Mother designated a whole category of Ancestral Beings to protect this new earth. These included her two daughters and a younger son to take over the responsibility for the Universe's three power symbols so that they may fertilise the universe. Their actions shaped the landscape, and performed marvelous acts, which both the Warlpiri and Wiradyuri still reenact today, indeed many Aboriginal groups reenact these in their ritual performances. Every spot visited by Mudyigali and her children became a sacred site, and every sacred site became part of a specific *Dreaming* itinerary or song cycle (also known as a song line) for both men and women. Some Ancestral Beings were created to live near the conception place of their emergence in the role of guardians, while Old Mother continued to travel through territories that belong to neighbouring Aboriginal groups. In her travels known to the Warlpiri people, Old Mother left behind "life forces" (Peterson, 1969, p. 27).

Each element of Dreaming has its own expression rules (language). There is Mother Earth Skin or Kinship language, Mother Earth Lore language, and Mother Earth Ceremony language. Skin language is when people change their speech to show respect towards each other. Lore language is all the Wiradyuri names for the land and things in it, but it is also the way Mother Earth can 'talk' to Wiradyuri people, such as their interpretation of seasonal indicators. Ceremony language is the way speech changes in ceremony. It is also the language of songs and the messages of sacred objects and designs. Each language supports the smooth function of Dreaming. Lore language is a person's ability to use all the other languages. There is always Lore in language because using the correct language inherently directs people and keeps them on the right path.

Many Aboriginal language groups identify three main kinds of life forces: in Wiradyuri they are Dyiny (jin), Dhinggaa (ding gaa) and Warrangan.

The Dyiny (jin), or the Mother Universe's Ceremony, was sung by her daughters into places to ensure the philosophy and understanding of honouring the inner purpose of all living things is adhered to and understood. Dyiny Ancestor enters the women through the sole of their foot, navel or uterus. Dyiny, however, has a more specific application: to the spirit or essence of the individual, a spirit that enters via the semen of the father and the egg of the mother and localises itself in the two kidneys. Male (father's semen) and female (mother's eggs) elements are thus accepted and held within every universal thing.

The Dhinggaa are the marks, signs, and designs mythical beings left behind, while the use of the word Dyiny invokes the unseen aspect of the Muraymin. Dhinggaa and Dyiny are complementary categories, with the former referring to the visible (and the latter to the invisible) traces of the Muraymin. Mother Universe's Lore is to ensure that living beings can see and feel her through visual geographical places, signs and symbols. These are our Wiradyuri people's Dhingaa. This is to express universal "Lore" in my mother's people (the Wiradyuri-weir rad you rhee Language speaking country) this is called Buyaa (boy yah), which is the creation Lore that comes from the Old Time.

The final key essence of the Muraymin of Wiradyuri, is called Warrangan, which is associated exclusively with the act of conception. Whereas the Warrangan is linked physically to the individual, wherever that individual may be, the essence or spirits associated with conception are all site-specific. While a mother is walking along, a Warrangan will penetrate her—through the womb, foot, or navel—in a fashion that animates the fetus. When the Warrangan emerges, that particular site becomes known as the conception site of the soon-to-be-born child. Later, the child will have special responsibilities and ritual obligations over the site and associated Ancestral Beings.

5.3.3 *Garramarra (garra mar ra) (to Take Or Hold)—Sky Father Power Symbol Responsibility Time*

The final Creation Phase refers to a specific segment of an ancestral itinerary at a given site and its vicinity. One or several songs, designs, and dance sequences are associated with a Song Cycle segment enacted during ritual ceremonies. The same basic elements of designs, songs, and dances performed by men and women are stylistically arranged according to age and gender that characterize a specific ritual activity.

The purpose of ceremony (in relation to Country) as follows: Ceremony is Dhadharra [re-enactment]. Copying and re-enacting the Creation Ancestors journeys across the universe. Remembering song, painting and dance, reminding us of who to be and how to work with Country, but it all comes back to our responsibility to our families' Dhingaa [grandmother's Country].

The songs are how we honour orally and remember the inner purpose of living things.

These simple rhythmic chants are the memory code of the inner purpose that Old Mother placed within each living thing and the power essence movements of all Ancestral family. Aboriginal people sing to honour the eternal giving of Old Mother.

Warlpiri Elder, Jerry Jangala, outlined the purpose of ceremony (in relation to Country) as follows: "Ceremony is Waya-parri-rni [re-enactment]. Copying and re-enacting the Jukurrpa (dya kar ba). Remembering song, painting and dance, reminding us of who to be and how to work with Country, but it all comes back to Warringiyi [grandfather and grandfather's Country]".

He goes on to say Ceremony is related to sustaining Country in five different ways:

1. learning knowledge
2. refreshing knowledge
3. celebrating knowledge
4. building relationships between people and Country
5. graduating people with all the skills to function on Country.

As Dussart and Warner (2000, p. 47) noted, '[l]ong before they are born and long after they die, the Warlpiri are directly and inexorably implicated in a complex repertoire of ceremonial activity'.

Yawulyu is the principal means by which Warlpiri women demonstrate their matrilineal clan identity as belonging to a defined clan estate. In my Wiradyuri Country each clan estate has one main Yawalanha (same as Yawulyu for Warlpiri), usually named for the principal Dhingaa (totem) of the estate ('country'). For example, Yawalanha ngadyang ('water/rain') is the main Yawalanha for a certain Country belonging to one of the Wiradyuri matrilineal Warrangan.

Aboriginal women understand these songs to be originally composed by ancestral beings in a timeless creational moment known in Wiradyuri as Muraymin, and these songs are intimately related to individual Wiradyuri women's identity, including their matrilineally inherited Country and associated ancestral stories. Yawulyu is multimodal, with complex interlocking components essential for proper performance. Within the Warlpiri people, singing is accompanied by dances, painted designs and other movements (Barwick & Turpin, 2013, p. 191).

Before dancing, women spend many hours painting one another's chests, breasts, and upper arms (and sometimes other body parts) with ochred designs, which link together the Songlines and Country with which they identify. Women play clapsticks, possum skin drums, and incorporate other percussive beats, including clapping against their thigh, and the thud of dancers' feet. There are no accompanying tonal instruments, and the pitch is determined in each performance by a lead singer who is, in most cases, the most senior owner of the particular Jukurrpa or the most proficient singer of a particular repertoire (Barwick & Turpin, 2013, p. 191).

Finally, Jukurrpa (Muraymin in Wiradyuri) refers to nocturnal dreams. When asleep, dreamers may have an Ancestral visit (dream) in which they "see and hear" Ancestral Beings. If the dreamers can remember information shared or song(s) sung by the Ancestral Beings as they wake, their dreams, after thorough examination by Jukurrpa experts, are usually integrated within an existing Jukurrpa itinerary. Through such dreams, the Warlpiri have learned about (and continue to learn about) the Jukurrpa. The Warlpiri maintain that nothing is new but simply forgotten. The Jukurrpa is believed to be immutable, whereas the reality of life is that cosmology and religious order are dynamic.

The songs come from the eternal Dreaming Lore Creation phase (Jukurrpa from Walpiri and Muraymin for Wiradyuri), a time-out-of-time in which Dreaming ancestors laid down the laws for humankind and formed the Country and all beings that live upon it. Most individual songs within a Song Cycle series are of unknown origin. However, others have come into the repertoire in living memory, usually through dreams in which songs (and associated dances, body designs, etc.) are revealed to the dreamer by an ancestral being (Dussart & Warner, 2000). Even in such cases, the songs are regarded as having always existed.

As previously mentioned, new songs may enter the repertory from time to time through revelatory dreams. Still, their source is attributed to the Dreaming ancestors rather than human agency, although those who receive the songs in dream may derive considerable social prestige through their composition (Dussart & Warner, 2000).

5.4 Conclusion

Across every Aboriginal-speaking Country, there are sacred sites that follow the journey and activities of the Universal Creation Ancestors of my Aboriginal people, Ancestors who hold fluid and interchangeable gender responsibility to Country.

In the permanent camps, females had their area (separate from the men) where they looked after the children and their own spiritual and earth-based business. All children (boys as well) were considered females until they went to ritual to be reborn as men between the ages of 12 to 14 years. Up until the age of five, there are no gender names for children in our system. We identified each other by key developmental cycles in a ceremonial sense that was obvious via appearance (when a child got hairs on their vagina or penis area, or a girl developed lumps known in English as breasts). Age was not a thing. If you were old, you had white hair. It was considered taboo for a male to marry until he had a full beard or before a female had been through the menstruation ceremony. The bloodline connection codes of families taught from a young age ensured that the bloodlines of each person were kept pure and right, as did the Mother-In-Law taboo, where a married man could never directly communicate with his wife's mother.

Then there are the topographical gender sites across Australia. Sacred sites such as the Three Brothers or the Three, Seven, Five and Two sister sites are incredibly important for our First People to play their part in ensuring the universe is balanced and these sites are refreshed and sung.

Many people believe that there is only one Three Sister site in the Blue Mountains in Sydney. Nothing could be further from the truth. The Sisters' journey is known in every language group across the entire Country, and I am now being sent Sisters' sites from around the world. As part of my ongoing responsibility research, which I have now been undertaking for over a decade to restore the ancient knowledge, location and ancestral genealogy of my people's Aboriginal Sacred sites, I have mapped approximately one hundred and twelve "Three Sister" sites, from the Torres Strait Islands to Tasmania and across the continent. As international awareness of my work grows, I am now also receiving Three Sister sites from all over the world. It seems to be the same teaching story, just in different oral language (expression) and locations. To undertake such research, I have collated hundreds upon hundreds of the earliest location maps of the colonists who renamed our places, removed all English place names, and started creating and translating maps my with my people's original Universal Responsibility names once again.

There is so much restoration work that needs to be done.

I dream of a day when the songs and ceremonies of Universal Creation are included once again singing and joyous across my people's land (all lands) according to Aboriginal Creation Lore. Celebrating the interconnectedness and genius of fertilisation that all living things share in. I sit quietly imagining a regular calendar of localised ceremonies actively occurring to refresh the universe and honouring all living things shared eternal ancestral journey. This is why Aboriginal (all Indigenous) people will continue to fight for our Aboriginal Creationism across each generation. Afterall, we are fighting to restore all living things responsibility and purpose to ensure the entire universe lives eternally in a cycle of mutualism and honour.

References

Barwick, L., & Turpin, M. (2013). *Central Australian women's traditional music: Yawulyu/Awelye. Indigenous music case study report for the sustainable futures for music cultures project. Revised version.* Griffith University Queensland Conservatorium Research Centre. https://doi.org/10.1093/9780190259075.003.0005

Berndt, R. M., & Burndt, C. H. (1964). *The world of the first Australians*. Ure Smith Pty Ltd.

Dussart, F., & Warner. (1937 [2000]). *The politics of ritual in an aboriginal settlement: Kinship, gender, and the currency of knowledge*. Smithsonian Institution Press. https://doi.org/10.1525/aa.2004.106.1.184

Meggitt, M. J. (1967). *Gadjari among the Walbiri Aborigines of Central Australia* (Oceania Monograph No 14). University of Sydney.

Mitchell, T. L. (1839). *Three expeditions in the interior of Eastern Australia with descriptions of the recently explored region of Australia Felix and of the Present Colony of New South Wales*. T. and W. Boone.

Mulvaney, J., Morphy, H., & Petch, A. (1997). *"My Dear Spencer": The letters of F. J. Gillen to Baldwin Spencer*. Hyland.

National Academy of Sciences. (1999). *Science and creationism: A view from the National Academy of Sciences* (2nd ed.). National Academies Press.

Oxley, J. (1820). *Journals of two expeditions into the interior of New South Wales undertaken by order of the British Government in the years 1817–18*. John Murray.

Peterson, N. (1969). Secular and ritual links: Two basic and opposed principles of Australian social organisation as illustrated by Warlpiri ethnography. *Mankind, 7*, 27–35. https://doi.org/10.1111/j.1835-9310.1969.tb00383.x

Stockton, K. (2021). *Gender(s)*. The MIT Press.

Part II
Unmaking Extraction: Toward Restoration and Repair

Chapter 6
The Root of The Matter: Forests and Colonial Histories in Aotearoa New Zealand

Anne Salmond and Manu Caddie

Abstract Forests, with their complex networks of relations among different life forms, have their own life histories, often intimately entangled with the lives of people. In Māori philosophy, the whakapapa or kin networks that trace the exchanges among land, forests, rivers, the atmosphere, the sea and human beings echo the patterns of these living systems. Around the world, richly diverse old growth forests have been felled for industrial uses and replaced by monocultures—pasture or tree plantations, for instance. This chapter explores the history of industrial forestry in Aotearoa New Zealand, underpinned by colonial ideas of 'property,' 'progress' and 'profit'. While industrial forestry has caused environmental harm in many parts of New Zealand, the impacts have been particularly severe in Tairāwhiti, the east coast of the North Island, which has some of the most erodible land in the world. In this article, we examine these dynamics, including the devastation caused by logs, fallen trees, forestry waste and sediment mobilised during Cyclone Gabrielle in 2023, the response initiated by Mana Taiao Tairāwhiti, a Māori-led community group, and a visionary programme, 'Recloaking Papatūānuku' that seeks to heal ravaged landscapes with indigenous afforestation.

Keywords Forests · Colonial · Māori · Papatūānuku · Relations

A. Salmond (✉)
University of Auckland, Auckland, New Zealand
e-mail: a.salmond@auckland.ac.nz

M. Caddie
Te Kotahi Research Institute, University of Waikato, Hamilton, New Zealand

Mana Taiao Tairāwhiti, Tūranganui-ā-Kiwa/Gisborne, New Zealand

M. Kennedy, E. O'Donnell (eds.), *People, Place and Nature in Indigenous-Settler Relations*, Indigenous-Settler Relations in Australia and the World 10, https://doi.org/10.1007/978-3-032-03600-1_6

6.1 Introduction: Thinking Through Forests

Forests, with their complex networks of relations among different life forms, have their own life histories, often intimately entangled with the lives of people. In contemporary scholarship, many forests once thought wild and pristine are found to have co-evolved with human communities over millennia; land, water, plants, animals and people engaged in reciprocal exchanges, shaping each other's futures (Hecht et al., 2014). In places like Amazonia, this entanglement between people and forests is so profound that shape-shifting forest animals (jaguars, for instance) and humans (shamans or warriors, for instance) are thought to share similar cultures, and change into each other (Vilaça, 2007; Viveiros de Castro, 1998). These and other Indigenous philosophies have helped to inspire a 'more-than-human' anthropology that investigates living communities including plants, animals and other life forms, as well as human beings, where forests may 'speak' and 'think' (e.g. Kohn, 2013). Even in the most remote parts of the world, however, forests have been destroyed or radically altered by global exchanges. As Anna Tsing describes, richly diverse old growth forests around the world—in Japan, China, the Pacific Northwest or Amazonia—have been felled for industrial uses and replaced by monocultures—pasture or tree plantations, for instance. In industrial forestry:

> Unwanted tree species, and indeed all other species, were sprayed with poison. Fires were absolutely excluded. Alienated work crews planted "superior" trees. Thinning was brutal, regular, and essential. Proper spacing allowed maximum rates of growth as well as mechanical harvesting. Timber trees were managed for uniform growth, without multispecies interference, and thinned and harvested by machines and anonymous workers. (Tsing, 2021, p. 41)

In this extractive process, Indigenous groups who live among forests are often displaced. As soils are exhausted or trees are stricken by disease or climate change, industrial plantations are abandoned as the foresters move on to other, more promising locations. Nevertheless, in these ravaged landscapes, forest remnants may still yield new forms of life, as processes of resurgence create new adaptive niches (Tsing et al., 2017).

In Aotearoa New Zealand, one of the last places on earth to be found and settled by people, these processes have a much shorter history than in many other parts of the planet. When the first star navigators from island Polynesia came ashore, perhaps just 800 years ago, they found a land of forests, with trees, plants and animals very different from those in their tropical homelands. The ancestors of those first settlers had crossed the Pacific, sailing from island to island and constantly adapting to new conditions. In the process, they evolved philosophies in which all life forms, including people and forests, are interconnected in vast kin networks, engaging with each other and endlessly creating new forms of life through reciprocal exchanges. In this article, we explore these ways of being, and the very different philosophies of the settlers who arrived from Europe and elsewhere from the late eighteenth century onwards, and how these have tangled and clashed in Aotearoa New Zealand.

Co-authors Anne Salmond and Manu Caddie are both from Tairāwhiti, the eastern part of the North Island, which features largely in this account: Salmond, a scholar in anthropology, Māori studies and environmental studies raised in the region, and Caddie, an Indigenous person who married into a local iwi (kin group) 26 years ago, and co-founded Mana Taiao Tairāwhiti, an environmental organisation led mostly by mana whenua (local Indigenous peoples). As on the marae (speaking ground), or in forests for that matter, we each have our own voices, shaping the way this article has been written.

6.2 Forests in Aotearoa New Zealand: Anne Salmond

6.2.1 Te Wao-nui-a-Tāne (Tane's Great Forest)

In the cosmological accounts taught in ancestral whare wānanga (schools of learning) in Aotearoa New Zealand, the emergence of the living world is often framed in the language of plant growth (Best, 1976, p. 62, quoting Hamiora Pio of Ngāti Awa):

Te Pu	*Origin, source, main stock of tribe,*
Te More	*Root cause, taproot*
Te Weu	*Hair, rootlet*
Te Aka	*Long thin roots, stem of climbing plant*
Te Rea	*Growth*
Te Wao-Nui	*Primeval Forest*
Te Kune	*Conception, pregnancy, growth*
Te Whe	*Noise, sound*
Te Kore	*Creative void, Nothingness*
Te Po	*Night*
Rangi	*Sky*
Papatuanuku	*Earth mother*

According to the nineteenth century *tohunga* (ritual expert), Te Matorohanga, everything in the living world is shaped by the same fundamental forces: "Nga mea katoa i tipu ai, i ora ai, i whai ahua ai, ahakoa rakau, kowhatu, manu, ngarara, ika, kararehe, tangata" ("All things are made to grow, live and take their form, whether trees, stones, birds, reptiles, fish, animals or people") (Smith, 1913, p. 13). The concept of tipu—to grow, increase; shoot, bud—is linked with the emergence of all living beings, including trees, stones, animals and people. It also gives rise to the word *tipuna*—ancestor, grandparent (literally, 'grown'). A founding ancestor might be spoken of as a take, or 'root ancestor,' as Edward Shortland remarked while describing a land dispute in the mid nineteenth century:

> "So and so is the root of the estate. Now then…" is the form of words in which they invariably commence. He then endeavours to prove that this root exercised some right of

> ownership undisputed by anyone and deduces, step by step, the descent of his clients from this ancestor or root. (Shortland, 1851, p. 96)

From these root ancestors, whakapapa (genealogy, descent) can be traced back to the beginnings of the cosmos. This is exemplified by Tāne-nui-a-Rangi, the ancestral house carved and decorated by Paki and Hinemoa Harrison at the University of Auckland. The house itself embodies Tāne, the shape-shifting ancestor who ascended the skies on a whirlwind to find the baskets of knowledge; and in his guise as Tāne Mahuta, is the ancestor of forests. In its carvings and paintings, this house tells the story of the beginnings of life on earth, as recounted in Te Rangikāheke's beautiful account Nga Tama a Rangi (Rangi's Children), written for Governor George Grey in 1849:

> E hoa, whakarongo mai! Kotahi ano te tupuna o te tangata maori, ko Rangi-nui e tu nei, ko Papa-tuanuku e takoto nei. Ki nga tangata maori, na Rangi raua ko Papa nga take o mua. [Friend, listen to me! There is only one ancestor of the Maori person, Rangi the great sky standing here, and Papa the earth lying here. According to Maori people, the take (root) of the past came from Rangi and Papa]. (Te Rangikaheke GNZMMSS 43:893)

Here, Earth and Sky are one ancestor, locked together in a primeval embrace. Between them, their children lie cramped and crowded, in utter darkness. In their frustration, Rangi and Papa's children talk about killing their parents, but after heated debate, decide to separate them instead. After one child after another tries and fails, Tāne-Mahuta, the forest ancestor, lies on his back and pushing up with his legs, thrusts his parents apart. As light floods into the world, Ranginui weeps for his wife, creating rivers and lakes, while Papatūānuku sends up mists of longing to greet him. Seeing his parents naked, Tāne-Mahuta cloaks his mother with forests, and his father with stars, while the wind ancestor Tāwhiri-matea, moved by their grief, attacks his brothers, unleashing hurricanes and cyclones. As storms hit land and sea, the trees, Tāne Mahuta's children, snap and fall; fish, the children of Tangaroa, ocean ancestor, dive into deep waters; while Haumia (fern root) and Rongo (kumara, sweet potato) hide in their mother's body.

Only Tūmatauenga, the ancestor of people, stands tall, defying the tempests. In revenge for his brothers' cowardice, he begins to harvest their children, catching Tangaroa's offspring with hooks and nets; pulling fern-root and kūmara out of the earth by their hair; felling some of Tāne Mahuta's children, the trees, and snaring the birds in nooses. When Tūmatauenga's descendants, the island star navigators, arrived in Aotearoa from their tropical homelands in about the thirteenth century, they set about clearing the forests for gardens, using stone tools and fire. It is estimated that by the time the first Europeans came ashore about five hundred years later, about a third of the North Island and much of the eastern half of the South Island had been cleared of forests, and were in grassland, fernland or shrubland (McGlone et al., 2022, p. 5).

As forest resources became depleted, rāhui (tapu restrictions) were placed to allow them to recover. Tohunga (ritual experts) called upon the ancestors of forests, gardens, rivers and the ocean in chants and seasonal rituals, making offerings to ask permission to harvest their offspring and enhance their life force (hau ora). If exchanges are balanced, the hau enters a state of ora (hau ora: wellbeing, health, good fortune, prosperity), but if they are unbalanced, the hau moves into a state of mate (hau mate: ill-being, ill-health, misfortune, death). This can apply to individuals, families, communities and ecosystems, including lakes, rivers or forests.[1]

6.2.2 Colonial Ideas About Forests

In early colonial times in Aotearoa New Zealand, the European creation stories were very different. Here, cosmological accounts were also powerful, for instance the biblical account in Genesis. Instead of a source of growth, God created heaven and earth, dividing light from the darkness and land from the seas, and all living creatures, including people:

> And God said, Let us make man in our image, after our likeness: and let them have dominion over the fish of the sea, and over the fowl of the air, and over the cattle, and over all the earth, and over every creeping thing that creepeth upon the earth.
>
> So God created man in his own image, in the image of God created he him; male and female created he them.
>
> And God blessed them, and God said unto them, Be fruitful, and multiply, and replenish the earth, and subdue it: and have dominion over the fish of the sea, and over the fowl of the air, and over every living thing that moveth upon the earth. (King James Bible, Genesis 1–28)

This creation story helped to shape philosophical musings about the relations between people and the wider world, for instance John Locke's theory of property:

> God, when he gave the world in common to all mankind, commanded man also to labour, and the penury of his condition required it of him. God and his reason commanded him to subdue the earth, i.e. improve it for the benefit of life, and therein lay out something upon it that was his own, his labour. He that in obedience to this command of God, subdued, tilled and sowed any part of it, thereby annexed to it something that was his property, which another had no title to, nor could without injury take from him. (Locke, 1690, Chapter 5, Section 32)

The same story also shaped the law of property in the early colonial period, as outlined by William Blackstone in his *Commentaries on the Laws of England* (Book 2, Chapter 1), for instance:

> In the beginning of the world, we are informed by holy writ, the all-bountiful creator gave to man "dominion over all "the earth; and over the fish of the sea, and over the fowl of "the air, and over every living thing that moveth upon the earth." This is the only true and solid foundation of man's dominion over external things, whatever airy metaphysical notions may have been started by fanciful writers upon this subject.

[1] Much of the basic research for this section of the paper was carried out as part of a Marsden research programme, 'Te Ao Tawhito: the Ancient Māori World' (2010–2015), with a team from Māori Studies at the University of Auckland—Anne Salmond, Hone Sadler, Robert Pouwhare, Jeni Curnow, Joe Te Rito and Jane McRae—exploring ancestral tikanga (ways of living and thinking) by studying and translating a large number of early Māori manuscripts, including the cosmological chants.

> The earth therefore, and all things therein, are the general property of all mankind, exclusive of other beings, from the immediate gift of the creator. And, while the earth continued bare of inhabitants, it is reasonable to suppose, that all was in common among them, and that every one took from the public stock to his own use such things as his immediate necessities required.

Rather than a nurturing mother, the earth was understood as inanimate, and created for human purposes. Ideas of 'improvement' also shaped scientific thinking, with models of social 'progress' from hunting and gathering ('savage') to agricultural ('barbaric') to industrial ('civilised') societies emerging at this time—for instance in this quote from Georg Forster, the young German naturalist who accompanied James Cook to New Zealand, in Dusky Sound in 1773:

> The superiority of a state of civilization over that of barbarism could not be more clearly stated, than by the alterations and improvements we had made in this place. In the course of a few days, a small part of us had cleared away the wood from the surface of more than an acre, which fifty New Zealanders, with their tools of stone, could not have performed in three months…
>
> Nor had science disdained to visit us in this solitary spot: an observatory arose in the centre of our works,… where the attentive eye of the astronomer contemplated the motions of the celestial bodies. The plants which clothed the ground.. in the forests likewise attracted the attention of philosophers, whose time was devoted to mark their differences and uses. In a word, all around us we perceived the rise of arts, and the dawn of science, in a country which had hitherto lain plunged in one long night of ignorance and barbarism!. (Forster, 2000, I,105-6)

Those who subdued the earth by clearing and planting it could claim it as property; and in the 'civilising' processes that followed, the land was cut up into blocks and sold as a commodity, uprooting and severing the sprawling, overlapping whakapapa networks, with the migratory, overlapping seasonal patterns of land use they fostered. Once sold, the land was generally cleared of forest, put in pasture and fenced, and protected by the laws of trespass.

In 1840, when Te Tiriti o Waitangi [The Treaty of Waitangi] was signed, about half of Aotearoa New Zealand's landscape was covered with indigenous forest. Over the next seventy years Te Wao-nui-ā-Tāne shrank to about a quarter of the landmass (Beattie & Starr, 2010, p. 195).

In ancestral *tikanga* (ways of being), the ruling idea of kinship among people and other life forms was interwoven with a strand of human control. Likewise in early colonial thinking, the ruling idea of human dominance was interwoven with a strand of kinship with Nature, and responsibility for its well-being. In his masterwork *Kosmos* (1845–62), for instance, the German naturalist Alexander von Humboldt wrote about a 'wonderful web of organic life' 'animated by one breath—from pole to pole, one life is poured on rocks, plants, animals, even into the swelling breast of man'—a description of the planet closely resonant with Māori ways of thinking (Von Humboldt quoted in Wulf, 2015, p. 120).

These different philosophical strands led to a colonial history in which the widespread clearance of indigenous forests alternated with attempts at sustainable harvesting, or their conservation (Beattie & Starr, 2010; McGlone et al., 2022). Thus while some settlers urged the need to clear the forests to establish a pastoral Eden, for instance the early Wesleyan missionary Schackenberg in 1841: "If you find your mind, your heart to be a wilderness, cultivate it in the same manner as you do your fields, cut down the bush, great and small—spare no sin" (Knight, 2009, p. 326), others lamented their destruction, for example William Pember Reeves in his closing poem in *Ao Tea Roa* (1898, in Kuzma, 2003, p. 454):

> Mighty are axe and fire, destroyers twain,
> Swift servants of the arch-destroyer, Man:
> And he is mighty as he hews amain,
> Bronzed pioneer of nations. Ay, but scan
> The ruined wonder never wrought again,
> The ravaged beauty God alone could plan!
> Bitter the thought: 'Is this the price we pay—
> The price for progress—beauty swept away?

This regret, plus a realisation that deforestation was having devastating impacts on flooding and erosion, led to efforts to conserve indigenous forests. In 1874, Julius Vogel's government passed a Forests Act, and a Chief Forest Conservator, Campbell-Walker, who had worked in India and was familiar with German forestry, was appointed. Quoting von Humboldt, he argued that indigenous forests should be managed for sustainable supplies of timber, flood protection, and their beneficial impacts on erosion and the climate (Beattie & Starr, 2010, p. 203). Campbell-Walker was opposed by those who regarded indigenous timber trees as inferior to Northern Hemisphere conifers, and heading for extinction—an idea that in the late nineteenth century, often mirrored ideas about the imminent demise of te iwi Māori, the Indigenous Peoples. According to John Sheehan MP, for instance:

> The same mysterious law which appears to operate when the white and brown races come into contact… by which the brown race, sooner or later, passes from the face of the earth—applies to native timber. (Beattie & Starr, 2010, p. 215)

Eventually, Campbell-Walker's critics won the day. In 1887 the position of Chief Conservator of Forests was disestablished, and the logging of indigenous forests resumed unabated. Once again, deforestation led to widespread flooding and erosion, and botanists including Leonard Cockayne, who helped to establish the Royal Forest and Bird Protection Society, fought to save them. By 1919 their efforts led to the founding of the State Forest Service, headed by a Canadian forester MacIntosh Ellis, who argued for a sustainable future for indigenous forests. By 1925, however, immense areas of exotic conifers were being planted, and conservationists began to advocate for indigenous forest reserves for wildlife and scenery, rather than sustainable timber production. This was controversial, for instance when the NZ Forest Service opposed the establishment of the Waipoua Forest reserve in 1948:

> To lock up 40,000 acres as a plant museum or 'tree cemetery' would be regarded by the Service and a large body of its supporters as fantastically wasteful of land and natural resources. (McGlone et al., 2022, p. 6)

After World War II, a shortfall in housing led to a surging demand for timber, and more indigenous forests were felled, including those on Māori land, and more exotic plantations were planted. At the same time the Forest Service, citing the risks of disease or climatic events to exotic trees, fought to preserve indigenous forests for future logging (McGlone et al., 2022 pp. 9–11). Loggers and sawmillers often circumvented these controls, however, and in response to their clearance, an assertive conservation movement emerged in the 1970s to fight for the preservation of natural old-growth forests. Still, tawa forests were being logged for chipping and wood pulp and indigenous forests were being clear-felled for exotic plantations.

While the Forest Service tried to study the sustainable management of indigenous forests, their efforts were small scale and sporadic, as their journal noted in 1960:

> The fact remains that planned experimentation as the basis of scientific study has been notoriously absent. In the whole of the country's indigenous forest estate there are probably not five sets of even small plots which have been studied and tended continuously over a period of the 40 consecutive years of departmental existence, let alone any longer. (McGlone et al., 2022 p. 11)

In 1987 the Forest Service was disestablished. Old growth indigenous forests were handed over to the newly formed Department of Conservation, and their logging was prohibited. At the same time, many exotic state plantations were sold to private investors, including those established as 'conservation forests' in response to widespread erosion. Since that time, the clear felling of pine plantations has dominated timber production in Aotearoa New Zealand, and the export of raw logs to markets like China and India.

While there has always been an interest in the sustainable management of indigenous forests in Aotearoa New Zealand, then, this has never happened at scale. Research and investment has overwhelmingly been channeled into industrial forestry with exotic monocultures, especially *pinus radiata*. This remains the case after concerns about climate change led to the creation of the Emissions Trading Scheme, and the use of production forests and then carbon farming with pine trees for carbon sequestration.

While these processes proved disruptive in many parts of Aotearoa New Zealand, the environmental impacts have been particularly severe in Tairāwhiti, the east coast of the North Island, which has some of the most erodible land in the world (Marden, 2009, p. 1). Here, concerns about the devastation caused by commercial forestry and carbon farming have come to a head in very recent times.

6.3 Forests in Tairāwhiti: Manu Caddie

6.3.1 Matemate-a-one

Wānangahia te atua o te rangi
O te rā, o te marama, o te whetū, o te kapua, o te hau
Tātaihia ko te kawa o te rangi,
Ko te nui o te rangi, ko te hua o te rangi
Ko te paki o te rangi, ko te pū o te rangi

Ko te mana o te rangi ka tau hā,
Whakatau ko te rangi e tū iho nei

Ka tau hā, whakatau
Ko te whenua e takoto ake nei
Takitakina ko te kawa o Papa
Ko te horanga ā-nuku, he atua!

He tipu te atua, he rākau te atua,
He ngārara te atua, he manu te atua
He oneone te atua e tangi ai te mapu

I te korowai ka takapau he kawa ora!

Let us invoke the atua of the sky
Of the sun, of the moon, the stars, the clouds and of the wind
Trace the kawa (ritual) of the sky
The expanse of the sky, the abundance of the sky
The glow of the sky, the origin of the sky
Welcome the mana of the sky
Invoke the sky that stands here

Let us invoke
the land that lies here
Trace the kawa of Papa
The spreading out of the earth, an atua!

The atua is growth, the atua is a tree
The atua is a reptile, the atua is a bird
The atua is the earth, weeping with indrawn breath

The cloak lain down as a kawa of life!
(Wayne Ngata, in Mana Taiao Tairāwhiti, 2023)

In this chant, Wayne Ngata summons up the ancestral powers of earth and sky. A former Māori Language Commissioner from Ūawa (Tolaga Bay), Tairāwhiti (the East Coast, on the North Island of Aotearoa New Zealand) Ngata recited this *karakia* (incantation) to formally open a meeting between representatives of Mana Taiao Tairāwhiti (literally, the mana of Tairāwhiti's living world), concerned Tairāwhiti residents, and the Ministerial Inquiry into Land Use. This Inquiry was established by the New Zealand government in the wake of two devastating cyclones in early 2023, to investigate the impacts of land use in the region and the causes and solutions for forestry slash (logs and other harvest waste from commercial pine plantations).

In his *karakia*, Ngata calls upon the power of *Papatūanuku*, the Earth Mother, who is weeping with indrawn breath. During Cyclones Hale and Gabrielle, the land was ravaged by storms that that stripped millions of tonnes of soil off the hills, ripping away the cloak of trees and leaving the earth naked and bare. Gouged by cut logs and toppled trees, local landscapes were scarred by more than one million landslides. As they moved downstream, rafts of logs formed dams in gullies, around bridges and on river bends that burst, creating mini-tsunamis. During the floods, forestry waste smashed through fences, roads and bridges, destroying paddocks, orchards, crops and vineyards, farm buildings and houses, choking streams and rivers, and smothering beaches and harbours with sediment and slash. More than 60 bridges in the region were destroyed or damaged. The impact on local landscapes and communities was devastating. Property, livelihoods and lives were lost, the link with land and sky was ruptured. Local people became fearful when it rained, and mourned the loss of familiar landmarks. For Māori, this kind of grief is known as *matemateāone*, literally 'the many deaths of the earth.' *Matemateāone* has a wide range of meanings, although these converge upon pain arising from relationships between people and the land. According to linguist Wharehuia Milroy:

> Matemateāone grows from within the group, knowing and getting to know each other. The physical cues such as trees, mountains, rivers and kainga [settlements] etc. are all factors that activate matemāteaone. (Milroy in Te Awekotuku & Nikora, 2003, p. 23)

In Tairāwhiti, this kind of grief has flowed in and out like the tide. Over the past ten years, it has surged as the region has experienced an accelerating series of catastrophic floods, caused in large part by the way land in the region has been used and abused during human occupation.

When the first humans arrived in Tairāwhiti about 800 years ago, local landscapes were cloaked in temperate rain forests. In coastal areas, where native forests were cleared for gardens and settlements on the foothills and river flats, forest stands were left for building timber, birds, bush foods, and fibres. (Coombes, 2001) In the hinterland, forest resources were seasonally harvested. Although about one third of Aotearoa was cleared of native forest prior to European arrival, mostly by burning, indigenous forests in Tairāwhiti remained relatively intact (except around coastal bays) until the late nineteenth century. At about that time, according to a European bushfeller in the region:

> The steep hills and river flats were bush covered right down to the beds of the rivers, which were hard and full of huge boulders. The water was clear and sweet, and it ran fast. Children swam in the clear pools, and there were eels, native trout and fresh water mussels. The native bush was beautiful. It was full of tawa, with plenty of totara, white pine and matai. There was beech forest at the higher levels. There were pongas and ferns of all sorts, and the undergrowth was thick and green. (Howard, 1976, p. 4)

After European settlement in Tairāwhiti, from the 1880s the land was surveyed into blocks, titles were awarded by the Native Land Court and land sales and clearance accelerated dramatically. In the early 1900s, the Royal Commission on Forestry stated that "No forest land, except if it be required for the special purposes of a climatic or scenic reserve and which is suitable for farm land, shall be permitted to remain under forest if it can be occupied and resided upon" (Royal Commission on Forestry, 1913, C12 XX).

Across the region, large areas of bush were felled and burned for pasture for sheep farming, including on Māori lands. Native trees were felled for houses, bridges, fences, jetties, farm and office buildings, and mills were set up to process the timber. This phase of forestry relied on native forests, but no attempt was made to manage them sustainably (Coombes, 2001). The erosion from this large-scale land clearance across the region was catastrophic. According to two geologists employed in 1920 to assess the impacts on the Waipaoa catchment, the results included:

> greatly increased sheet-washing of the soils; great increase in the number of slips, slumps and rain-gullies; aggradation of the stream-beds; wandering of the streams over valley-bottoms; lateral erosion of the river-banks; burying of culverts and bridges; filling-in of the Gisborne Harbour; and more severe and frequent floods. (Henderson and Ongley, 1920, p. 29)

Although Macintosh Ellis, the first Director of Forests in New Zealand (1920–1928), suggested the restoration and sustainable harvesting of native forests, government initiatives including Land Clearance grants, Stock Retention schemes, Land Development and Marginal Lands Loans schemes encouraged further clearance of land in the region that had reverted to shrublands since the 1930s. These provided subsidies for land, fertiliser grants, reduced loans, and guaranteed minimum livestock prices (Fuller et al., 2023, p. 429). It was not until the late 1950s that afforestation schemes were proposed as a way forward in Tairāwhiti. The afforestation schemes involved mass plantings of Pinus radiata, an exotic conifer, initially as conservation plantations, and targeted planting of willow and poplar on erosion-prone hillsides and gullies. (Gundry, 2012, p. 222) A timely opportunity to explore sustainable approaches to native forestry was squandered.

Even so, remaining stands of forest near Māori settlements were still extensively used for food and other resources. As *kaumātua* (elder) Rūtene Irwin reported:

> When I was a kid the Pipiwhakao forest provided for many of my family's needs: It was our supermarket, you see. I'd go there and trap birds—kereru especially. We'd set up traps by bending branches and tying them down with string ... Wait, wait, wait some more, then ... dinner. At other times of the year, we'd collect kiekie leaves—the inner ones which are whiter, just like you'd go for the heart of a good lettuce. We call that tawhara.
>
> I'd collect tutu and drain the juices out so my mother could make a dessert out of it with karengo to set it. There were all sorts of berries and fruits which could be collected there. Even in the 1960s, those foods were important as [a] supplement to shop food. How much more important those forests must have been before there were shops. (Irwin in Coombes, 2001)

On farms owned by Māori, erosion was often blamed on poor land management, although farms owned by Pākeha (European New Zealanders) suffered similar damage. According to a Department of Lands and Survey report in 1964, for instance:

> The reasons usually given for the quite evident deterioration in Maori land in this part of the region include the difficult nature of the land itself, the unfavourable climate, the extensive erosion, the granting of leases to Europeans without right of compensation for improvements and without rights of renewal and the difficult title position of most of the land. All these reasons are valid but the most important of all is the personal factor.
>
> With some notable exceptions the Maori has yet to become a good farmer under present day conditions ... Many Maoris seem to lack some essential attribute for a business of this type. They do not generally show the needed ability to plan ahead and budget for essential requirements such as maintenance, topdressing, and stock replacement. They are usually good workers but not good managers. (Department of Lands and Survey, 1964, p. 83)

In 1979, the Raukūmara Forest Park was established, 110,000 hectares of mature indigenous forest extensively grazed by deer, goats, pigs, possums and other pests; and in 1987 the Park was handed over to the newly created Department of Conservation. In that same year, the New Zealand Forest Service was disestablished, the pine plantations that had been established for soil conservation were sold by the state to private interests, and wholesale harvesting began once again. As John Ruru, who had just become the district manager of the New Zealand Forestry Corporation in Tairāwhiti, lamented (in Gundry, 2012, p. 223):

> We had gone through the stage of acquiring the land, planting the trees and establishing the forest, and it had got to the stage of setting up the forests for harvesting. When the decision was finally made, it really cut up a lot of people, myself included, because I felt let down and cheated by the government.
>
> We were told a whole lot of lies. They had an ulterior motive and that was to sell the forests. My argument was that the forests couldn't be sold because of the reasons for which they were established—that is, erosion control… Fronting up to the people was the worst part, as virtually all of them lost their jobs when the forests were sold.

Pastoral farming on steep hillsides was still causing severe erosion, with the formation of gullies and sedimentation in rivers, most spectacularly during Cyclone Bola in 1988. In response to this and the declining profitability of sheep and beef farming, many local farms were sold to forestry companies. As Colin Williams, a local farmer, remarked (Gundry, 2012, p. 191):

> One of the big tragedies is that there's so much beautiful strong land on the Coast that's gone into trees. It would have been much better to keep it in farming, which is more profitable for people and communities like Tokomaru Bay, Ruatoria and Tolaga Bay. Now any profits from forestry—apart from local trucking firms and the like—goes overseas and is not spent locally. The profit from farming was always spent locally, providing employment for so many people. It's a green desert now.

After Cyclone Bola in 1990, Basil Graeme, North Island Conservation Officer for Forest & Bird, wrote about forestry in Tairāwhiti:

> There is a role for commercial forestry, enhancing the regional economic base, and for catchment management on the better hills close to the plains. However the most cost- effective catchment protection for a huge area between these different plantation types is to allow the bush to return.
>
> All this land needs is retirement and pest control; nature will do the rest. It is here in the naturally revegetating lands that conservationists can enhance the process with judicious planting of seed source trees that are regionally scarce. The manuka and scrub species are already there, in pockets and gullies, waiting to do their job.
>
> If it is appropriate for this region to return to subsidised pseudo-commercial forestry, then it is essential to first zone land suitable for commercial forestry and zone out with a green line those lands requiring permanent tree cover. We have enough to pay for without replacing one cycle of inappropriate land use with another.
>
> The nation does owe a debt to the East Cape. We have had 100 years of wrenching export earnings from these hills. We will all have to pay if the category 3 lands are to be purchased, replanted and retired. We will all have to pay if most of the category 2 lands are to be purchased and retired. Anything less than this commitment to a "green line" is a decision to let the region die. (Graeme, 1990, p. 39)

In 1992, local iwi Ngāti Porou set up its own commercial forestry operation in partnership with Hansol, a Korean company, expanding its holdings over the next twenty years and retaining jobs and profits in the region (Gundry, 2021, p. 224). Once again, however, a combination of pasture on erodible steeplands and the clear-fell harvesting of pine plantations led to severe flooding, now aggravated by forestry waste in the floodwaters, culminating in floods in Ūawa (Tolaga Bay) in 2018 that battered roads, fences, bridges and houses with tangled masses of forestry logs, and provoked a national outcry.

In the latest phase, carbon farming with pine trees, funded by the Emissions Trading Scheme, is expanding in the region, with highly erodible landscapes being sprayed with defoliant (Smale, 2023) and then mass planted with relatively short-lived, shallow rooting, highly flammable monocultures of pine trees (BDO Spicer, 2021). Given the risks of fire, disease and increasing storm damage, this is not a credible form of long-term carbon sequestration in a time of climate change. For local communities, pest- and weed-ridden plantations of ageing and dying pine trees across vast parts of most catchments are the most likely legacy of this latest central government policy, conceived at a distance from the region but impacting local landscapes throughout Tairāwhiti and beyond.

In September 2021 a group of local Māori organised an event to inform landowners about alternatives to carbon farming with pine trees, including indigenous afforestation. At this event, ‘Recloaking the Whenua,’ more than 300 people joined online sessions featuring land use researchers, central and local government officials, ecologists, farmers, entrepreneurs and conservationists. Following this event, a regional initiative called Te Weu Tairāwhiti was formed to focus on the impacts of carbon farming, climate change and land use in the region. This group began to interview local leaders and families working in farming and forestry sectors about the likely impacts of the rapid transition to ‘permanent pine’ that would no longer be harvested, and organised educational resources, events and roadshows to share information and facilitate feedback from residents.

Soon afterwards, a study examining the impacts of deforestation on five major catchments in Tairāwhiti noted that, in one of them, the Waiapū river is the highest sediment yielding river in New Zealand, and one of the worst in the world, with severe hillslope erosion and river bed aggradation that have accelerated sharply since human arrival (Fuller et al., 2023). After the cyclones, Manaaki Whenua Landcare Research, a Crown Research Institute, used satellite imagery to analyse how East Coast landscapes under different kinds of vegetation cover had fared. They found that the land under indigenous forests held the land together five times better than pine plantations, and ten times better than harvested plantations. (McMillan et al., 2023).

6.3.2 Indigenous Reforestation in Tairāwhiti

Alongside the expansion of commercial *pinus radiata* plantations in Tairāwhiti, and the experience of widespread pre-planting and post-harvest erosion, various projects to restore these fragile landscapes into indigenous forest emerged from about 2000 onwards, although compared with industrial plantations, most of the early projects were relatively small in scale. These include the Whinray Ecological Trust, formed to restore Whinray Reserve in 1999 (DoC, 2024); Waikereru on the Waimatā River, led by Dame Anne and Jeremy Salmond from 2000 (Waikereru, 2024); Te Kurī a Paoa / Young Nick's Head and around Te Wherowhero Lagoon south of Gisborne city, initiated by Ngai Tamanuhiri with Geoff Park in 2001, and by landowner John Griffin (Park, 2001; Wikipedia, 2024); the Uawanui Sustainability project led by Te Aitanga-a-Hauiti in collaboration with the Allan Wilson Centre for Research Excellence from 2011 (Massey University, 2024); the restoration of indigenous forests for carbon farming on Nuhiti Q station in Anaura, led by the Ferris whānau from 2011 (Mercer, 2018); and the restoration of Waikanae Stream led by Ngai Tāwhiri and Gisborne District Council in Gisborne city from 2011 (Gisborne District Council, n.d.).

More recently, other indigenous reforestation projects have been initiated in the region, including Whaia Titirangi, the restoration of Titirangi, the ancestral hill that stands alongside the port in Gisborne city, led by Ngāti Oneone since 2018 (Trees that Count, 2024); Raukūmara Pae Maunga, initiated by Ngāti Porou and Te Whānau-a-Apanui with the Department of Conservation, a massive project to restore 150,000 hectares of indigenous forest in the Raukūmara Forest Park which attracted a government grant of $34 million in 2020 (Raukūmara Pae Maunga, 2024); a range of iwi-led projects funded by the Jobs for Nature programme from 2021–24 (Gizzy Local, 2021); and the Toha project, launched in 2023, which offers Tairāwhiti biodiversity projects for investment (Toha, 2024). It is striking that the majority of these projects have been initiated by local hapū and iwi (kin groups), with much of the work of planting, pest and weed control being carried out by young Māori. Ancestral philosophies of care for the whenua remain strong, and these helped to inspire local responses to the cyclones.

6.3.3 Cyclone Gabrielle, the Mana Taiao Tairāwhiti Petition and the Ministerial Land Use Inquiry

Shortly after Cyclone Hale hit Tairāwhiti on 6 January 2023, a largely iwi-led group called Mana Taiao Tairāwhiti was formed and drafted a petition asking Gisborne District Council to support an independent inquiry into forestry and land use in the region, to tighten regional land use regulations and monitoring, and to back a community-led Just Transition planning process with central government support. Within two weeks, nearly 10,000 people signed the petition. The evening before the petition was presented to Council, a young boy was killed by a freshly cut log washed down from a harvest site on the Waimata River onto Waikanae Beach, where the river meets Te Moananui-a-Kiwa, the Pacific Ocean. After a highly charged meeting, the Gisborne District Council supported the petitioners, and formally asked the Government to establish an inquiry into forestry and land use in the region.

Three weeks later, on 12 February 2023, a second, more devastating cyclone hit the East Coast. During Cyclone Gabrielle, ten people were killed in Hawkes Bay and one in Tairāwhiti, and hundreds of roads, bridges, homes and crops were destroyed, resulting in damage estimated by Treasury at $9 to 14.5 billion (Pullar-Strecker, 2023). Within a fortnight, the Minister for the Environment and Minister for Forestry had announced a Ministerial Inquiry into Land Use in Tairāwhiti and Te Wairoa (the district to the south of Tairāwhiti). A former Cabinet Minister, Hekia Parata from Tairāwhiti, was appointed to chair the Panel, and two experts in resource management and forestry joined her to investigate historic and contemporary land use in the region, consult with the public, and report back to Ministers within a two-month timeframe (Parata et al., 2023).

Like other locals, Mana Taiao Tairāwhiti had very little time to organise their submission to the Inquiry. They sent out an online call for volunteers, and within 48 hours, over 110 researchers and writers from Aotearoa New Zealand and overseas responded. A dedicated team was set up to manage the process, and distinguished researchers from institutions including the London School of Economics and Cambridge University joined local experts and lay people, deciphering decades-old science, summarizing community, and land-based insights, and distilling several hundred research papers into key points and summaries. In this remarkable crowd-sourced initiative, the volunteer researchers were sorted into seven groups, each addressing a particular inquiry theme, holding on-line meetings and rapidly accumulating a vast document repository. Another team organised the primary material with a spreadsheet, while a separate group developed technology to allocate readings and manage summaries and reviews. This infrastructure was improvised on the fly, with Mana Taiao Tairāwhiti overseeing the coordination.

One week before the deadline, a group of readers and reviewers formed a dedicated writing team that distilled the work of the researchers into a 29-page, four-part submission. This made three key recommendations, asking the government to:

1. Order the immediate cessation of clear-felling on erosion-prone land;
2. Support the immediate native reforestation of harvested and erosion-prone grasslands;
3. Order the cessation of planting pine trees for harvest or carbon farming on erosion-prone land.

The Mana Taiao Tairāwhiti submission also included a presentation deck, a comprehensive 58 page thematic summary, a searchable annotated database of reviewed documents, and five papers written by subject experts, in addition to the main document, all prepared in less than a month (Mana Taiao Tairāwhiti). Alongside this detailed work, in April 2023 Renee Raroa, a young Maori leader from Tairāwhiti, presented an eloquent plea to the UN Permanent Forum on Indigenous Issues in New York on behalf of Mana Taiao Tairāwhiti, outlining the devastation in the region and stating, 'The only acceptable solution is an immediate moratorium in clear-felling, and the restoration of biodiverse native cover' (Raroa, 2023). The Inquiry received more than 600 submissions, and the Panel's report contained 49 recommendations (Parata et al., pp. 10–41). All of Mana Taiao Tairāwhiti's recommendations were included in the Inquiry Panel's final report to government (Ministerial Inquiry into Land Uses in Tairawhiti and Wairoa, 2023).

6.3.4 Recloaking Papatūānuku

The seed for a new national initiative, 'Recloaking Papatūanuku,' was sown at the 'Recloaking the Whenua' conference organised by the group that would become Mana Taiao Tairāwhiti in September 2021, and at a later conference 'O Tātou Ngāhere' sponsored by Pure Advantage (a group of business leaders dedicated to a sustainable future for Aotearoa New Zealand, see Pure Advantage, 2023) and Tāne's Tree Trust (researchers into indigenous forests), held in Wellington on 27–28 October 2022. Informed and guided by *mātaraunga* Māori (Māori traditional knowledge), "Recloaking Papatūānuku" is designed and supported by a consortium including Pure Advantage and Tāne's Tree Trust; a range of Māori entities and landowners with deep connections with ancestral lands, including Mana Taiao Tairāwhiti; and leading environmental NGOs, among others.

'Recloaking Papatūānuku' is a nature-based plan based on *mātauranga* and leading-edge science and practice. It aims to reforest and rehabilitate 2.1 million hectares of indigenous forests around waterways and on highly erodible land across Aotearoa New Zealand over the next decade. The protective cloak of forests will serve as permanent carbon sinks and habitat for many rare and endangered species, enrich and stabilise soils, enhance the quality of waterways, create jobs and weave climate and ecological resilience into landscapes across New Zealand. The Recloaking Papatūānuku' business plan has been rigorously tested by leading analysts, working on a voluntary basis. Landowners would be paid by the New Zealand government in perpetuity to protect existing indigenous forest and/or retire low productivity farmland, and funds from the Nationally Determined Commitments for net emission reductions under New Zealand's Paris Agreement would pay for initial pest control, fencing and planting and ongoing management of the forests.

While the New Zealand government proposes to meet much of its Paris Agreement commitment by purchasing offshore credits, this strategy does not enjoy wide public support. Schemes for planting trees to offset emissions, particularly exotic monocultures, have been harshly criticised for allowing polluters to continue to pollute, while causing additional environmental damage. The costs of international credits are high (potentially $24 billion for New Zealand's commitment), their ongoing availability is uncertain, and some of these schemes have been revealed as fraudulent.'Recloaking Papatūānuku' offers an alternative with multiple environmental and social benefits. It provides a pioneering example of a nature-based, Indigenous-inspired solution to address the crises of climate, biodiversity and culture loss, with the urgency and ambition that Papatūānuku (Earth Mother) deserves. The success of this initiative could render Aotearoa New Zealand the first nation in the world to achieve a nature-positive status.

With the support of multiple stakeholders across Aotearoa New Zealand, this transformative initiative is well placed to literally change the face of the country, and so far the signs are positive. ‘Recloaking Papatūanuku’ has been profiled by its stakeholders at COP28 in Dubai in December 2023, and presented to the incoming New Zealand government. Most recently, a documentary about the project, *Think like a Forest,* featuring environmental practioners from Tairāwhiti, was launched at Parliament by the Minister for Climate Change, Simon Watts (https://www.youtube.com/watch?v=PuS6G_Rbwl). Working with Papatūānuku, Ranginui (Sky Father) and their children to recloak the *whenua* with the cloak of Te Wao Nui ā Tāne is a daunting challenge, for the Tairāwhiti region and for the nation. For the sake of future generations, our responses must be active, courageous and strategically astute, not ill-considered, evasive and lazy.

Mauri tū, mauri ora.
Mauri noho, mauri mate.
Stand up, and live.
Sit down, and die.

6.4 Conclusion

From the Land Wars of the 1860s to the Land March and occupations in the 1970s, and since the establishment of the Waitangi Tribunal in 1975, Māori kin groups have fought fiercely for their *whenua* (lands). Despite past and ongoing ravages of landscapes across the country, the ancestral idea of land, forests, rivers, the ocean and people as living beings interconnected in a vast kin network remains potent, generating world-leading environmental initiatives. In 2017, for instance, the Whanganui River became the first river in the world to be recognised as a legal person, with its own life and rights. A fight to recognise the rights of Hine-moana, the ocean, is also under way. ‘Recloaking Papatūānuku’ is another example of this philosophy; with the proliferation of Indigenous reforestation initiatives across Tairāwhiti testifying to the flax-roots strength of these commitments.

Many of the rising generation, taught in *kohanga reo* (preschool language ‘nests’), *kura kaupapa* (Māori immersion schools) and *whare wānanga* (Māori tertiary institutions) and schooled in *Te Reo* (Māori language) and *tikanga* (ancestral ways of being), are deeply committed to caring for Papatūānuku, with her forests, mountains, rivers and oceans. These young people are prepared to fight for te *tino rangatiratanga* (the independence) of their *whenua*, *kāinga* (dwelling places) and *taonga* (treasures), promised to their ancestors by Queen Victoria in Te Tiriti o Waitangi in 1840. During 2024, attempts led by a small libertarian political party in the coalition government to erase that Treaty promise triggered protests even larger than the 1975 Land March.

The perennial struggle for the wellbeing of the *whenua*, rivers, forests and rights of Māori continues into a new generation. In a time of collapsing ecosystems and climate change, these young leaders are taking their fight to the United Nations, and their message to the world is powerful. As Renee Raroa (2023) urged the UN Permanent Forum on Indigenous Issues in New York: "We call on the Forum to encourage all member states to center the rights and wellbeing of Indigenous Peoples, indigenous ecosystems, land and waters as the only true pathway to climate stability. In this there is hope for all."

References

BDO Spicer. (2021). *Report on the impacts of permanent carbon farming on Te Tairawhiti region: Stepping into the shoes of our Mokopuna.*

Beattie, J., & Starr, P. (2010). Global influences and local environments: Forestry and forest conservation in New Zealand, 1850s–1925. *British Scholar, 3*(2), 191–218.

Best, E. (1976). *Maori religion and Mythology Part I.* A.R. Shearer.

Coombes, B. (2001). *Ecological impacts and planning history, Gisborne Inquiry, Wai 814 A20.* Waitangi Tribunal.

Department of Conservation. (2024). https://www.doc.govt.nz/parks-and-recreation/places-to-go/east-coast/places/whinray-and-motu-scenic-reserves/whinray-scenic-reserve/

Department of Lands and Survey. (1964). *The land utilisation survey of Gisborne – East Coast Region.* Department of Lands and Survey.

Forster, G. (2000). *Voyage Round the World* (ed Thomas, N. & Berghof, O.) University of Hawai'i Press.

Fuller, I. C., Brierley, G. J., Tunnicliffe, J., Marden, M., McCord, J., Rosser, B., Hikuroa, D., Harvey, K., Stevens, E., & Thomas, M. (2023). Managing at source and at scale: The use of geomorphic river stories to support rehabilitation of Anthropocene riverscapes in the East Coast Region of Aotearoa New Zealand. *Frontiers in Environmental Science, 11*, 1162099.

Gisborne District Council. (n.d.). https://www.gdc.govt.nz/__data/assets/pdf_file/0018/7281/gdc-waikanae-stream.pdf

Gizzy Local. (2021). Jobs for nature. https://www.gizzylocal.com/post/jobs-for-nature

Graeme, B. (1990). *Forest & Bird Magazine*, November.

Gundry, S. (2012). *A splendid isolation Gisborne: East Coast 1950–2012.* H.B. Williams Memorial Library Book Trust.

Gundry, S. (2021). The Waimata River: Settler history post 1880, Research Report 3, Te Awaroa. https://www.waikereru.org/assets/documents/WaimataReport3.pdf

Hecht, S. B., Morrison, K. D., & Padoch, C. (2014). *The social life of forests: Past, present and future of Woodland resurgence.* University of Chicago Press.

Henderson and Ongley. (1920). *The geology of the Gisborne and Whatatutu subdivisions, Raukumara division.* NZGS.

Howard, G. (1976). *Erosion al Mangatu.* Government Printer.

Knight, C. (2009). The paradox of discourse concerning deforestation in New Zealand: A historical survey. *Environment and History, 15*(3), 323–342.

Kohn, E. (2013). *How forests think: Toward an anthropology beyond the human.* University of California Press.

Kuzma, J. (2003). New Zealand landscape and literature, 1890–1925. *Environment and History, 9*(4), 451–461.

Locke, J. (1690). *Second treatise of civil government.*

Mana Taiao Tairāwhiti. (2023). Taiao Mātāmua Mana Taiao Tairāwhiti submission to the Ministerial Inquiry on Land Use in Tairāwhiti, Tūranganui-a-Kiwa and Te Wairoa. https://manataiao.wordpress.com/wp-content/uploads/2023/04/taiao-matamua-mana-taiao-tairawhiti-submission_corrected.pdf

Marden, M. (2009). Gullies supply most sediment to major East Coast river systems. *Farm Forestry New Zealand, 2009*, 1.

Massey University. (2024). https://www.massey.ac.nz/research/research-impact-stories/massey-uawa-partnership-delivers-environmental-restoration-and-safe-local-food-sources/

McGlone, M., Bellingham, P., & Richardson, S. (2022). Science, policy, and sustainable Indigenous forestry in New Zealand. *New Zealand Journal of Forestry Science, 52*(8). https://doi.org/10.33494/nzjfs522022x182x

McMillan, A., Dymond, J., Jolly, B., Shepherd, J., & Sutherland, A. (2023). *Rapid assessment of land damage – Cyclone Gabrielle*. Manaaki Whenua Landcare Research.

Mercer, L. (2018). *Testing the New Zealand ETS to facilitate native forest regeneration on Māori land: The ETS and Māori landowners: The Nuhiti-Q case study*. https://waronz.wordpress.com/wp-content/uploads/2018/12/The-ETS-and-M%C4%81ori-landowners-the-Nuhiti-Q-case-study-405423

Ministerial Inquiry into Land Uses in Tairawhiti and Wairoa. (2023).*Outrage to Optimism: Report of the Ministerial Inquiry into land uses associated with the mobilisation of woody debris (including forestry slash) in Tairāwhiti / Gisborne District and Wairoa District.*. https://environment.govt.nz/assets/Outrage-to-optimism-superseded.pdf

Parata, H., et al. (2023). *Outrage To Optimism: Report of the Ministerial Inquiry into land uses associated with the mobilisation of woody debris (including forestry slash) and sediment in Tairawhiti/Gisborne District and Wairoa*. Ministry for the Environment.

Park, G. (2001). *An ecological plan for restoring the indigenous ecosystems of Te Kuri a Paoa (Young Nick's Head) and Te Wherowhero, Muriwai*, prepared for Ngai Tamanuhiri by Geoff Park, National Library MS-Papers-9392-096.

Pullar-Strecker, T. (2023). Repair bill from cyclone and Auckland floods at least $9b, Treasury estimates. *Stuff.co.nz*, April 27. Retrieved December 1, 2024, from www.stuff.co.nz/business/131883544/repair-bill-from cyclone-and-auckland-floods-at-least-9b-treasury-estimates

Pure Advantage. (2023). *Recloaking Papatūānuku* on Pure Advantage website. Auckland. Retrieved December 1, 2024, from https://pureadvantage.org/recloaking-papatuanuku/

Raroa, R. (2023). Speech to UN permanent forum on Indigenous issues. https://www.youtube.com/watch?v=U8DVrGypDlY

Raukūmara Pae Maunga. (2024). Raukūmara Pae Maunga. https://www.raukumara.org.nz/ (accessed 12 September 2025).

Royal Commission on Forestry. (1913). *The report of the Royal Commission on Forestry*.

Shortland, E. (1851). *The Southern Districts of New Zealand*. Longman, Paul, Green and Longmans.

Smale, A. (2023). Tairāwhiti's Trauma, Honouring the Future. *Newsroom*. https://newsroom.co.nz/2023/10/07/honouring-the-future/

Smith, S. P. (1913). *The Lore of the Whare Wananga, or the Teachings of the Maori College*. Thomas Avery for the Polynesian Society.

Te Awekotuku, N., & Nikora, L. W. (2003). *Nga Taonga o Te Urewera. A Report prepared for the Waitangi Tribunal's Urewera District Inquiry*, August (Wai 894, doc B6).

Toha. (2024). Toha network, how it works. https://toha.network/#howitworks

Trees That Count. (2024). https://treesthatcount.co.nz/profiles/whaiatitirangi?offset=0&page=1&limit=12

Tsing, A. L. (2021). *The mushroom at the end of the world: The possibility of life in capitalist ruins*. Princeton University Press.

Tsing, A. L., Bubant, N., Gan, E., & Swanson, A. (2017). *Arts of living on a damaged planet: Ghosts and monsters of the anthropocene*. University of Minnesota Press.

Vilaça, A. (2007). Cultural change as body metamorphosis. In C. Fausto & M. Heckenberger (Eds.), *Time and memory in Indigenous Amazonia: Anthropological perspectives* (pp. 169–193). University Press of Florida.

Viveiros de Castro, E. (1998). Cosmological deixis and Amerindian perspectivism. *Journal of the Royal Anthropological Institute, 4*, 469–488.

Waikereru. (2024). https://www.waikereru.org/

Wikipedia. (2024). Young Nick's Head. https://en.wikipedia.org/wiki/Young_Nick%27s_Head

Wulf, A. (2015). *The invention of nature: Alexander von Humboldt's new world*. Vintage Press.

Chapter 7
Cultural Waters: Integrating Indigenous Knowledge and Governance for Justice and Sustainability in Central Queensland

Clint J. Hansen and Matthew J. Currell

Abstract This chapter examines the cultural significance of water for the Iman People of Central Queensland, focusing on the impacts of Coal Seam Gas (CSG) extraction. By documenting the views and lived experiences of Iman People, it examines how these values might shape regenerative futures and advance restorative justice for Indigenous Peoples, who are often marginalized in extractive industry agreements. The analysis emphasizes the necessity of incorporating Indigenous Cultural Values into water management plans, both above and below ground, to ensure sustainability and justice. This study highlights the transformative potential of integrating cultural and scientific approaches to water management, with implications for governance reform and sustainability.

Keywords Indigenous water values · Cultural significance · Water management · Sustainability · Ecological stewardship

7.1 Introduction

Water is fundamental to life, deeply woven into the cultural, spiritual, and ecological systems of First Nations communities globally (Dixon & Morgan, 2023; Kelsey Leonard et al., 2023; Laborde & Jackson, 2022; Pueblo Action Alliance, 2021). For the Iman People of Central Queensland, water is not merely a resource but a sacred

C. J. Hansen (✉)
Faculty of Engineering, Monash University, Clayton, VIC, Australia
e-mail: clint.hansen@monash.edu

M. J. Currell
School of Engineering, Royal Melbourne Institute of Technology, Melbourne, VIC, Australia

The Australian Rivers Institute, Griffith University, Brisbane, QLD, Australia

M. Kennedy, E. O'Donnell (eds.), *People, Place and Nature in Indigenous-Settler Relations*, Indigenous-Settler Relations in Australia and the World 10, https://doi.org/10.1007/978-3-032-03600-1_7

entity sustaining life, culture, and identity (Hansen et al., 2024; Marshall, 2017). This paper examines the cultural values of water in the context of Coal Seam Gas (CSG) activities on Iman lands, and explores how these values might guide the creation of regenerative futures, while advocating for justice for Indigenous communities globally.

Indigenous worldviews often regard water as a living being, essential to the well-being of communities and ecosystems (Hurlbert et al., 2024; A. Poelina et al., 2021). For the Iman People, water embodies both spiritual and physical sustenance, intertwining with the health of the People, Land, and environment (Nenquimo & Anderson, 2024). This understanding starkly contrasts with industrial paradigms, which often view water as an economic resource to be extracted, stored, and distributed for commercial gain. In the context of Coal Seam Gas (CSG) extraction, industrial actors prioritize water extraction for fracking and production processes, leading to significant depletion of groundwater reserves and contamination of vital waterways. These processes disregard the cultural and ecological significance of water, posing existential threats to First Nations communities who rely on these waters for their spiritual, cultural, and physical well-being. The exploitation of water in these industries is particularly harmful as it disrupts Indigenous governance systems, erodes cultural connections to water, and accelerates environmental degradation that disproportionately affects Traditional Owners and local ecosystems (Mercer et al., 2014; Osborne et al., 2024; Poelina et al., 2023a).

7.1.1 Iman Country: Land, Water, and Cultural Significance

Iman Country is located in Central Queensland (see Fig. 7.1), covering a diverse landscape of rivers, floodplains, groundwater basins, and dryland ecosystems. The Dawson River (Wardingarri) is the lifeblood of Iman Country, playing a crucial role in cultural ceremonies, songlines, and ecological sustainability.

Key Water Systems at Risk from CSG Mining

- Dawson River (Wardingarri): A major river system central to Iman cultural identity and spiritual practice. It provides drinking water, supports aquatic totems, and sustains wetland ecosystems threatened by CSG development.
- Fitzroy Basin: The largest river basin draining into the Great Barrier Reef, heavily impacted by industrial water extraction for mining and agriculture.
- Groundwater and Artesian Springs: The Great Artesian Basin (GAB) underlies Iman Country, feeding underground water seeps, mound springs, and creeks that hold cultural significance.
- Seasonal Wetlands and Billabongs: Temporary water bodies provide habitat for native fauna, totemic species, and medicinal plants, all of which face disruption from CSG-related groundwater drawdowns.

Water sources within Iman Country hold profound ancestral and cultural significance, serving as custodians of heritage and sites for ceremonies, totemic

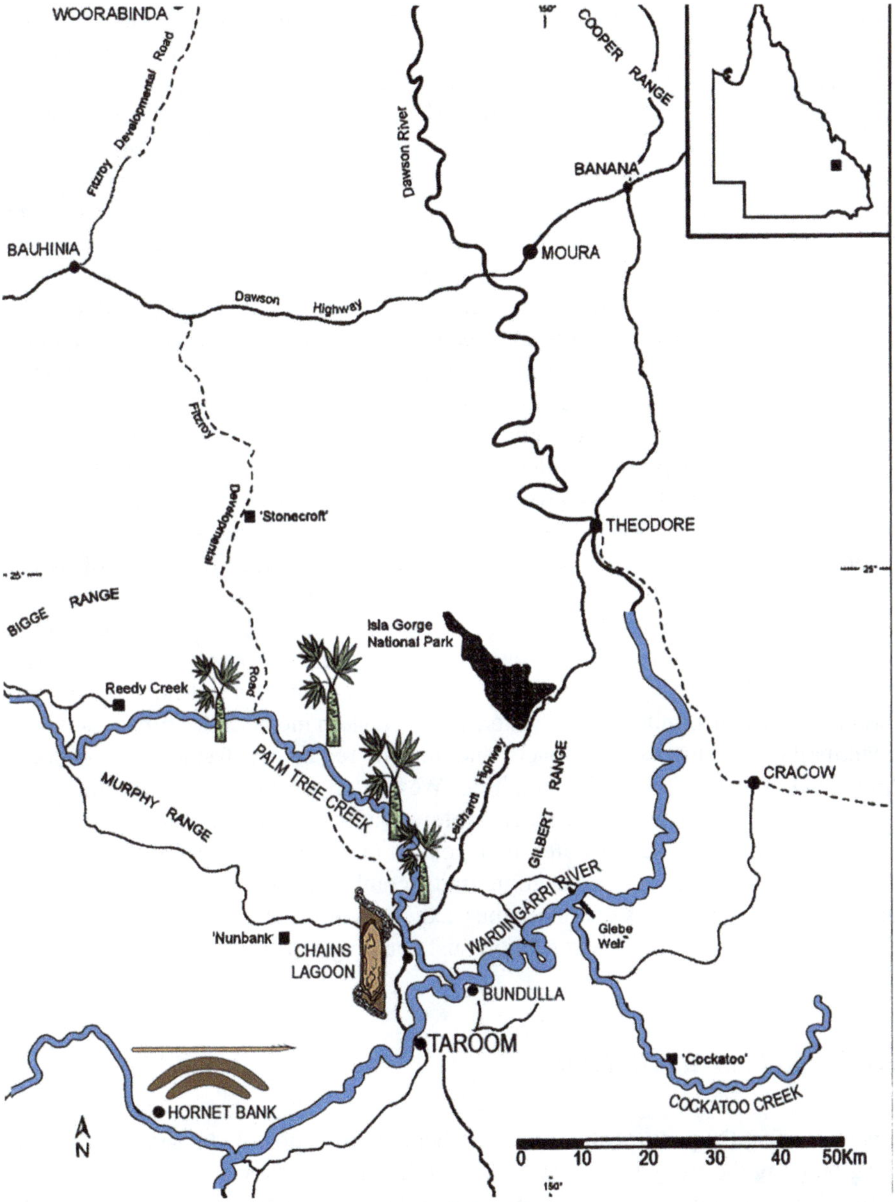

Fig. 7.1 Map adapted from L'Oste-Brown, (1995) showing some of the culturally significant places on Country along with the, Bundulla Mission (Previously known as Taroom Aboriginal Reserve), Chains lagoons, Hornet bank and Palm tree creek

connections, and the sustenance of flora and fauna central to the Iman identity. Any threat to these waters imperils not only the environment but also the cultural and spiritual integrity of the Iman People (Currell et al., 2022). This perspective contrasts with conventional scientific approaches, focusing on empirical assessments of water's physical and chemical properties (Jackson & Palmer, 2012; Weir, 2009). While these methods are vital for understanding environmental processes and impacts, they often overlook the interconnectedness and sacredness emphasized in Indigenous Knowledge systems (Moggridge et al., 2019). Bridging these paradigms through an integrative approach can enhance water management by fostering strategies that are both scientifically robust and culturally respectful (Berkes, 2018).

The extraction of CSG in Central Queensland has raised critical concerns among the Iman People, particularly regarding the contamination and depletion of vital water resources, both above and below ground (Campbell et al., 2024; Hansen et al., 2024). CSG operations extract considerable quantities of groundwater (OGIA, 2021) and frequently produce water – termed 'co-produced water' by the industry – containing pollutants that risk groundwater and surface water quality and threaten the survival of culturally significant species (Khan & Kordek, 2014; Hadwen Georgy, 2023). The Iman People's fears reflect the broader implications of industrial activity on ecosystems and Indigenous cultural heritage.

Integrating Indigenous scientific and Western scientific methods of water quality assessment provides a path toward sustainable management practices (Johnson et al., 2016). Indigenous approaches prioritize long-term stewardship and the cultural and spiritual significance of water, while Western methods contribute detailed, quantitative data to assess impacts and inform regulatory frameworks (López-Maldonado et al., 2024). By aligning Western hydrological monitoring with Indigenous perspectives, this research explores pathways for integrating Indigenous and scientific approaches to water management on Iman Country. Drawing on survey data collected from the Iman community, combined with collaborative methodologies, this chapter seeks to honour the sacredness of water, uphold cultural values, ensure ecological integrity, and support sustainable governance frameworks.

7.1.2 Reflexivity Statement

Understanding the positionality of researchers in Indigenous-led research is crucial for grounding the study in ethical and culturally appropriate methodologies. As co-authors of this chapter, we bring distinct but complementary perspectives to this work.

Hansen: As a First Nations Iman scholar with a background in engineering, I bring a unique dual perspective to this research. My technical expertise enables rigorous quantitative analysis, while my lived experience as an Iman Countryman informs qualitative and culturally sensitive methodologies. This duality supports a balanced, respectful approach that honours both scientific rigour and Indigenous Knowledge.

Currell: As a non-Indigenous scholar trained in water science and engineering, taught overwhelmingly from Western perspectives, I approached this research primarily from the perspective of seeking to listen and learn from First Nations peoples. I am looking for ways in which my training may help assist and support Iman peoples' aspirations for empowerment and greater involvement in decisions affecting the waters of their Country.

By bringing these perspectives together, this research aims to reflect both Indigenous and scientific approaches to water management while advocating for knowledge-sharing, cultural respect, and governance reforms that centre Indigenous voices

7.1.3 Objectives and Research Hypothesis

The central research hypothesis is that integrating Western hydrological methods with Indigenous Knowledge/Science will enhance the effectiveness and sustainability of water management. However, further evidence is needed to evaluate how such integration influences outcomes in practice. For example, assessing changes in ecological health, cultural recognition, and governance processes before and after the meaningful inclusion of Indigenous Knowledge. Such evidence will take considerable time to collect and analyse and will require long-term commitment, cooperation, and shared decision-making between Indigenous knowledge holders, water managers, policymakers, and scientific researchers. At this stage, it was determined that a key step was to further understand Iman peoples' views and knowledge regarding water impacts of CSG development, and the level of collaboration and consideration of Indigenous perspectives in water management and governance, as identified by their lived experiences.

By bringing together the rigorous observational methods of Western science with the deep, relational understanding of water embedded in Indigenous Knowledge systems, this research asserts that water management practices can be strengthened through a more holistic and just approach. Indigenous Knowledge, which has sustained ecosystems and communities for millennia, offers sophisticated understandings of hydrological systems, ecological interdependencies, and cultural governance structures. Integrating these approaches fosters more comprehensive, culturally grounded, and environmentally sustainable water management strategies. This synergy is expected to enhance the protection of natural water sources, amplify recognition of water's cultural and spiritual significance, and lead to more just and informed decision-making in water governance.

The findings of this research aim to inform and advocate for changes in water management practices by emphasizing the importance of incorporating Indigenous perspectives into hydrological monitoring and assessment. Additionally, the outcomes will contribute to educational efforts by highlighting the cultural significance of water sites from an Indigenous viewpoint and promoting best practices that integrate both scientific data and Indigenous Knowledge. Ultimately, this integrative approach aspires to support the development of sustainable, culturally respectful, and inclusive water management strategies in Central Queensland and beyond.

7.1.4 *Significance of the Study*

The significance of this study lies in its exploration of how Indigenous perspectives on water, particularly those of the Iman People in Queensland, can be integrated into future water management practices and (potentially) treaty negotiations (Creamer et al., 2024). By highlighting the cultural, spiritual, and ecological importance of water to Indigenous communities, the study provides a framework for more inclusive and sustainable management strategies that respect and uphold Indigenous rights. The research calls attention to the need for a holistic approach to water management that goes beyond Western scientific methods, incorporating Indigenous Knowledge to ensure the long-term health of water resources. This study not only contributes to the academic understanding of Indigenous water values but also has practical implications for policy-making, providing insights that help to promote more equitable and culturally sensitive governance structures in natural resource management. This will be of great importance both locally and globally – for example, providing a stronger basis to inform treaty negotiations in Queensland/Australia (where the research is focussed), and learnings and insights for other Indigenous Peoples seeking ways to safeguard the health of waters and country for intergenerational benefit and justice.

7.1.5 *Coal Seam Gas (CSG) Mining: Methods, Impacts, and Scale in Australia*

Coal Seam Gas (CSG) extraction is a form of unconventional gas mining that involves extracting methane trapped in underground coal seams. Unlike conventional gas reserves, which exist in porous rock formations, CSG is held within coal seams by water pressure. The extraction process typically involves:

- Drilling Wells: Wells are drilled into underground coal seams, often several hundred meters deep, to reach gas reserves.
- Hydraulic Fracturing ("Fracking") or Dewatering: Depending on the seam's permeability, gas is released either by pumping out groundwater (depressurization method) or through hydraulic fracturing (fracking), which injects water, sand, and chemicals at high pressure to break the coal seam and release gas (de Rijke, 2018).

- Produced Water Management: The extracted water, known as produced water, can contain heavy metals, salts, and other contaminants, posing risks to surrounding ecosystems (Currell et al., n.d.).

7.1.6 Impacts of CSG Mining on Land, Water, and Ecosystems

The environmental and cultural consequences of CSG extraction are significant, particularly in areas with deep cultural and ecological value, such as Iman Country. Key risks include:

- Water Contamination and Groundwater Drawdown: The removal of large volumes of groundwater can lower the water table, affecting springs, wetlands, and cultural water sites. Contaminants such as sodium, boron, and cadmium have been detected in CSG-affected waterways (Davies et al., 2015).
- Land Degradation and Fragmentation: The expansion of CSG infrastructure leads to land clearing, habitat destruction, and fragmentation of Country, disrupting wildlife corridors and traditional land use.
- Threats to Cultural and Spiritual Waterways: Iman cultural sites linked to waterways, aquifers, and floodplains face permanent ecological damage from water extraction and contamination, undermining traditional custodianship practices.
- Air Pollution and Greenhouse Gas Emissions: CSG operations release methane, a potent greenhouse gas, contributing to climate change and regional air quality issues.
- Expansion and Scale of CSG in Australia: CSG production is concentrated in Queensland, particularly in the Surat and Bowen Basins, with over 7,000 wells drilled and extensive infrastructure networks (Mallants et al., 2017, 2018).

These impacts are particularly concerning for the Iman People, given their long-standing cultural, spiritual, and environmental responsibilities over the water systems in their Country.

7.2 Methodology

7.2.1 Mixed Methods

This research aims to help protect culturally significant water resources in Central Queensland, deepen understanding of Indigenous perspectives on water management, and foster collaboration between Indigenous communities and scholars. To achieve these goals, a mixed-methods approach was employed, combining quantitative and qualitative data collection and analysis. This approach integrates Indigenous Standpoint Theory with technical and scientific practices, creating a novel framework for water management research (Arsenault et al., 2019; Castleden et al., 2017).

The mixed-methods design was selected to address the multifaceted nature of the research topic, which intersects engineering, politics, and social systems. Quantitative methods provided empirical data to assess Iman views and experiences of water-related issues, while qualitative methods captured the cultural and spiritual significance of water for the Iman People. This approach ensured a comprehensive analysis that respects and incorporates Indigenous Knowledge alongside Western scientific practices.

The research was conducted from March 2021 to November 2024, focusing on:

- Areas impacted by CSG activities within the Fitzroy Basin and Dawson River catchment.
- Participants with direct connections to Iman lands and cultural practices.

Criteria for material selection included cultural significance, alignment with research goals, and relevance to water governance frameworks. The primary sources of data upon which the research drew encompassed:

Surveys: Structured surveys were distributed to the Iman community during the 2024 NAIDOC celebrations on Iman Country. The full questionnaire is available on request from the corresponding author. Surveys included:

- Likert-scale questions: A common survey method that measures attitudes and perceptions by asking respondents to rate their level of agreement or disagreement with a statement on a multi-point scale, typically ranging from 'strongly agree' to 'strongly disagree.'(Duignan et al., 2022). These were used to gauge Iman peoples' perceptions and experiences of water management, contamination risks, and governance.
- Yes/no questions: For binary insights on specific issues.
- Open-ended questions: To capture detailed, qualitative responses reflecting cultural and personal significance.**Secondary Data Collection:**

- Literature Review: A review of existing studies on Coal Seam Gas (CSG) activities, Indigenous water governance, and cultural water values informed the study framework and contextualized findings.
- Policy Analysis: Relevant policies, such as the Fitzroy Basin Water Plan and Indigenous water reserves, were critically examined.

7.2.2 *Relationship to Existing Methodologies and Unique Aspects*

This research integrates Indigenous Standpoint Theory, decolonial yarning methodologies, and mixed-methods approaches to examine the water governance challenges posed by Coal Seam Gas (CSG) mining on Iman Country. While elements of this methodology align with existing frameworks in Indigenous research and environmental governance, it is also uniquely tailored to the specific cultural, ecological, and governance concerns of the Iman People.

7.2.2.1 Alignment with Existing Methodologies

This research builds upon methodologies commonly used in Indigenous water governance studies, including:

- Decolonial Yarning Methodologies: The use of yarning as a research tool aligns with frameworks used in First Nations-led studies, prioritizing oral storytelling, relational accountability, and knowledge-sharing (Bailey & Clark, 2024; Dawn Bessarab & Bridget Ng'andu, 2010).
- Indigenous Standpoint Theory: The reflexive approach taken in this study is consistent with Indigenous standpoint methodologies that center First Nations lived experiences, cultural authority, and resistance to colonial knowledge systems (Maddison & Nakata, 2019; Nakata, 2007).
- Community-Based Participatory Research (CBPR): The study follows CBPR principles by actively involving Iman community members in research design, data collection, and interpretation, ensuring that findings are culturally meaningful and actionable (Rigney, 1999).

7.2.2.2 Unique Aspects Tailored to the Iman People & CSG Mining

While this study draws from established Indigenous research methodologies, it is uniquely adapted to the Iman People's specific experiences and governance concerns regarding CSG activities. Key distinctions include:

Culturally Grounded Water Survey Design:

- The survey structure was co-developed with Iman knowledge holders to ensure it accurately reflected cultural values, governance concerns, and lived experiences related to CSG extraction.
- Unlike many standardized environmental surveys, this study integrates questions on cultural responsibilities, totemic species, and spiritual relationships with water, ensuring that the methodology reflects Iman perspectives.

Focus on Coal Seam Gas-Specific Water Risks:

- Unlike broader Indigenous water governance studies, this research is directly focused on the impacts of CSG on water systems, including groundwater depletion, contamination risks, and policy failures.
- The inclusion of independent water sampling, eDNA analysis, and field research is specific to this study, strengthening its ability to assess scientific and Indigenous perspectives side by side.

Reciprocity & Indigenous-Led Knowledge Integration:

- The research process involved ongoing knowledge exchange with Iman community members, including field-based learning, cultural advisory roles, and knowledge-sharing sessions.
- Unlike purely extractive research approaches, this study ensures that research findings are returned to the community in accessible forms, such as educational brochures and community presentations.

By combining Indigenous epistemology, scientific research, and community-led governance insights, this methodology not only aligns with established research approaches but also ensures that both the research process and its outcomes remain culturally relevant and actionable for the Iman People.

7.2.3 Integration of Traditional Ecological Knowledge (TEK) in Research Methods

This study integrates Traditional Ecological Knowledge (TEK) frameworks alongside Indigenous Standpoint Theory and mixed methods approaches to provide a holistic understanding of water governance and environmental stewardship in the context of CSG mining on Iman Country. TEK is defined as the cumulative body of knowledge, practices, and beliefs passed down through generations of Indigenous Peoples regarding the relationships between living beings, ecosystems, and resource management (Berkes, 2012, 2018; Kinch, 2022). Unlike conventional Western ecological frameworks, TEK emphasizes interconnected relationships, long-term environmental observations, and ethical responsibilities toward land and water (Moggridge et al., 2022).

7.2.3.1 Alignment with TEK Frameworks

This study aligns with key TEK-based research approaches commonly used in Indigenous environmental governance:

- Relational Understanding of Water: TEK recognizes water as a living entity with cultural, spiritual, and ecological significance, rather than viewing it as a commodity (Jackson & Barber, 2013).
- Longitudinal Ecological Knowledge: The study incorporates multi-generational Iman knowledge regarding changes in water systems, aligning with TEK's emphasis on historical environmental observations and place-based management strategies (Leonard et al., 2013).
- Embedded Cultural Governance: This research follows TEK methodologies that prioritize Indigenous-led governance structures, laws, and protocols, particularly in water management (Barber & Jackson, 2012).

- Reciprocity and Ethical Responsibility: TEK emphasizes reciprocity between humans and the environment, which aligns with the study's commitment to community engagement and knowledge-sharing as a form of ethical research practice (Moggridge et al., 2019, 2022; Moggridge & Thompson, 2021).

7.2.3.2 TEK in Practice: How It Was Applied in This Study

The application of TEK in this research was not merely theoretical but actively incorporated into the research design, data collection, and interpretation:

- Guidance from Cultural Advisors: Water sampling and survey design were conducted with Iman cultural advisors present, ensuring Indigenous governance principles informed research activities.
- Yarning as a Research Method: Yarning sessions document oral histories and cultural connections to water, providing qualitative insights that complement scientific data (Bessarab & Ng'andu, 2010).
- Environmental Indicators & eDNA Analysis: The study integrates Indigenous observations of water health and quality with Western scientific methods such as eDNA testing, demonstrating a two-way knowledge exchange (Moggridge et al., 2019).
- Use of Cultural Landscapes as Water Governance Frameworks: The Wardingarri (Dawson) River and associated waterways were examined not only as hydrological systems but as cultural landscapes, where water health is inseparable from Indigenous rights and custodianship (Heinrichs & Rojas, 2022; Robinson et al., 2016).

By weaving TEK into each stage of the research process, this study ensures that Indigenous knowledge is not just acknowledged but actively centered in discussions of water governance, ecological sustainability, and CSG impacts.

7.2.4 Survey Design and Implementation

The primary data for this research were collected through surveys designed to explore the Iman People's perceptions of water-related issues and Coal Seam Gas (CSG) activities. These surveys provided Iman participants with a confidential platform to express their views using Likert scale, yes/no, and open-ended questions. Among participants, two groups were identified/surveyed:

1. Group 1 – Participants with no prior education on CSG and water issues.
2. Group 2 – Participants who received targeted educational materials designed to bridge knowledge gaps about CSG and its impacts.

7.2.4.1 Development and Purpose of the Educational Brochure

An educational brochure was provided to Group 2, which was created as part of an Honours research project, which both authors of this paper supervised. The fieldwork was conducted by the authors alongside Honours students, with water sampling undertaken on Iman Country under the guidance of an Iman cultural advisor, ensuring that Traditional Knowledge informed the research process. The field campaign, water testing, and associated travel costs—including cultural advisory support—were funded through the Lynn Brake Scholarship, awarded to lead author Clint Hansen.

The brochure incorporated key findings from ecotoxicological water sampling, environmental DNA (eDNA) analysis, and prior fieldwork (Currell et al., 2022) to connect scientific research with real-world implications for water governance. This data was selected because it originated from independent third-party research, ensuring an unbiased representation of CSG-related impacts. Unlike data published by mining companies, which may prioritize industry narratives, this information was gathered through rigorous scientific methods and aligned with Indigenous perspectives on water stewardship.

Key topics covered in the brochure included:

- Produced Water: Explanation of its chemical composition (e.g., sodium, boron, cadmium) and associated contamination risks.
- Groundwater Drawdown: Analysis of CSG's effects on the Great Artesian Basin and the ecological consequences for connected surface waters and culturally significant wetlands.
- Cultural Significance: Acknowledgment of Iman cultural values related to water resources, such as the Wardingarri (Dawson) River, emphasizing the spiritual and governance importance of maintaining water integrity.

Two key principles guided the selection of data for inclusion in the brochure:

1. Scientific Integrity – Ensuring that findings were independent, peer-reviewed, and methodologically sound, derived from field-based water sampling campaigns.
2. Community Relevance – Prioritizing information that was meaningful to the Iman community, focusing on the direct impacts of CSG extraction on culturally significant water sources and emphasizing both risks and Indigenous-led solutions for sustainable water management.

By incorporating findings from Indigenous-led investigative research rather than industry reports, the educational materials aimed to empower the Iman People with evidence-based insights, reinforcing community agency in decision-making regarding water governance and CSG development.

7.2.4.2 Development of the Survey and recruitment of participants

The survey was developed through a consultative process involving Indigenous community members and academic experts. This ensured that the questions were culturally sensitive, relevant to the Iman People's lived experiences, and aligned with the research objectives. Pilot testing with a small group of Iman individuals conducted prior to finalisation confirmed the clarity and appropriateness of the questions, before broader distribution. Feedback from this process informed revisions, ensuring the questions were respectful, accessible, and effective in capturing participants' perspectives on the key topics identified in the research objectives.

Participants were recruited during the 2024 NAIDOC celebrations on Iman Country, chosen for its cultural significance and high attendance. Recruitment occurred through face-to-face engagement, where the research team explained the study's purpose, assured confidentiality, and emphasized the voluntary nature of participation. This culturally safe environment fostered trust and encouraged community engagement. Convenience sampling was employed during the NAIDOC celebrations to maximize community participation. Participants were recruited from diverse demographics within the Iman People to ensure representation of various perspectives.

7.2.4.3 Ethical Considerations

This research adhered to the National Health and Medical Research Council's (National Health and Medical Research Council & Commonwealth of Australia: Canberra, 2018) (NHMRC) *Keeping Research on Track II Guidelines*, ensuring alignment with the six core values:

Spirit and Integrity: This research is designed to benefit past, present, and future generations of Iman Peoples. It honors the memory of Elders like my Uncle John Lawson, who fought for Land rights, and engages current generations in protecting our Water, Country, and Culture from the threats posed by the extractive industry (Langton, 2013).

Cultural Continuity: The methodology supports better management and protection of water-dependent assets in the Great Artesian Basin (GAB), including groundwater-dependent ecosystems and culturally significant sites (Barber & Jackson, 2012). It engages First Peoples of Central Queensland in defining values, aspirations, and priorities for ongoing water management.

Equity: This research aims to develop a framework for Indigenous water management that empowers First Nations People, providing them with greater decision-making power over how waterways are used and managed. This framework has the potential to be a model for other First Peoples communities (Barber & Jackson, 2012).

Reciprocity: This research is grounded in mutual respect and knowledge sharing, reflecting the ethical responsibility of Indigenous scholars to uphold community obligations beyond academic work. As an Iman researcher, I incorporate reciprocity by actively contributing to cultural and environmental initiatives that support my community. Beyond academic responsibilities, Iman author Hansen has engaged in community-led efforts to sustain cultural practices and advocate for water rights. This includes:

- Creating Traditional craft kits for cultural ceremonies, ensuring the continuation of cultural knowledge-sharing among younger generations.
- Serving on the Bundulla Mission Site Erosion Committee, where I work alongside community members to address environmental challenges affecting culturally significant areas.
- Assisting Iman community members in understanding water rights, translating complex legal and policy frameworks into culturally relevant terms, and supporting engagements with government agencies and industry stakeholders in a way that centers Indigenous governance perspectives.

These commitments demonstrate ongoing reciprocity beyond the research project itself, reinforcing the importance of Indigenous scholars actively participating in both knowledge creation and community advocacy (Rose, 2005).

Respect: Trust and cooperation between the research team and the Iman community are crucial for the success of this project. Ongoing consent and negotiation of research activities and outcomes will be maintained throughout the project (O'Bryan, 2019b).

Responsibility: The authors accept full responsibility for the design and conduct of this research and are committed to maintaining the health of Iman Country and its People throughout the project and beyond, upholding the values of caring for Country and community in all aspects of life (O'Bryan, 2019a).

7.2.5 Collation and Analysis of Data

Survey responses were collected in both paper and digital formats based on participant preferences. Open-ended responses were transcribed and analysed thematically, while quantitative data from Likert scale and yes/no questions were coded and analysed using statistical software. This mixed-methods approach enabled a holistic interpretation of the data, blending statistical rigor with qualitative depth.

7.2.5.1 Quantitative Data Analysis

Following the compilation of sixteen surveys completed by Iman community members, Likert scale responses were recorded in a database (MS Excel workbook), and descriptive statistics were calculated to identify patterns and trends. For interpretation, the Likert scale responses were assigned numerical values ranging from 1 to 5, where:

1 = Strongly Agree
2 = Agree
3 = Neutral
4 = Disagree
5 = Strongly Disagree

Mean scores were used to summarize overall trends, with lower scores indicating stronger agreement (greater concern or dissatisfaction) and higher scores reflecting weaker agreement (less concern or dissatisfaction). A score of 3 represented a neutral stance, providing a "break-even" value for assessing trends.

7.2.5.2 Qualitative Analysis Process

Open-ended survey responses underwent a systematic coding and thematic analysis process, detailed as follows:

Familiarization with the Data: The lead researcher thoroughly read all responses to understand participants' perspectives, taking notes to capture initial impressions and identify potential patterns.

Development of a Coding Framework: An iterative coding framework was developed, guided by the research objectives and emerging themes from the data. Key topics included the cultural significance of water, perceptions of CSG activities, and governance concerns.

Coding of Data: Responses were uploaded into MaxQDA software, where each response was systematically coded. Adjustments to the framework were made as new themes emerged.

Identification of Themes: Codes were grouped into broader themes to highlight overarching patterns. For example, responses related to water's cultural and spiritual values were categorized under the theme "Cultural Significance of Water." This thematic grouping structured complex narratives into actionable insights. Thematic coding identified key areas of concern, including environmental risks, cultural significance of water, dissatisfaction with governance, and interest in community-led monitoring roles.

Triangulation and Verification: To ensure reliability, themes were cross-checked by multiple researchers in the team. Discrepancies in interpretation were resolved through discussion, ensuring accurate representation of participants' views.

Integration with Quantitative Data: Qualitative findings were integrated with quantitative data to provide a comprehensive understanding of community perspectives. For instance, statistical trends in Likert scale responses were contextualized with illustrative quotes from the qualitative responses. The integration of qualitative and quantitative findings provided a comprehensive understanding of the participants' perspectives (whereas reliance on one or the other type of data may have missed the full richness of insight available through the participants' responses).

7.2.5.3 Measures to Eliminate Bias

To ensure objectivity and cultural integrity:

- Community Engagement: Surveys and focus groups were co-designed with Iman community members to reflect their priorities and avoid external biases (Khan et al., 2024).
- Reflexive Practice: Regular reflection sessions were conducted to mitigate potential biases arising from my dual role as an Iman scholar and researcher (Russell-Mundine, 2012).
- Data Validation: Cross-validation was performed by involving Indigenous Knowledge holders in the interpretation of qualitative themes (Wright et al., 2016).

7.2.5.4 Innovation

The research utilized Water B(l)ack,[1] a methodological tool designed for Indigenous-led water research. This tool integrates cultural perspectives with technical analysis, aligning with Indigenous Standpoint Theory. By prioritizing culturally grounded insights, Water B(l)ack ensures that both qualitative and quantitative data reflect the values and priorities of the Iman People (Hansen et al., 2024; Kelsey Leonard et al., 2023).

7.3 Results

7.3.1 Group Comparisons

There were significant and meaningful differences between Group 1 (participants without prior education on CSG-related issues) and Group 2 (participants with prior education on CSG-related issues), as described in Table 7.1. Further quantitative analysis of these differences is presented in Figs. 7.2 and 7.3.

[1] Hansen: Waterb(l)ack is a methodological framework I have coined to center Indigenous sovereignty, resistance, and decolonial water governance, integrating Traditional Knowledge with scientific analysis to challenge extractive industries and reclaim cultural water rights on Iman Country.

Table 7.1 Comparison of perceptions on CSG-related issues between participants with and without prior education

	Group 1: Participants without prior education on CSG-related issues	Group 2: Participants With prior education on CSG-related issues
Yes/No questions		
Familiarity with CSG Co-produced water	Responses to this question were mixed, with several participants unfamiliar with CSG co-produced water. Reviews in academic scholarship and policy documents on CSG document the potential risks associated with the release of such waters and have noted co-produced water management to be one of the major issues relating to ground and surface waters in areas of CSG development (Khan & Kordek, 2014; Mallants et al., 2018).	Group 2 participants demonstrated a higher familiarity level, suggesting that some form of prior education (see Sect. 7.2.1) effectively bridged knowledge gaps about CSG's environmental impacts.
Likert scale questions		
Adequacy of policy and governance measures for CSG	Responses leaned toward dissatisfaction, with a mean score of 4.3 and a median of 4.5 (Fig. 7.2). Group 1 participants generally feel that current governance and policies inadequately protect Iman Country and its cultural resources.	Responses leaned toward strong dissatisfaction, with a mean score of 4.6 and a median of 4.7. Based on comparison with Group 1, education likely deepened participants' awareness of governance shortcomings (Fig. 7.3).
Support for Continued CSG Development	Most responses leaned toward dissatisfaction, with a mean score of 4.3 and a median of 4.5 (Fig. 7.2). Group 1 participants generally feel that current governance and policies inadequately protect Iman Country and its cultural resources.	Opposition to further development was similarly strong among Group 2 (mean: 4.4; median: 4.5), with similar scores evident among the groups (Fig. 7.3).
Awareness of water's cultural, spiritual, and ecological values	High awareness was indicated, with a mean score of 1.9 and a median of 2.0 (Fig. 7.2). This underscores water's intrinsic cultural value as an ingrained perspective, reflecting deep, shared cultural understanding.	High awareness was confirmed (mean: 1.8; median: 2.0), reflecting shared cultural understanding across groups (Fig. 7.3).
Risk to clan group totems and native fauna	Participants expressed high concern (mean and median scores of 4.5) about the impact of CSG on fauna and clan totems, reflecting heightened ecological awareness, particularly regarding species of spiritual and cultural significance.	Consistently high concern (mean and median: 4.5), with education reinforcing the importance of protecting culturally significant species.
Desire for enhanced water protection	Overwhelming agreement was reflected, with a mean score of 1.5 and a median of 1.5 (mid-way between Strongly Agree and Agree) (Fig. 7.2). Participants expressed dissatisfaction with current protections and a strong desire for better safeguards to ensure long-term sustainability.	Strong support for enhanced water protections was indicated (mean: 1.3; median: 1.5), consistent with Group 1 findings (Fig. 7.3).

(continued)

Table 7.1 (continued)

	Group 1: Participants without prior education on CSG-related issues	Group 2: Participants With prior education on CSG-related issues
Satisfaction with the footprint of gas infrastructure	Scores indicated strong dissatisfaction (mean: 4.6; median: 5.0, see Fig. 7.2), particularly regarding encroachment of gas infrastructure on traditional lands and visible impacts on sacred sites (as elaborated in qualitative responses later – see below). These concerns were tied to the degradation of culturally significant spaces.	Strong dissatisfaction (mean: 4.6; median: 4.5), with concerns centered on cultural and ecological impacts, similar to Group 1 (see Fig. 7.3).

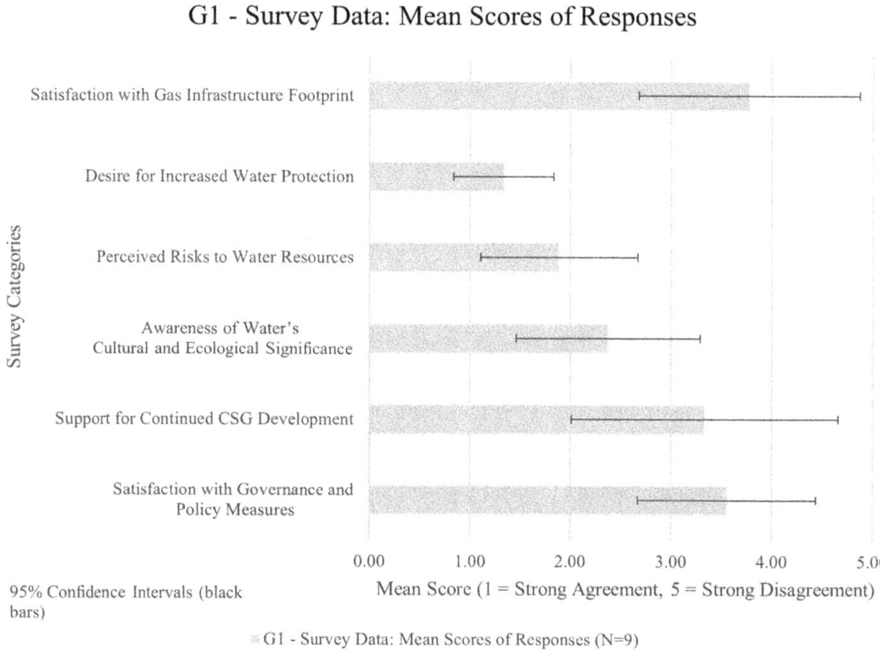

Fig. 7.2 G1 - Survey data: Mean scores of responses

7.3.2 *Thematic Insights*

Participants identified significant risks related to water contamination, environmental degradation, and declining water quality (Table 7.2), which align with independent scientific findings.

- Water Contamination: Concerns were raised about the presence of toxic chemicals in CSG-produced water. Water sampling conducted as part of this research confirmed elevated levels of sodium, boron, and cadmium, substances known to cause soil degradation, aquatic toxicity, and potential human health risks (Currell et al., 2022).

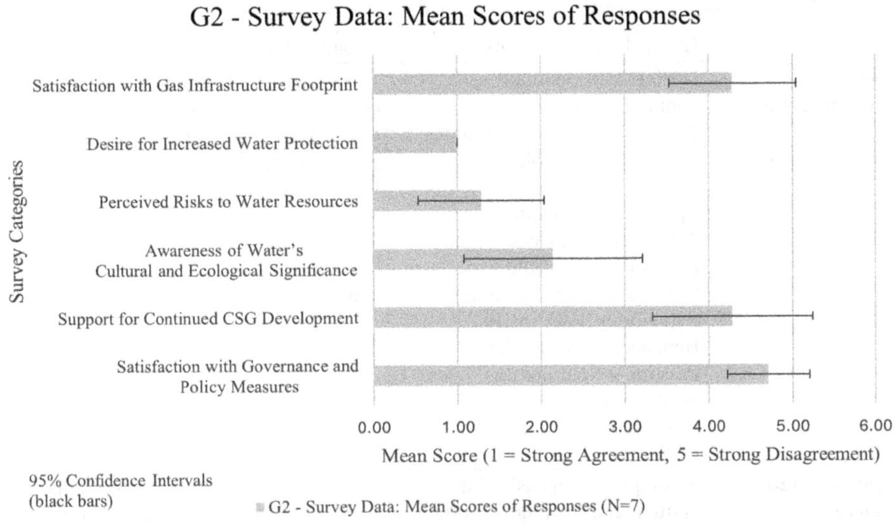

Fig. 7.3 G2 - Survey data: Mean scores of responses

- Loss of Aquatic Species: eDNA analysis identified a decline in native aquatic species, with specific reductions in culturally significant fish populations. This supports concerns raised by participants about disruptions to totemic species and broader ecosystem changes (Moggridge et al., 2019).
- Cumulative Environmental Damage: Community members expressed frustration with long-term, compounding contamination issues, emphasizing that "mistakes happen regardless of policy or governance". These concerns align with scientific evidence that CSG infrastructure leads to cumulative water pollution over time, particularly affecting groundwater recharge areas and wetland habitats (Navi et al., 2015).

These findings reinforce the importance of Indigenous-led environmental monitoring, ensuring that both scientific data and Traditional Knowledge inform governance decisions on water protection.

7.3.3 Statistical Analysis: Welch's t-Test

Further analysis of the quantitative data was conducted using Welch's t-test, a statistical method used to determine whether there are significant differences between the means of two independent groups when variances are unequal (Ruxton, 2006). Unlike the traditional Student's t-test, Welch's t-test is more robust in cases where sample sizes are different or variance homogeneity cannot be assumed, making it particularly suitable for this study given the differences in responses between the

Table 7.2 Comparison of thematic concerns related to water, culture, and CSG development

Theme	Group 1: Key concerns	Group 2: Key concerns
Water contamination and environmental risks	Participants identified contamination of sacred water, destruction of flora and fauna, and long-term ecological damage (Ali et al., 2017). Specific concerns included: "There are many contaminants in this water. I would expect tight governance guidelines and regulations… testing for Benzene, Mercury, Carcinogens, Poisons—Neurotoxins." "Poor drinking water quality, loss of water supply, and potential health issues."	Participants highlighted specific concerns, such as: "Contamination construction ongoing over time; when it gets capped, they leave CSG, and we are left to deal with consequences—even unknown ones at the time." "Kills animals and people at worst, makes them sick, causes allergies from contaminated water."
Cultural and spiritual connection to water	Participants emphasized the cultural and spiritual importance of water (de Rijke, 2018). "Water connects people and communities to land and each other… waterways, creeks, rivers are central to spiritual and cultural practices." "Water is sacred, a part of our ceremonies and identity. Without it, we lose a part of who we are."	Participants emphasized the cultural and spiritual connection to water: "To me, water's cultural value is in how it connects to cultural and spiritual history, stories, and meaning."
Interest in Becoming a Water Ranger	A strong interest in water ranger roles was expressed, reflecting a desire for community-led monitoring and protection of water resources: "Water Rangers and Iman people need to be back on Country, monitoring and protecting our land."	Like Group 1, participants expressed a strong interest in water ranger roles: "If we had a say, we would prevent harm before it gets worse."
Concerns about industrial development and infrastructure	Participants expressed frustration with visible ecological and cultural degradation (Mercer et al., 2014). "Our old people would not allow this. The land does not look the same, with gas stations everywhere."	Similar concerns about ecological and cultural degradation were noted (Hadwen Georgy, 2023). "Gas operations pollute water and destroy habitats of our totems."

two participant groups (1 and 2). Data, including the mean difference and *p*-value are summarised in Table 7.3 and Fig. 7.4.

The Welch's t-test result for the question on Adequacy of Governance Measures (see Table 7.3 and Fig. 7.4) was statistically significant, with a *p*-value of 0.0053. This indicates a clear difference in perception between the two groups regarding the adequacy of governance measures for CSG activities. Group 2 (educated

Table 7.3 Statistical comparison of responmses on CSG governance, cultural values, and environmental risks

Variable	Group 1 Mean	Group 2 Mean	Mean difference	*p*-value	Significance
Adequacy of governance	3.56	4.67	−1.11	0.005	**Yes**
Support for continued CSG development	3.33	4.17	−0.79	0.239	No
Awareness of water's cultural values	2.14	2.33	0.19	0.846	No
Perceived risks to water resources	1.75	1.33	0.42	0.341	No
Desire for increased water protection	1.33	1.00	0.33	0.170	No
Satisfaction with gas infrastructure	3.88	4.33	−0.45	0.395	No

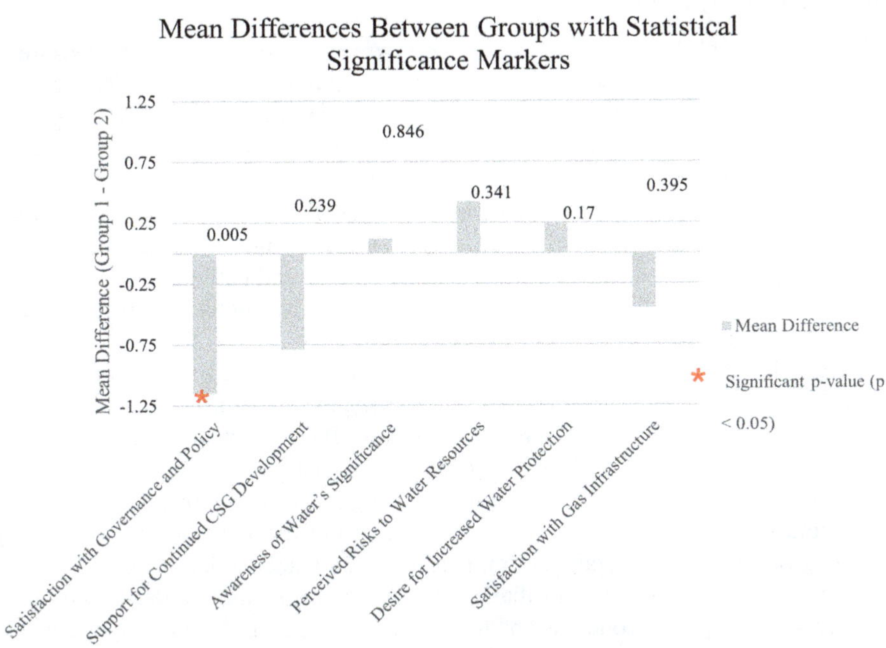

Fig. 7.4 Mean differences between groups with statistical significance markers (*p* value for Welch's t-test indicated above each bar)

participants) rated governance more critically than Group 1, suggesting that their additional knowledge about CSG-related issues may have increased their awareness of policy shortcomings.

The statistical significance in these results highlights the importance of education in shaping critical perspectives within the community, particularly around governance measures. Group 2's more informed stance likely stems from the nuanced awareness facilitated by prior educational exposure. While the next largest difference in means between the two groups was for the question regarding 'support for continued CSG development', this was not a statistically significant difference (Table 7.3).

7.4 Discussion

7.4.1 Connecting Scientific Research to Real-World Implications

The findings from this study highlight the direct environmental and health risks associated with CSG-related water extraction and contamination. Scientific research, including ecotoxicological water sampling and eDNA analysis, has provided tangible evidence of these risks, reinforcing the lived experiences and concerns of Iman community members.

- Human Health Implications: Participants expressed concerns about exposure to contaminants such as benzene, cadmium, and other neurotoxins, which are present in CSG-produced water (Ali et al., 2017; Davies et al., 2015). Long-term exposure to these substances has been linked to carcinogenic risks, respiratory diseases, and endocrine disruption (Currell et al., 2022).
- Wildlife and Ecosystem Health: eDNA analysis of water sources near CSG operations revealed declining populations of culturally significant aquatic species. Participants emphasized the impact on clan totems and native fauna, supporting scientific findings that contamination and groundwater depletion disrupt fragile wetland ecosystems (Moggridge et al., 2019).
- Water Levels and Sustainability: Groundwater monitoring has confirmed drawdown effects in areas affected by CSG activity, exacerbating stress on the Great Artesian Basin and connected river systems (Baublys et al., 2021; Campbell et al., 2025; Post et al., 2020). Reduced water availability threatens cultural water practices, such as ceremonies conducted along the Wardingarri (Dawson) River.
- Cumulative Contamination Risks: Evidence from independent water sampling suggests that trace metals and organic pollutants accumulate in sediment and water over time, posing risks that extend beyond immediate contamination incidents. Participants' concerns about long-term ecological damage align with scientific warnings about irreversible changes in water chemistry that impact aquatic biodiversity and agricultural use (Ali et al., 2017).

By integrating scientific and Indigenous knowledge, this study underscores the urgent need for stronger regulatory oversight, culturally informed water management, and community-led monitoring initiatives.

7.4.2 The Cultural Significance of Water

This study reaffirms the profound cultural importance of water for the Iman People. Water is not merely a physical resource but a sacred entity central to the spiritual and cultural identity of the community. Both quantitative and qualitative findings reveal that water contamination is perceived as a threat to the community's physical and spiritual well-being. These findings align with broader Indigenous perspectives, emphasizing water as a life-giving force deeply interconnected with cultural practices (Hill et al., 2022; Marshall, 2017; Poelina et al., 2023b).

Short-answer responses emphasized the inseparability of water, cultural practices, and ecological health. Participants frequently cited threats to totems and native fauna as critical risks, reinforcing the intertwined cultural and ecological dimensions of water. For instance, one respondent from Group 1 noted:

> *The land has spiritual, physical, social, and cultural significance to Aboriginal People. Waterways, creeks, and rivers are important to spiritual and cultural practices as well as environmental management and food production. Water connects people and communities to land and to each other.*

Similarly, a Group 2 participant explained:

> *To me, the cultural value of water, as I put it on my face and in my mouth to connect, is about living in harmony. Water, the sun, and the moon are all life—it is all deeply connected.*

> *Water connects people and communities to land and each other… waterways, creeks, rivers are central to spiritual and cultural practices.*

These insights align with Likert-scale results where participants demonstrated high awareness of water's cultural values (Mean = 1.9).

This cultural perspective contrasts with Western scientific approaches that often treat water as a quantifiable resource that can be allocated to economic sectors (e.g., the use of terms such as 'consumptive pool'[2] and 'available' water within management plans and frameworks) (Maclean, 2015; Jackson et al., 2021). While empirical methods focus on water quality, availability, and ecosystem impacts, they frequently overlook the relational and spiritual dimensions central to Indigenous worldviews (Hartwig et al., 2018). These findings highlight the critical need to integrate both perspectives to create holistic and culturally respectful water management frameworks.

---

[2] Bureau of Meteorology, "Consumptive Pool," Australian Water Information Dictionary, accessed December 5, 2024 (Bureau of Meteorology, 2024)

7.4.3 Implications for Regenerative Futures

The study reveals the Iman People's active role in envisioning regenerative futures. They strongly support stewardship roles, such as Water Rangers, to safeguard their water resources (Gergan & Curley, 2023). Likert-scale responses, where participants overwhelmingly supported stronger water protections (Mean = 1.54), align with qualitative insights advocating for active community involvement in water management.

The results indicate a knowledge gap regarding CSG co-produced water among Group 1 participants. This lack of familiarity may have influenced the way they expressed environmental concerns in subsequent responses, as their perspectives tended to be more general rather than focused on specific risks associated with CSG operations. The implications of this gap are significant, as an understanding of co-produced water and its risks is essential for meaningful engagement in water governance discussions. This finding highlights the importance of targeted environmental education and accessible public information to ensure that the Iman community is well-equipped to advocate for water protections.

Thematic analysis identified specific practices—such as controlled burns and Traditional Ecological Knowledge (TEK)-based conservation methods—as central to regenerative efforts. These practices reflect a sophisticated understanding of how Indigenous Knowledge can complement scientific assessments to enhance ecological resilience (Moggridge et al., 2019; Steffensen, 2020). Respondents emphasized the need for integrating these approaches to address the impacts of CSG activities on water sources. A Group 1 participant highlighted the importance of this approach:

> *We need our waters to be natural and pure. The practices above ground support ways of living for our people, including fishing and maintaining natural marine life.*

Likewise, a Group 2 participant emphasized collaborative stewardship:

> *I would like all stakeholders to come together in a yarning circle to co-design sustainable systems. Through respect for culture, Iman People can preserve our country in its natural state.*

The significant difference observed in governance perceptions highlights the transformative impact of education on community engagement with policy frameworks. Beyond raising awareness, education equips individuals with the skills to critically analyze policy measures, advocate for Indigenous perspectives, and demand accountability from policymakers and industry stakeholders (McCarty & Lee, 2014). This connection emphasizes the need for tailored educational initiatives to empower Indigenous communities to propose alternative, culturally sensitive solutions and advocate for more inclusive governance frameworks.

A key insight from the comparative analysis between Group 1 and Group 2 is the apparent impact of education on participants' perceptions of governance. Group 2, who had prior exposure to educational materials on CSG impacts, expressed greater dissatisfaction with governance frameworks compared to Group 1. This suggests that education deepened participants' awareness of governance shortcomings, equipping them with a more critical lens through which to evaluate regulatory measures. This aligns with broader research on environmental governance, where increased access to knowledge often results in stronger advocacy for reform and accountability. The findings highlight the importance of continued education and community engagement in strengthening Indigenous participation in decision-making processes related to water governance (Figs. 7.1 and 7.2).

Another key consideration is the potential for self-directed exploration by participants following their engagement with the research materials. The information provided in the brochure, which detailed the environmental, cultural, and governance-related impacts of CSG extraction, likely encouraged participants to further investigate case studies, reports, and scientific data on the topic. This suggests that access to targeted educational materials may play a role in shaping perspectives and fostering deeper critical engagement with environmental governance issues. While this assumption is not directly measurable in the current study, it highlights the importance of knowledge-sharing mechanisms and the potential for ongoing self-education within Indigenous communities when provided with accessible and relevant information. Future research could explore the extent to which such informational resources influence long-term perspectives and advocacy efforts related to water governance.

The integration of quantitative and qualitative findings further allows the following conclusions to be elucidated:

- Governance Concerns: Dissatisfaction with governance and policy protections correlated with dissatisfaction with CSG infrastructure (−0.612). Qualitative responses reinforced this critique, highlighting calls for transparent, culturally inclusive policies.
- Cultural and Environmental Risks: Perceived risks to totems and native fauna were consistently emphasized, with a positive correlation noted between these concerns and dissatisfaction with infrastructure (0.432).
- Support for Water Protections: Strong support for enhanced water protections in Likert-scale responses was echoed in qualitative calls for stricter regulations, oversight, and community-led monitoring.

These findings demonstrate the interconnectedness of governance, cultural values, and environmental sustainability, emphasizing the need for integrated approaches. Regenerative futures, as envisioned by the Iman People, offer a blueprint for blending Indigenous and scientific strategies to restore ecosystems while upholding cultural integrity (Gibbs et al., 2024).

7.4.4 Justice for Indigenous Peoples

The findings highlight the broader struggles of Indigenous communities worldwide in confronting extractive industries. Dissatisfaction with governance structures (Mean = 4.8) and frustration over non-transparent decision-making processes were recurring themes in both qualitative and quantitative data. Respondents emphasized the exclusion of Indigenous voices and the inadequacy of current governance measures to address community concerns. One Group 1 participant questioned governance accountability and transparency:

> *I don't have a copy of governance measures (regulations) of CSG water extraction. Do all extraction companies follow these regulations, and are they documented and forwarded to Indigenous groups? Any failure to comply must result in penalties.*

Similarly, a Group 2 participant critiqued enforcement:

> *Disagree—while there are regulations around such matters, enforcement of the regulations is insufficient, and penalties do not discourage breaches of the regulations.*

This likely enabled Group 2 participants to articulate criticisms more effectively, as they had greater context for assessing and challenging governance structures. Research in Indigenous and environmental policy supports this trend. Studies show that knowledge-sharing initiatives and community education programs empower Indigenous groups to engage critically with governance frameworks. Yua et al. (2022) found that informed communities are more likely to articulate environmental critiques and demand accountability from policymakers in resource governance settings. Similarly, Schlosberg and Carruthers (2010) highlight that increased environmental literacy among Indigenous groups enhances their ability to advocate for sustainable governance practices and challenge exploitative policies.

Respondents called for co-management structures that respect Indigenous sovereignty and incorporate cultural values into policy-making. Locally, this means empowering the Iman People to co-manage water resources, while nationally and globally, systemic reforms are needed to address historical marginalization and ensure equity in resource governance (Adams & Sarvestani, 2024).

7.4.5 Local, National, and Global Implications

The integration of short-answer responses with Likert-scale findings paints a vivid picture of a community deeply concerned about the cultural and environmental impacts of CSG activities. Locally, the Iman People's calls for stronger water protections and inclusive governance demand immediate policy changes. One Group 1 participant suggested:

> *There needs to be policies in place to legally uphold lore established by Iman Peoples. Governance measures should always include input from Traditional Owners and Elders.*

A Group 2 participant echoed this sentiment:

> *I believe there needs to be more involvement of Iman People in the decision-making governance of our natural resources. More weight should be given to Iman values when protecting biodiversity.*

Nationally, these findings resonate with broader movements for Indigenous water rights and environmental justice in Australia (Moggridge, 2021). Globally, the Iman People's experiences mirror those of Indigenous groups worldwide, underscoring the need for systemic changes that center Indigenous voices in governance, resource management, and environmental stewardship.

Global examples provide a compelling context. The Standing Rock Sioux Tribe's opposition to the Dakota Access Pipeline and the Cofán People's legal battles in Ecuador to protect rainforest from extractive industries, highlight the transformative potential of centering Indigenous perspectives in governance frameworks (Burnside, 2022; Nelson & Shilling, 2018)). Integrating Indigenous and scientific approaches can address historical inequities while advancing sustainability. These cases demonstrate that integrating Indigenous Knowledge not only upholds Indigenous rights but also promotes broader ecological and cultural justice (Lianne, 2022).

7.4.6 Challenges Faced by Indigenous-Led Water Management Projects

Despite their significant contributions, Indigenous-led water management initiatives face numerous challenges:

- Legal and Political Barriers: Limited recognition of water rights and fragmented governance frameworks hinder effective management.
- Resource Constraints: Insufficient funding and technical capacity limit the scalability of projects.
- Environmental and Climate Pressures: Climate change exacerbates vulnerabilities, complicating restoration efforts.
- Social and Cultural Pressures: Marginalization, loss of Traditional Knowledge, and intergenerational tensions threaten the continuity of Indigenous management practices.

Despite their significant contributions, Indigenous-led water management initiatives face numerous challenges that limit their effectiveness and long-term sustainability. These challenges arise from systemic legal and political barriers, resource constraints, environmental and climate pressures, and social and cultural tensions that impact Traditional Knowledge transmission and governance structures.

7.4.6.1 Legal and Political Barriers

Limited legal recognition of Indigenous water rights and fragmented governance frameworks continue to hinder effective water management. In many jurisdictions, Indigenous groups must navigate complex regulatory systems that fail to grant them full decision-making authority over water resources (Marshall, 2017). Australia's

Native Title system, for example, recognizes land rights but does not automatically confer substantive water rights, restricting Indigenous governance in this domain (Barber & Jackson, 2012; Jackson & Langton, 2011). The exclusion of Indigenous voices from key water policy discussions further reinforces systemic inequities, making it difficult for communities to implement sustainable water stewardship programs.

7.4.6.2 Resource Constraints

Indigenous water management projects often face significant financial and technical barriers. Many initiatives operate with limited funding, relying on short-term grants rather than sustained investment, which affects their ability to scale up or maintain long-term monitoring programs (Maclean, 2015; Robinson et al., 2016). Additionally, the technical expertise required for integrating Indigenous Knowledge with contemporary water management frameworks often depends on external consultants or university collaborations, rather than being community-driven. This reliance on outside expertise can create power imbalances and limit the autonomy of Indigenous water governance (Morgan et al., 2004; Weir, 2009).

7.4.6.3 Environmental and Climate Pressures

Climate change exacerbates existing water vulnerabilities, further complicating Indigenous water management efforts. Rising temperatures, prolonged droughts, and extreme weather events disrupt freshwater systems, making it increasingly difficult to restore and protect water sources (Moggridge et al., 2019). Additionally, industrial activities such as coal seam gas (CSG) extraction and large-scale agriculture continue to deplete and contaminate water supplies, disproportionately affecting Indigenous communities who depend on these waters for cultural, spiritual, and economic purposes (Hartwig et al., 2020, 2021). Without strong environmental protections and legal recognition, Indigenous-led restoration efforts face ongoing challenges in mitigating these risks.

7.4.6.4 Social and Cultural Pressures

Beyond environmental and legal barriers, Indigenous water governance is also shaped by social and cultural challenges. The marginalization of Indigenous voices in policy decisions has led to the erosion of Traditional Knowledge systems, particularly as younger generations face limited opportunities to engage in cultural water practices (S. Jackson, 2018). Intergenerational tensions arise as communities navigate the balance between modern water management approaches and ancestral stewardship methods (O'Donnell et al., 2024). The loss of language and cultural frameworks associated with water knowledge further threatens the continuity of

Indigenous-led governance, underscoring the urgency of revitalization programs and policy support.

These intersecting challenges highlight the systemic barriers Indigenous communities face in reclaiming water sovereignty. However, despite these constraints, many Indigenous groups continue to assert their rights, engage in grassroots advocacy, and develop innovative water management models that blend Traditional Knowledge with contemporary science.

A Group 1 participant voiced concerns over governance and transparency:

> *Mistakes happen regardless of policy or governance. Is safety compromised? Do companies really tell the truth about outcomes, or does profit rule decisions*

A Group 2 participant expressed frustration with inadequate oversight, articulating a greater level of experience/knowledge with respect to governance of the CSG industry (in line with the findings of the Queensland Audit Office, 2020):

> *Not until there is more control over the industry, which the QLD Government appears reluctant to enforce, can we ensure the safety of our water. Current governance is insufficiently controlled and lacks Iman input.*

Addressing these challenges requires systemic reforms, including legally binding water rights, capacity-building support, and equitable co-management arrangements. For example, New Zealand's (Te Awa Tupua (Whanganui River Claims Settlement) Act, 2017), which granted legal personhood to the Whanganui River, exemplifies the potential for transformative reforms that recognize Indigenous cultural and spiritual connections to water while creating collaborative governance frameworks. Similarly, the Murray-Darling Basin Plan in Australia has introduced cultural water entitlements for Indigenous groups, but challenges persist in ensuring that these rights are adequately funded and implemented to reflect Traditional Knowledge and community needs (Hartwig et al., 2023). Globally, the experience of the Standing Rock Sioux Tribe in the United States underscores the difficulties of challenging extractive industries within existing governance systems, as seen in their opposition to the Dakota Access Pipeline (Friedman, 2020).These examples illustrate both the opportunities and the barriers to creating equitable governance structures that integrate Indigenous leadership and values in water management.

7.5 Conclusion

The cultural values of water held by the Iman People of Central Queensland provide invaluable insights into rethinking water management in the face of industrial development. These values highlight the sacredness of water, its intrinsic connection to community identity, and the necessity of governance systems that prioritize ecological and cultural sustainability. The findings highlight the urgency of integrating Indigenous Knowledge systems into water management to honour water's cultural, spiritual, and ecological significance.

By advocating for community-led stewardship and stronger protections for water resources, this research offers a pathway toward regenerative futures that benefit not only Indigenous communities but also contribute to the global effort to safeguard water resources for all (Biekart et al., 2023). It illustrates how Indigenous perspectives can inform innovative solutions, blending traditional practices with modern approaches to create sustainable and equitable water governance.

Ultimately, learning from and collaborating with Indigenous communities presents an opportunity to build a future that respects the rights of all Peoples, values the sacredness of water, and ensures the long-term sustainability of our planet's most vital resource.

References

Adams, M. M., & Sarvestani, D. B. (2024). Indigenous peoples' rights discourse: Toward Hemispheric indigenous climate and environmental justice. *Environmental Justice*. https://doi.org/10.1089/env.2023.0076

Ali, A., Strezov, V., Davies, P., & Wright, I. (2017). Environmental impact of coal mining and coal seam gas production on surface water quality in the Sydney basin, Australia. *Environmental Monitoring and Assessment, 189*(8), 408–408. https://doi.org/10.1007/s10661-017-6110-4

Arsenault, R., Bourassa, C., Diver, S., McGregor, D., & Witham, A. (2019). Including indigenous knowledge systems in environmental assessments: Restructuring the process. *Global Environmental Politics, 19*(3), 120–132. https://doi.org/10.1162/glep_a_00519

Bailey, N., & Clark, Y. (2024). Exploring bonding and attachment in Aboriginal families. *Australian Journal of Psychology, 76*(1). https://doi.org/10.1080/00049530.2024.2356117

Barber, M., & Jackson, S. (2012). Indigenous engagement in Australian mine water management: The alignment of corporate strategies with national water reform objectives. *Resources Policy, 37*(1), 48–58. https://doi.org/10.1016/j.resourpol.2011.12.006

Baublys, K. A., Hofmann, H., Esterle, J. S., Cendón, D. I., Vink, S., & Golding, S. D. (2021). Geochemical influences on methanogenic groundwater from a low rank coal seam gas reservoir: Walloon Subgroup, Surat Basin. *International Journal of Coal Geology, 246*. https://doi.org/10.1016/j.coal.2021.103841

Berkes, F. (2012). Context of traditional ecological knowledge. In *Sacred ecology*. Routledge.

Berkes,F.(2018).*Sacredecology*(Vol.1,4thed.).Routledge.https://doi.org/10.4324/9781315114644

Bessarab, D., & Ng'andu, B. (2010). Yarning about yarning as a legitimate method in Indigenous research. *International Journal of Critical Indigenous Studies, 3*(1), 37–50. https://doi.org/10.5204/ijcis.v3i1.57

Biekart, K., Camfield, L., Kothari, U., Melber, H., Melber, H., Biekart, K., Kothari, U., & Camfield, L. (2023). Rethinking development and decolonising development studies. In *Challenging global development*. Palgrave Macmillan. https://doi.org/10.1007/978-3-031-30308-1_1

Bureau of Meteorology. (2024). *Australian water information dictionary: Consumptive pool*.

Burnside, W. R. (2022). Gregory Cajete (1994). In *Foundations of socio-environmental research* (pp. 627–633). https://doi.org/10.1017/9781009177856.052

Campbell, A. G., Cartwright, I., Cendón, D. I., & Currell, M. J. (2024). Multiple isotope tracers reveal the sources of water sustaining ecologically and culturally significant springs, and their vulnerability to mining development. *Journal of Hydrology (Amsterdam), 132078*. https://doi.org/10.1016/j.jhydrol.2024.132078

Campbell, A. G., Cartwright, I., Webb, J. A., Cendón, D. I., & Currell, M. J. (2025). Using geochemical and geophysical data to characterise inter-aquifer connectivity and impacts on shallow aquifers and groundwater dependent ecosystems. *Applied Geochemistry, 178*, 106217. https://doi.org/10.1016/j.apgeochem.2024.106217

Castleden, H. E., Hart, C., Harper, S., Martin, D., Cunsolo, A., Stefanelli, R., Day, L., & Lauridsen, K. (2017). Implementing Indigenous and western knowledge systems in water research and management (Part 1): A systematic realist review to inform water policy and governance in Canada. *International Indigenous Policy Journal, 8*(4), 1–33. https://doi.org/10.18584/iipj.2017.8.4.7533

Creamer, J., Buchanan, C., Ingram, I., Vonda, M., Atkinson, R., & State of Queensland. (2024). Truth-telling and Healing Inquiry, Report 1: The story of Queensland connects us all. www.truth-telling-qld.com.au

Currell, M., Hansen, C., Nicholson, R., Esmond, M., & Dooley, K. (2022). Designing a monitoring program and conceptual models to protect ecological and cultural values of waterways vulnerable to coal seam gas impacts. In *Hydrology and Water Resources Symposium, HWRS 2022*, 912–923.

Currell, M., Hansen, C., Nicholson, R., Esmond, M., & Dooley, K. (n.d.). *Designing a monitoring program and conceptual models to protect ecological and cultural values of waterways vulnerable to coal seam gas impacts*. https://doi.org/10.3316/informit.916866948672904

Davies, P. J., Gore, D. B., & Khan, S. J. (2015). Managing produced water from coal seam gas projects: Implications for an emerging industry in Australia. *Environmental Science and Pollution Research International, 22*(14), 10981–11000. https://doi.org/10.1007/s11356-015-4254-8

Dawn Bessarab, & Bridget Ng'andu. (2010). Yarning about yarning as a legitimate method in Indigenous research [Article]. *International Journal of Critical Indigenous Studies, 3*(1), 37–50. https://doi.org/10.5204/ijcis.v3i1.57

de Rijke, K. (2018). Produced water, money water, living water: Anthropological perspectives on water and fracking. *Wiley Interdisciplinary Reviews: Water, 5*(2). John Wiley and Sons Inc. https://doi.org/10.1002/WAT2.1272

Dixon, R. D., & Morgan, T. (2023). Yinbarnini Ngukunginyi (Singing of water). *Oceania*. https://doi.org/10.1002/ocea.5378

Duignan, S., Moffat, T., & Martin-Hill, D. (2022). Be like the running water: Assessing gendered and age-based water insecurity experiences with Six Nations First Nation. *Social Science & Medicine, 1982(298)*, 114864–114864. https://doi.org/10.1016/j.socscimed.2022.114864

Friedman, L. (2020). Standing Rock Sioux Tribe wins a victory in Dakota access pipeline case. *New York Times*.

Gergan, M. D., & Curley, A. (2023). Indigenous youth and decolonial futures: Energy and environmentalism among the Diné in the Navajo Nation and the Lepchas of Sikkim, India. *Antipode, 55*(3), 749–769. https://doi.org/10.1111/anti.12763

Gibbs, M. C., Parker, L. M., Scanes, E., & Ross, P. M. (2024). Recognising the importance of shellfish to first nations peoples, indigenous and traditional ecological knowledge in aquaculture and coastal management in Australia. *Marine and Freshwater Research, 75*(4). https://doi.org/10.1071/MF23193

Hadwen Georgy. (2023). *Report: Climate concerns and impact of coal seam mining in Australia.*

Hansen, C. J., Currell, M. J., & Flynn, E. (2024). Iman Peoples Water Sovereignty: Extractive Industries in Central Queensland. *The Extractive Industries and Society, 20*, 101560. https://doi.org/10.1016/j.exis.2024.101560

Hartwig, L. D., Jackson, S., & Osborne, N. (2018). Recognition of Barkandji water rights in Australian settler-colonial water regimes. *Resources (Basel), 7*(1), 16. https://doi.org/10.3390/resources7010016

Hartwig, L. D., Jackson, S., & Osborne, N. (2020). Trends in Aboriginal water ownership in New South Wales, Australia: The continuities between colonial and neoliberal forms of dispossession. *Land Use Policy, 99*, 104869. https://doi.org/10.1016/j.landusepol.2020.104869

Hartwig, L. D., Jackson, S., Smart, J. C. R., & Osborne, N. (2023). Water trading by Aboriginal organisations in NSW, Australia. *Journal of Rural Studies, 100*, 102997. https://doi.org/10.1016/j.jrurstud.2023.03.005

Hartwig, L. D., Markham, F., & Jackson, S. (2021). Benchmarking Indigenous water holdings in the Murray-Darling Basin: A crucial step towards developing water rights targets for Australia. *Australian Journal of Water Resources, 25*(2), 98–110. https://doi.org/10.1080/13241583.2021.1970094

Heinrichs, D. H., & Rojas, R. (2022). Cultural values in water management and governance: Where do we stand? *Water (Basel), 14*(5), 803. https://doi.org/10.3390/w14050803

Hill, R., Harkness, P., Raisbeck-Brown, N., Lyons, I., Álvarez-Romero, J. G., Kim, M. K., Chungalla, D., Wungundin, H., Aiken, M., Malay, J., Williams, B., Buissereth, R., Cranbell, T., Forrest, J., Hand, M., James, R., Jingle, E., Knight, O., Lennard, N., et al. (2022). Learning Together for and with the Martuwarra Fitzroy River. *Sustainability Science, 17*(2), 351–375. https://doi.org/10.1007/s11625-021-00988-x

Hurlbert, M., Acharibasam, J. B., Datta, R., Strongarm, S., & Starblanket, E. (2024). Decolonizing indigenous drinking water challenges and implications: Focusing on indigenous water governance and sovereignty. *Water (Basel), 16*(5), 748. https://doi.org/10.3390/w16050748

Jackson, S. (2018). Water and Indigenous rights: Mechanisms and pathways of recognition, representation, and redistribution. *Wiley Interdisciplinary Reviews: Water, 5*(6), e1314.

Jackson, S., & Barber, M. (2013). Recognition of Indigenous water values in Australia's Northern Territory: Current progress and ongoing challenges for social justice in water planning [Article]. *Planning Theory & Practice, 14*(4), 435–454. https://doi.org/10.1080/14649357.2013.845684

Jackson, S., Carmody, E., & Hartwig, L. (2021). Treading water on Indigenous water rights: The serious deficiencies of water allocation planning and management in NSW under the Murray Darling Basin Plan. *Pandora's Box, 27*, 72–97.

Jackson, S., & Langton, M. (2011). Trends in the recognition of indigenous water needs in Australian water reform: The limitations of "cultural" entitlements in achieving water equity. *Journal of Water Law, 22*, 109–123.

Jackson, S. E., & Palmer, L. R. (2012). Modernising water: Articulating custom in water governance in Australia and East Timor. *International Indigenous Policy Journal, 3*(3), 7. https://doi.org/10.18584/iipj.2012.3.3.7

Johnson, J. T., Howitt, R., Cajete, G., Berkes, F., Louis, R. P., & Kliskey, A. (2016). Weaving Indigenous and sustainability sciences to diversify our methods. *Sustainability Science, 11*(1), 1–11. https://doi.org/10.1007/s11625-015-0349-x

Khan, S., & Kordek, G. (2014). Coal Seam Gas: Produced Water and Solids Prepared for the Office of the NSW Chief Scientist and Engineer (OCSE). https://www.chiefscientist.nsw.gov.au/__data/assets/pdf_file/0017/1088/OCSE-Final-Report-Stuart-Khan-Final-28-May-2014.pdf

Khan, N., Keck, L., Sykes, C., Rowden, C., Simister, J., Fenlon, S., Mccallum, E., Bell, M., Whiting, D., Shelton, C., Wells, G., Giles, R., Howard, J., & Peckham, S. (2024). Diversity, community engagement and co-design in research: A rapid review. *BMJ Leader*. https://doi.org/10.1136/leader-2024-001046

Kinch, R. A. (2022). *Indigenous storytelling, Cherokee traditional ecological knowledge, and place-based education*. https://doi.org/10.18666/JOREL-2022-11601

L'Oste-Brown, S. (1995). *Living under the Act: Taroom Aboriginal Reserve 1911–1927* (Luke Godwin, Gordon Henry, Ted Mitchell, & Vera Tyson, Eds.). Dept. Environment and Heritage, Queensland.

Laborde, S., & Jackson, S. (2022). Living waters or resource? Ontological differences and the governance of waters and rivers. *Local Environment, 27*(3). https://doi.org/10.1080/13549839.2022.2044298

Langton, M. (2013). *The quiet revolution: Indigenous people and the resources boom*. HarperCollins Publishers Australia.

Leonard, K., David-Chavez, D., Smiles, D., Jennings, L., Alegado, R. A., Tsinnajinnie, L., Manitowabi, J., Arsenault, R., Begay, R. L., Kagawa-Viviani, A., Davis, D. D., van Uitregt, V., Pichette, H., Liboiron, M., Moggridge, B., Carroll, S. R., Tsosie, R. L., & Gomez, A. (2023). Water back: A review centring rematriation and Indigenous Water research sovereignty. *Water Alternatives, 16*(2), 374–428.

Leonard, S., Parsons, M., Olawsky, K., & Kofod, F. (2013). The role of culture and traditional knowledge in climate change adaptation: Insights from East Kimberley, Australia. *Global Environmental Change, 23*(3), 623–632. https://doi.org/10.1016/j.gloenvcha.2013.02.012

Lianne, C. L. (2022). *Serpent river resurgence: Confronting uranium mining at Elliot Lake*. University of Toronto Press.

López-Maldonado, Y., Anstee, J., Neely, M. B., Marty, J., Mastracci, D., Ngonyani, H., Ogashawara, I., Salyani, A., Sharma, K., & Sims, N. C. (2024). The contributions of Indigenous people's earth observations to water quality monitoring. *Frontiers in Water, 6*. https://doi.org/10.3389/frwa.2024.1363187

Maclean, K. (2015). Crossing cultural boundaries: Integrating Indigenous water knowledge into water governance through co-research in the Queensland Wet Tropics, Australia. *Geoforum, 59*, 142–152. https://doi.org/10.1016/j.geoforum.2014.12.008

Maddison, S., & Nakata, S. (2019). *Questioning indigenous-settler relations* (Vol. 1). Springer Singapore Pte. Limited. https://doi.org/10.1007/978-981-13-9205-4

Mallants, D., Bekele, E., Schmid, W., Miotlinski, K., & Bristow, K. (2017). *Literature Review: Identification of Potential Pahways to Shallow Groundwater of Fluids Associated with Hydraulic Fracturing, Project Report Prepared by the Commonwealth Scientific and Industrial Research Organisation*. CSIRO) as Part of the National Assessment of Chemicals Associated with Coal ….

Mallants, D., Jeffrey, R., Zhang, X., Wu, B., Kear, J., Chen, Z., Wu, B., Bekele, E., Raiber, M., Apte, S., & Gray, B. (2018). Review of plausible chemical migration pathways in Australian coal seam gas basins. *International Journal of Coal Geology, 195*(June), 280–303. https://doi.org/10.1016/j.coal.2018.06.002

Marshall, V. (2017). Overturning Aqua Nullius: Securing aboriginal water rights. In *Overturning Aqua Nullius*. Aboriginal Studies Press.

McCarty, T. L., & Lee, T. S. (2014). Critical culturally sustaining/revitalizing pedagogy and indigenous education sovereignty. *Harvard Educational Review, 84*(1), 101–124. https://doi.org/10.17763/haer.84.1.q83746nl5pj34216

Mercer, A., De Rijke, K., & Dressler, W. (2014). Silences in the boom: Coal seam gas, neoliberalizing discourse, and the future of regional Australia. *Journal of Political Ecology, 21*(1), 279–302. https://doi.org/10.2458/v21i1.21137

Moggridge, B. J. (2021). Indigenous water knowledge and values in an Australasian context. *Australian Journal of Water Resources, 25*(1), 1–3. https://doi.org/10.1080/13241583.2021.1935919

Moggridge, B. J., Betterridge, L., & Thompson, R. M. (2019). Integrating aboriginal cultural values into water planning: A case study from New South Wales, Australia. *Australasian Journal of Environmental Management, 26*(3), 273–286. https://doi.org/10.1080/14486563.2019.1650837

Moggridge, B. J., & Thompson, R. M. (2021). Cultural value of water and western water management: An Australian Indigenous perspective. *Australian Journal of Water Resources, 25*(1), 4–14. https://doi.org/10.1080/13241583.2021.1897926

Moggridge, B. J., Thompson, R. M., & Radoll, P. (2022). Indigenous research methodologies in water management: Learning from Australia and New Zealand for application on Kamilaroi country. *Wetlands Ecology and Management, 30*(4), 853–868. https://doi.org/10.1007/s11273-022-09866-4

Morgan, M., Stelein, L., & Weir, J. (2004). Indigenous rights to water in the Murray Darling Basin In support of the indigenous final report to the living Murray Initiative. Canberra: AIATSIS. https://aiatsis.gov.au/sites/default/files/research_pub/morgan-strelein-weir-dp14-indigenous-water-rights-murray-darling-basin_0_3.pdf

Nakata, M. N. (2007). *Disciplining the savages, savaging the disciplines* (1st ed.). Aboriginal Studies Press.

National Health and Medical Research Council, & Commonwealth of Australia: Canberra. (2018). *Keeping research on track II: A companion document to Ethical conduct in research with*

Aboriginal and Torres Strait Islander Peoples and communities: Guidelines for researchers and stakeholders. NHMRC. www.nhmrc.gov.au/guidelines-publications/ind3

Navi, M., Skelly, C., Taulis, M., & Nasiri, S. (2015). Coal seam gas water: Potential hazards and exposure pathways in Queensland. *International Journal of Environmental Health Research, 25*(2), 162–183. https://doi.org/10.1080/09603123.2014.915018

Nelson, M. K., & Shilling, D. (2018). *Traditional ecological knowledge: Learning from indigenous practices for environmental sustainability* (M. K. Nelson & D. Shilling, Eds.). Cambridge University Press.

Nenquimo, N., & Anderson, M. (2024). *We will be Jaguars: A memoir of my people*. ABRAMS PR.

O'Bryan, K. (2019a). *Indigenous rights and water resource management: Not just another stakeholder* (1st ed.). Routledge. https://doi.org/10.4324/9781351239820

O'Bryan, K. (2019b). Victorian Indigenous rights legislation, environment and land management legislation and the management of water resources. In *Indigenous rights and water resource management* (1st ed., pp. 121–143). Routledge. https://doi.org/10.4324/9781351239820-8

O'Donnell, E., Clark, C., & Killean, R. (2024). Rights and relationality: A review of the role of law in the human/water relationship. *Water Alternatives, 17*(2), 207–238.

OGIA. (2021). Underground Water Impact Report for the Surat Cumulative Management Area. OGIA, Brisbane. https://www.dlgwv.qld.gov.au/__data/assets/pdf_file/0008/1584728/uwir-2021-report.pdf

Osborne, T., Cifuentes, S., Dev, L., Howard, S., Marchi, E., Withey, L., Rocha, S., & da Silva, M. (2024). Climate justice, forests, and Indigenous Peoples: Toward an alternative to REDD + for the Amazon. *Climatic Change, 177*(8), 128. https://doi.org/10.1007/s10584-024-03774-7

Poelina, A., Brueckner, M., & McDuffie, M. (2021). For the greater good? Questioning the social licence of extractive-led development in Western Australia's Martuwarra Fitzroy River region. *The Extractive Industries and Society, 8*(3), 100827. https://doi.org/10.1016/J.EXIS.2020.10.010

Poelina, A., Toussaint, S., & Muecke, S. (2023a). 'Why can't we speak up for ourselves…?' Water futures and ethnographic provocations. *Oceania, 93*(3), 366–376. https://doi.org/10.1002/ocea.5374

Poelina, A., McDuffie, M., & Perdrisat, M. (2023b). Martuwarra Fitzroy River Watershed: One society, one river law. *PLOS Water, 2*(9), e0000104. https://doi.org/10.1371/journal.pwat.0000104

Post, D. A., Crosbie, R. S., Viney, N. R., Peeters, L. J. M., Zhang, Y. Q., Herron, N. F., Wilkins, A., Janardhanan, S., Karim, F., Aryal, S. K., Pena-Arancibia, J., Lewis, S., Evans, T., Vaze, J., Chiew, F. H. S., Marvanek, S. P., Henderson, B., Schmidt, B., & Herr, A. (2020). Impacts of coal mining and coal seam gas extraction on groundwater and surface water. *Journal of Hydrology (Amsterdam), 591*, 125281. https://doi.org/10.1016/j.jhydrol.2020.125281

Pueblo Action Alliance. (2021). *Water back*. https://www.puebloactionalliance.org/water-back

Queensland Audit Office. (2020). *Managing coal seam gas activities (Report 12—2019–20)*. https://www.qao.qld.gov.au/reports-resources/reports-parliament/managing-coal-seam-gas-activities

Rigney, L. I. (1999). Internationalization of an Indigenous Anticolonial Cultural Critique of Research Methodologies: A Guide to Indigenist Research Methodology and Its Principles. *Wicazo Sa Review, 14*(2), 109–121. https://doi.org/10.2307/1409555

Robinson, C. J., Maclean, K., Hill, R., Bock, E., & Rist, P. (2016). Participatory mapping to negotiate indigenous knowledge used to assess environmental risk. *Sustainability Science, 11*(1), 115–126. https://doi.org/10.1007/s11625-015-0292-x

Rose, D. (2005). An indigenous philosophical ecology: Situating the human. *The Australian Journal of Anthropology, 16*(3), 294–305. https://doi.org/10.1111/j.1835-9310.2005.tb00312.x

Russell-Mundine, G. (2012). Reflexivity in indigenous research: Reframing and decolonising research? *Journal of Hospitality and Tourism Management, 19*(1), 85–90. https://doi.org/10.1017/jht.2012.8

Ruxton, G. D. (2006). The unequal variance t-test is an underused alternative to Student's t-test and the Mann–Whitney U test. *Behavioral Ecology, 17*(4), 688–690. https://doi.org/10.1093/beheco/ark016

Schlosberg, D., & Carruthers, D. (2010). Indigenous struggles, environmental justice, and community capabilities. *Global Environmental Politics, 10*(4), 12–35. https://doi.org/10.1162/GLEP_a_00029

Steffensen, V. (2020). *Fire country: How indigenous fire management could help save Australia.* Hardie Grant Travel.

Te Awa Tupua (Whanganui River Claims Settlement) Act. (2017). https://www.legislation.govt.nz/act/public/2017/0007/latest/whole.html

Weir, J. K. (2009). *Murray River country an ecological dialogue with traditional owners* (1st ed.). Aboriginal Studies Press.

Wright, A. L., Wahoush, O., Ballantyne, M., Gabel, C., & Jack, S. M. (2016). Qualitative health research involving indigenous peoples: Culturally appropriate data collection methods. *Qualitative Report, 21*(12), 2230–2245. https://doi.org/10.46743/2160-3715/2016.2384

Yua, E., Raymond-Yakoubian, J., Aluaq Daniel, R., & Behe, C. (2022). A framework for co-production of knowledge in the context of Arctic research. *Ecology and Society, 27*(1), 34. https://doi.org/10.5751/ES-12960-270134

Chapter 8
The River is an Island: (Re)imagining a More-than-human Future for the Waimatā Catchment

Megan Thomas, Billie Lythberg, Dan Hikuroa, and Gary Brierley

Abstract The emergence of an era of more-than-human relations in the academy marks a response to contestations around human-centric assertions of the Anthropocene. More-than-humanism extends beyond the dichotomy of 'humans vs Nature', encompassing a spectrum of relationships between the human and non-human. This chapter gazes through a more-than-human lens in seeking to understand riverscapes, focusing upon the Waimatā River in Gisborne, Aotearoa New Zealand, from the perspective of the river itself. Our approach interprets two representations—geomorphic evolution and ecopoetics—of this riverscape, demonstrating the multiple ways riverscapes, their stories, and in this case, a 'mountains to the sea' lens, can be expressed. Art, culture, and science come together as storied knowledge through the disciplinary vehicle of geography—in particular geoethics and geographies of affect. While this approach may be viewed as divisive in some circles, we feel that a radical reorganisation of process is required to keep up with the radical reorganisation of landscapes currently occurring across the globe, including in the Waimatā catchment.

Keywords More-than-human · Ecopoetics · Riverhood · Relationship · Geoethics

M. Thomas (✉) · G. Brierley
School of Environment, University of Auckland, Auckland, New Zealand
e-mail: megan.thomas@auckland.ac.nz

B. Lythberg
Business School, University of Auckland, Auckland, New Zealand

D. Hikuroa
Faculty of Arts and Education, University of Auckland, Auckland, New Zealand

M. Kennedy, E. O'Donnell (eds.), *People, Place and Nature in Indigenous-Settler Relations*, Indigenous-Settler Relations in Australia and the World 10, https://doi.org/10.1007/978-3-032-03600-1_8

8.1 Introduction

The era of more-than-humanism in the academy marks a response to the human-centric dichotomy of the Anthropocene (Thomas et al., 2024). The epoch characterised by 'human' dominion over 'nature', in which the two are separate and the latter modifies at will, with little consideration for the consequence, has brought us to a state of rupture (Head & Gibson, 2012). The 2019 Intergovernmental Science-Policy Platform on Biodiversity and Ecosystem Services (IPBES) Biodiversity and Ecosystem Services Report noted that 75% of the earth's surface has been significantly altered by human action, a proportion that continues to increase. Recognising that humans have worked with earth systems for millennia, what may be perceived by some as 'natural' landscapes and ecosystems are often humanised and culturalised. This is exemplified by the firing of landscape or cultural burning in Australia by Indigenous people, shaping species distribution and co-constructing ecosystems (Fletcher et al., 2021; Reardon-Smith, 2023).

Following the industrial revolution and the impacts of colonisation, human activities have altered the pace of natural earth system processes at an alarming rate (Syvitski, 2012). Direct, intentional responses to management actions have been accompanied by indirect responses to anthropogenically-induced climate change, and increases in the frequency and magnitude of extreme weather events (Brierley & Fryirs, 2022; IPCC, 2023). Reparative responses to such changes require a profound rethinking of human actions and relationships. More-than-humanism extends beyond the arguably Eurocentric dichotomy of 'human vs. nature', encompassing a spectrum of relatedness between the human and non-human (Abram, 1996) that directly acknowledges the non-human as both existing and having causal influence of its own (Rogers et al., 2013). To view a landscape in this way dissolves the boundary between nature and society/ culture, instead considering their mutual co-constitution and the coherence and holism of the entire system.

This chapter offers a glimpse through the burgeoning academic more-than-human lens in an effort to understand riverscapes (channel and surrounding catchment) from the perspective of the river itself. We draw on Boelens et al.' (2023) concept of 'Riverhood[1]' and three of the associated riverscape ontologies to contextualise this pursuit. The riverscape ontologies approach sees "river-as-ecosociety", wherein systems co-evolve through myriad interactions between environment, climate and culture; "river-as-territory", where different actors construct landscape differently; and "river-as-subject", the idea that rivers are living entities. To put this approach into practice, we interpret two narratives that demonstrate the conduct and messaging of riverscape stories.

We utilise Critical Physical Geography (Lave et al., 2014) and its consideration that landscapes are the products of physical and social factors to reimagine the ways landscapes are known and expressed. Towards the aim of the Let the River Speak (hereafter LTRS) Project (Salmond et al., 2019), of which the authors are a part, we

[1] The unifying state of being a river. Thinking in this way can guide the ways rivers are imagined, defined and lived with (Boelens et al., 2023).

look to give the river itself a 'voice'. A "geoethical future" (Sharp et al., 2022), where environmental ethics (the moral implications of human action towards the environment) is embedded into research and other human–non-human interactions, situates place from a non-human perspective. Our approach builds upon geomorphic work by Fuller et al. (2023), who define a river story as an account of evolutionary trajectory. We expand their definition, challenging the assumption that river stories are merely recounts of human occupation. We believe river stories must also encompass the lived histories and experiences of the riverscape, from the perspective of the river itself, and in any medium that befits the content.

We contend that adopting a more-than-human lens broadens both the breadth and depth of understandings of river systems, for stories extend through time much further than human memory and perhaps beyond human futures. From this perspective, river stories might be better defined as expressions of the river translated for general human consumption. In this way, river stories, much like riverscapes themselves, are agents of connectivity and relationality. They invoke a sense of care, of stewardship and of empathy. Given their inherent trans-disciplinarity, geographers are well positioned to weave these storied knowledges together (Koppes, 2022), thereby representing the interrelationships of more-than-human landscapes.

Across much of the world, the state of rivers reflects how people think of them (Wooltorton et al., 2022) and their (inter)actions with them. In turn, human well-being is often directly dependent on rivers. Hence, the multi-media scripting and sharing of river stories is of ever-increasing importance. Intertwined social relationships with degraded riverscapes affects ways of life, health and cultural well-being, altering opportunities to interact with water for food, art, recreation and cultural activities (Parsons & Fisher, 2020). Our first narrative in this contribution presents an account of the geologic and geographic evolution of the Waimatā River (Fig. 8.1),

Fig. 8.1 (**a**) Minor areas of remnant bush in headwaters, (**b**) Evidence of erosion and slash in flume-like section, (**c**) Pine plantations on hillslopes, (**d**) Regenerating bush at Waikereru Ecosanctuary

located in the Gisborne region of New Zealand's North Island. Following this, our second narrative conveys understandings of relations to riverscapes through a more artistic and differently-scientific representation in the form of a poem by Māori writer Hone Peneamine Anatipa Te Pona Tuwhare (Ngāti Korokoro, Ngāti Tautahi, Te Popoto, Te Uri-O-Hau), 'The River is an Island'.

We posit that recognising, respecting and working with multiple ways of knowing the same landscape/ riverscape through a transdisciplinary or pluralist approach can enhance social cohesion by creating and expressing a shared sense of identity (Lyver et al., 2016), thereby supporting efforts to foster social resilience (Lubinski & Thoms, 2018). We challenge readers to consider how river stories that comprise geologic, geographic, Indigenous and poetic perspectives can generate affective responses.

8.2 Geomorphic Evolution

Although the Waimatā story begins tens of millions of years ago, the East Cape region, within which the Waimatā is situated, located in the East Coast Basin (ECB) (Fig. 8.2), has a much more ancient history. The origins of the ECB begin 105 million years ago with cessation of the collision and mountain building along the Gondwana margin, during which the basement rocks of New Zealand were created (Ballance, 2017; Coates & Cox, 2002). For the next 80 million years, prior to the initiation of the subduction of the Hikurangi subduction margin, the ECB resided on a passively subsiding continental margin (Coates & Cox, 2002). Erosion processes of land coupled with the lack of subduction allowed the formation of a seaward sloping sediment wedge (Ballance, 2017). As the adjacent landmass wore down the initially coarse-grained sediments were progressively broken down, and clay became a significant proportion of the wedge. Under its own weight and compression between the Australian and Pacific, the wedge lithified to form weak sedimentary layers of the ECB, which were crumpled and thrust atop the continental landmass (Ballance, 2017). Emplacement of the layers broke up the previously coherent Taitai Sandstone, creating distinctive remnant mountains in the Gisborne area that are comprised of the largest of these ripped up sandstone blocks (Mike Marden and Colin Mazengarb, *personal communication*, 2023). Landscapes in the region retain a selective memory of past geologic disturbance events (Brierley, 2010).

Gisborne is in the forearc zone between the active volcanic zone (Taupo Volcanic Zone) and the subduction trench (Hikurangi Trough; Fig. 8.2). Within the forearc are three different parallel zones; the frontal arc, the forearc sedimentary basin and the accretionary complex (Ballance, 2017). The Raukumara Ranges, in which the Waimatā has its headwaters, are located within the frontal arc as part of the North Island Axial Ranges. Given the catchment lies adjacent to a major tectonic boundary, faults are abundant, earthquakes are common and uplift rates are high. This suite of processes created the highly erosive landscapes that explain the high sediment yields experienced by the catchment today (Marden et al., 2008).

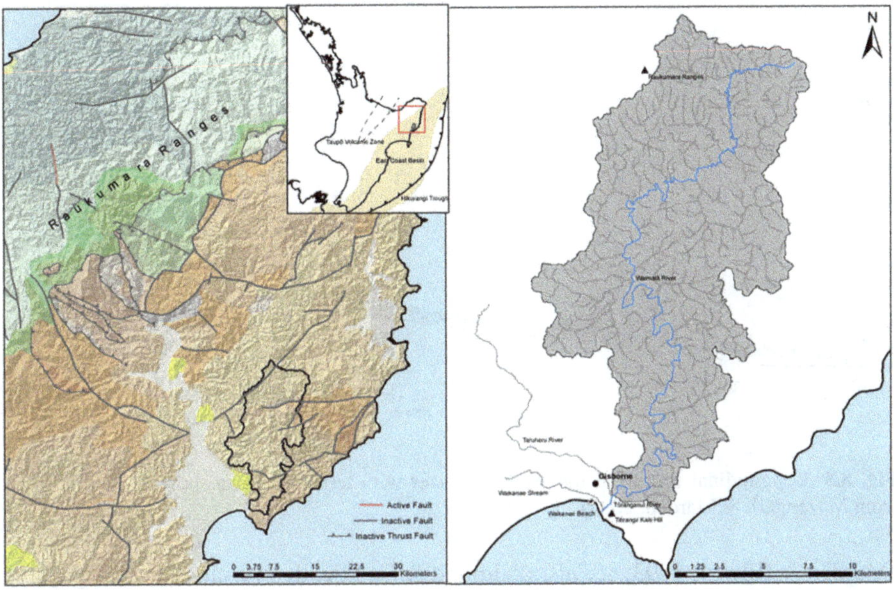

Fig. 8.2 Location of the Waimatā catchment and broad geologic setting

While the geologic history of the area is a pertinent chapter in the Waimatā riverscape story, it does not dictate the entire plot. Climatic fluctuations are imprinted on this geologic history. For example, profound changes have occurred since the Last Glacial Maximum ~16,000 years ago when the sea level was ~130m lower than presently (Ballance, 2017). Associated with rising and falling sea levels, four main terrace levels are present within the Waimatā catchment (Fig. 8.3). The Waipaoa-1 terrace surface signals the ceasing of aggradation at the end of the Last Glacial Maximum (Marden et al., 2008). Subsequent warmer conditions led to increased vegetation cover, more rainfall, and reduced sediment loads, thereby instigating incisions that produced a series of degradational terraces (Litchfield & Berryman, 2005). The terraces constrain the through-put zone[2] of the river (Fig. 8.2), limiting lateral migration of the channel and space to store sediment within the channel margins (Harvey et al., 2021). As a result, much of the trunk stream of the Waimatā acts as a flume, rapidly flushing water and associated materials from the mountains to the sea (Fuller et al., 2023). Hence, landscapes in the Waimatā catchment also retain a strong climatic memory, with past climate and geology shaping boundary conditions—collectively, these conditions influenced the impacts, consequences and memory of past human activities, thus forming distinctive legacy effects (Brierley, 2010; Salmond et al., 2022).

European settlement in the late 1800s led to extensive deforestation to facilitate pasture for grazing (Poverty Bay Herald, 1895). Subsequently, to meet the national demand for timber, a trio of planting booms occurred through the twentieth century

[2] Section of a river where sediment is flushed through and channel eroded during high flows.

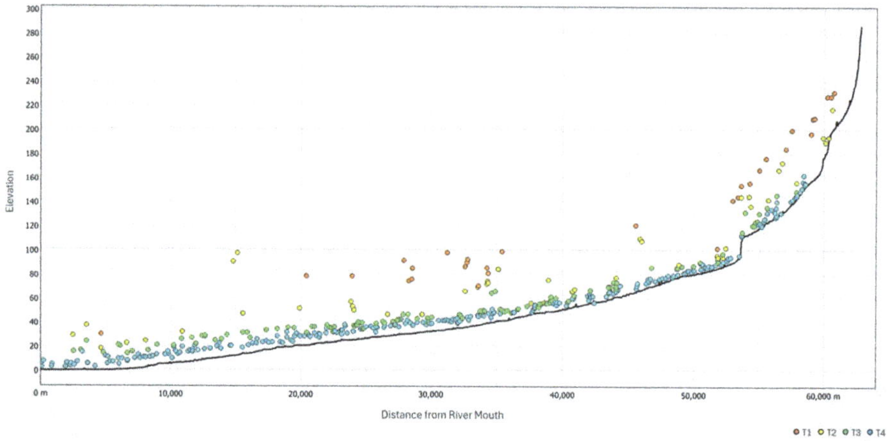

Fig. 8.3 Longitudinal Profile of the Waimatā River with terrace heights shown. (Data sourced from Mazengarb & Marden, 2024)

(Marden & Rowan, 1993; see also Salmond & Caddie, this volume). Planting *Pinus radiata* was encouraged due to fast growth rates and potential economic success. Little thought was given to the consequences of planting shallow rooting and highly flammable trees in the erodible landscapes of the Waimatā (Marden & Seymour, 2022). Today, only 10–20% of the catchment remains under indigenous forest cover. This, along with the repeated harvesting of pine trees, has led to large volumes of fine-grained sediment and slash entering the channel. These changing boundary conditions, coupled to larger and more frequent storm events (e.g., cyclones Gabrielle and Hale in 2022 and 2023; Green, 2022; Hunt, 2023), have altered the distribution, rate and effectiveness of geomorphic processes in the region.

'Normal' is being altered, changing the time scales over which geomorphological units 'come into and out of being' (Massey, 2001). Here, the nature of space-time is important. Depending on the spatio-temporal context, the same phenomena may have different properties and behavioural traits (Massey, 2001). Given the importance of spatio-temporal context, geologic and geomorphic river stories gain traction (Fuller et al., 2023). While universal laws and generic science principles fashion baseline processes, on their own they cannot accurately express the complete nature of a riverscape. In efforts to Let the River Speak, catchment-specific knowledges reflect multiple knowledges and different ways in which understandings are expressed (Salmond et al., 2019).

8.3 Ecopoetics

Nature writing, or as discussed more specifically here, ecopoetics, is an example of a different way in which these understandings can be expressed. When considering how society can engage with a profoundly changing environment, and thus human–non-human relations, ways of thinking that are affective rather than only scientific or political (Lidström & Garrard, 2014) are of equal importance. However, arbitrary disciplinary boundaries often prevent moves towards this direction. Instead, transcending these boundaries, rejecting anthropocentrism and accepting the benefits of affectual thinking can help in making sense of a profoundly changing world. Ecopoetry is poetic expression that questions the ideals of the environment only having value in relation to humans. It acts to affirm the intrinsic value of the non-human (Lidström & Garrard, 2014).

In Moana Oceania narratives, the cosmos is in a continuous state of movement. Star gods map the skies in star canoes, and their descendants haul land from the ocean (Royal, 2006). The north island of Aotearoa, Te-Ika-a-Maui, is a fish caught by the adventurer-ancestor Maui with his grandmother's jawbone. Sliced into by Maui's brothers, the fish takes on the contours of mountains, valleys, steep cliffs and deep ravines (Te Rangihaeata in Grey, 1971). Ki uta ki tai—from the mountains to the sea—rivers course through the land and land embraces water. In his 1970 poem, Hone Tuwhare, one of the foremost eco-poets on Aotearoa, addresses one such Aotearoa river in the terms of its complex momentum, dynamics and relationships: a river in and as island (see also Hikuroa et al., 2025):

The river is an island[3]
You are river. This way and that
and all the way to sea two escorts
shove and pull you. Two escorts
in contention.
Left bank or right bank, how can
you be a river without either?
Thus are U bends made. Thus are
S bends made. Your direction
is assured and sometimes running
perfectly and quite straight.
A low bank on your left holds your
laughing stitches in. On our right
side skips another hushing your
loud protests.
You are river. Joy leaping down
a greenstone stair-way: anger cradled
in a bed of stones.
You're a harbour; a lake; an island
only when your banks lock lathered
arms in battle to confine you: slow-
release you.
Go river, go. To ocean seek your
certain end. Rise again to cloud;
to a mountain—to a mountain
drinking from a tiny cup.
Ah, river
you are ocean: you are island.
—Hone Tuwhare

[3] First published as 'The River Is An Island.' An Anthology of Twentieth Century New Zealand Poetry. Ed. Vincent O'Sullivan. Auckland, NZ: Oxford UP, 1970. 181–182.

Tuwhare describes a river that cannot be itself without its banks, 'two escorts in contention', pushing and pulling. 'You are river,' he proclaims, 'a harbour; a lake; an island.' Calling us to visualise the building blocks of language, Tuwhare traces sibilance and assonance, soundscapes and cadence. 'Laughing stitches' held by low banks evoke 'aho' or genealogical threads (see Tapsell, 1997). As opposing banks 'skip' alongside, 'hushing… loud protests,' the land is fleet-footed, and the remonstrating water is gently subdued.

Tuwhare writes river and banks in co-constitutive contention. Emotive connections and relations accompany biophysical realities. River cascades playfully down pounamu (greenstone) stairs; elsewhere, in turmoil, it finds land's cradle. He offers the ocean as river's end yet reminds us of water's constant cycling, rising from river to cloud to mountain to sea and back again—ki uta ki tai. The river becomes an island when its banks demand it be so: confining it, slow-releasing it. This dynamic relationship inverts the sense that water carves its way through the land, instead asserting a relationship with land channelling water, holding it, containing it, partaking in mutual transformation—a confluence.

Without being unduly literal, it's interesting to contemplate the geomorphic type of river evoked in Tuwhare's poem. An island suggests divided flow—braided or anabranching? Flow shoves and pushes in contention—a moderate energy system (not a sluggish stream)—the balance of impelling and resisting forces, or secondary currents generated along concave banks at the outside of tight meander bends? Banks constrain channelised flow, so this is not a discontinuous watercourse or a swamp. The U and S bends indicate meandering traits—perhaps reflecting wandering gravel-bed rivers that are intermediary between braided and active meandering streams (there are lots of these rivers in New Zealand, but it's only in recent years that they have been scientifically described).

> One of the defining characteristics of such anabranching systems is that they have variable sinuosity—some channels can be very wiggly, others have low sinuosity 'running perfectly and quite straight'. An asymmetrical channel cross section is inferred, with the low bank on the left the convex slope on the inside of a bend, where 'laughing stitches' are perhaps products of depositional processes and patterns. In contrast, 'loud protests' on the right (outer bank) reflect erosion and lateral migration of the channel. It's a gravel-bed system ('a bed of stones'), with stepped features on the channel bed (a 'greenstone stairway' may be a riffle, perhaps more likely a step-pool sequence or a cascade—bedrock is likely close to the surface—the distribution of pounamu indicates a west coast, South Island location). And then the river transitions to the lower slopes adjacent to the sea, characterised by the 'slow release' of flow and sediment. While the river is conveyed as a living, indivisible entity, it is part of something bigger—driven by processes of the hydrological cycle that connect clouds, mountains and oceans.Ah, river
> you are ocean: you are island

While the river retains its integrity in its own right, it is innately part of something much bigger.

Every river is geomorphologically distinctive (perfect, unique; Brierley et al., 2013; Phillips, 2007). Although various aspects of the flume-like, terrace-constrained, largely fine-grained bed of the Waimatā River differ from characteristics expressed in Tuwhare's poem, distinct bigger-picture echoes are clear in the sentiment of diversity, connection, linkage, integrity and relationality.

Two years later, in Wind Song and Rain,[4] Tuwhare (1972) celebrates rippling water 'wind-huffed' with poetry, but eschews still water as 'a poem winded,' and a 'mirrored distortion'—stories silenced. He writes of the generative power of turbulence, of letting waters speak, of rivers in dynamic relationship with land, with weather, with emotions, with purpose. In the words of John Huria (Ngāi Tahu, Muaūpoko, Ngāti Apa ki te Rā Tō, Ngāti Rangi) (1993, p. 335, cited in Newman & Sullivan, 2024, p. 23), Tuwhare 'relegates the notion of "personification" to the rapidly filling basket of Eurocentric redundancies—how can you personify a person? But Tuwhare does more than transcend anthropomorphism. He evokes dynamic whakapapa networks that bind all life forms together, whereby rivers and people are kin and rivers are recognised as more ancient and powerful than persons. This lens offers the prospect to begin to envision and enact a profound reorganising of things.

8.4 Discussion

The above two narratives offer a complex, interwoven story abound with clear differences but also striking similarities. Tuwhare teaches us much about rivers in an artful form that affects as it informs and expands our sense of what rivers are, can be, and can do. Returning to the geologic and geographic languages of the first narrative's Waimatā, could we also imagine their poetics?

I was seaward sloping
          A wedge
                    w   e a r i   n   g

D
  O
    W
      N

My coarse grains *fining* into clay

Between Australian and Pacific plates
I crumpled
Became sedimentary, igneous
(A layer cake, a magma mass)

Lithified and congealed, I thrust upwards...

The potential relationship between prose and poem is undeniable. These are not typecast relationships that are set in stone; they are emergent—dynamic and

[4] First published as 'Wind Song and Rain.' Sap-wood & Milk. Hone Tuwhare. Dunedin, N.Z.: Caveman, [1972].

ever-evolving. Such multiple expressions of riverscapes as living, indivisible entities that operate from the 'mountains to the sea' yet connected to much bigger realities (e.g. hydrological cycles) demonstrate how a 'river' can never be *one* thing (Cohen et al., 2023). This variety is born out of the co-evolution of water and society, resulting from interactions over varying spatial and temporal scales (Linton, 2010). River-as-ecosociety describes such ontological conceptualisations of riverhood (Boelens et al., 2023).

In Aotearoa New Zealand, a Whanganui elder said during a Treaty of Waitangi claim lodgement for long-term damage to the Whanganui River—"I am the river, the river is me"[5] (Salmond, 2020). Cultural identity, social relationships and knowledge are cultivated through experiences with and along riverscapes (Anderson et al., 2019) through time. Meanings of water become known as a result of these relationships, morphing notions of riverscapes from externalised natural entities to sites of human and more-than-human flows (Krause & Strang, 2016; Jackson et al., 2022) and identity (Muru-Lanning, 2016). A sort of 'socio-cultural landscape' is formed from what initially may appear to be 'only' natural in its constitution.

Building on Carl Sauer's work on 'cultural geographies' and the 'Morphology of Landscape' (1925) more recent work defines landscape in itself as a cultural image (Cosgrove, 2000). Both interpretations posit the making of landscapes as a human endeavour, prompting co-evolutionary investigations that re-animate the 'missing matter of landscapes' (Whatmore, 2006). Herein, a relational approach extends the spectrum of human and more-than-human entities that populate the millennia-old co-constitution of humans with and 'within' nature (Milgin et al., 2020). Such personal and varied relationships are inherently tied to emotional connections.

Here, the crossover from the potentially abstract to the concrete may be observed. Fostering emotional connections and utilising a relational approach can cultivate social resilience. Historically, emotions have been thought of as primitive (Ahmed, 2004) and void of rationality, especially in the 'hard' sciences of physics and mathematics. In this instance of physical geography, knowledge and feelings are inseparable in many Indigenous ontologies (Mathews, 2022; Dell, 2021). In many Australian Indigenous traditions, one calls to 'River' to seek an intimate relationship (Poelina et al., 2023). Similarly, Twance (2019) observes the lack of separation between science and spirituality in Anishinaabe ways as represented in the changing meanings of traditional rock art through time from continued conversations with landscape. Each art piece means something different to each person at different times as it is born out of their relationship with the land it represents. Landscapes and all they encompass are fabricated and experienced through emotion (Pile, 2010). In turn, emotions themselves become a way of knowing/ knowledge. Thus, the concept of incorporating science and art or culture finds footing.

Affective geographies investigate emotional and subjective experiences of humans in relation to landscapes (O'Grady, 2018). The nature of emotion is not assumed or taken for granted, as bodily responses are also considered. Research

---

[5] Tiirama Thomas Hawira, Waitangi Tribunal, 1999, p. 56.

undertaken through the lens of affective geography would not simply ask, 'how does this make you feel?' It would also include the observation of people in space (McCormack, 2003) or, in this instance, riverscapes. This approach extends beyond Anthropocene notions that restrict the ability of humans to interact with more-than-human relations, embracing a 'world of feeling' (Albrecht, 2019). Landscapes, from which modern society are often distanced, can become de-spirited through the restrictions placed upon them. Affectual repair can lead to physical repair (RiverofLife et al., 2022). To achieve this reweaving of the human and more-than-human requires looking at and listening to landscapes carefully; becoming more observational prompts more care (Poelina et al., 2022). Indigenous worldviews recognise landscape as kin rather than an external 'thing' (Cohen et al., 2023), where humans are both related and responsive to it (RiverofLife et al., 2022). In this instance, kin does not refer to birth, but rather to reciprocal relationships that contribute to life and well-being (Ingold, 2000; Cohen et al., 2023). As relationships are restored and once again reciprocal, a 'poetic ecology' (Poelina et al., 2023) is produced, where humans feel the more-than-human and recognise being felt by it. Mathews (2009) observes these poetic meanings as contributing to a riverscape's living agency. In becoming observational, caring and reciprocal, mutually beneficial engagement with riverscapes is enhanced. Notions often associated with human-human relationships, such as fairness, obligation and stewardship, begin to be applied to riverscape relationships.

Solastalgia, a sense of grief, is an important example of an emotional connection to riverscapes (Albrecht et al., 2007). It refers to the loss of loved places resulting from environmental decline, born from closeness and deep meaning. Collective grief, or solastagia, can help to create togetherness while exposing the reciprocal nature of human and more-than-human relations. This can create opportunities for change. Given the openings solastalgia presents, it should be spoken about more often. Sharing riverscape stories through art, literature, poetry, and music can help speak the often unspoken (Cunsolo Willox, 2012; Santos & Wainwright, 2024). Bringing riverscape stories into social spaces of everyday life is a therapeutic practice for coping with ever-changing times. It allows for hydrophilic (water-loving) encounters to occur, fostering social resiliency (Kitson et al., 2023). Similarly, river stories put the idea of a "geoethical future" (Sharp et al., 2022) into practice. Coproduction of multiple, relational knowledges helps to situate place from a non-human perspective. Seeing riverscapes as living and part of embedded entanglements with humans and more-than-humans changes the approach to living with them entirely. A river-as-subject ontology (Boelens et al., 2023) emerges in which rivers themselves are living (subject) with their own rights, agency and significance rather than objects to be controlled. Simultaneously, a river-as-territory ontology (Boelens et al., 2023) is highlighted through the myriad ways humans and non-humans construct meanings of place.

In Aotearoa New Zealand, Māori worldviews express concerns for reciprocity (the interconnected nature of humans and non-humans and affectiveness (emotional connections between the human and non-human. In Te Ao Māori, the Māori world, people, their ancestors, and landscapes are intertwined and indistinguishable.

Humans exist in a kin-based relationship with the universe and everything in it (Hikuroa, 2017), and rivers are ancestors. Here, kinship and place (riverscapes) are linked through reciprocal relationships (Hikuroa et al., 2022). This relational understanding situates humans as one part of a much larger network of linked kin (Salmond, 2014) and teina, or juniors, to rivers and land.

How could true connection to place be developed across human and more-than-human worlds. What other lenses might be used to gaze through to explore these trans-local relations (Boelens et al., 2023)?

In the same way that the imperial past, irretrievably entangled with the development of academic disciplines and colonialism, can indicate what not to replicate, it can also provide insight and inspiration through innovators and their ideas. Building upon Wulf (2015), we look to the scholarship of Alexander von Humboldt (1769–1859), Ernst Haeckel (1834–1919) and John Muir (1838–1914). Their various works, knowingly or otherwise, offer examples of how fostering emotional connections and relational approaches to landscapes can cultivate social change; examples we might learn from today. These scholars resisted disciplinary divides and the separation of affect from science.

German polymath, geographer, naturalist and explorer Friedrich Wilhelm Heinrich Alexander von Humboldt was a proponent of romantic philosophy and science. His work built upon eighteenth and nineteenth century expressions of the Enlightenment—an age of global maritime history, scientific exploration and imperial expansion in which many scientists were involved in systematic classification of taxa (Mohan & Tamma, 2021). While Humboldt emphasised critical contributions derived from scientific observation and analysis, he believed that emotions should also be an integral part of interpretations of the Earth as a living organism (Wulf, 2015). In a similar vein, Kant (1783/2001) noted the 'thing-in-itself' can never be known as reality is always filtered through the perspective of the viewer. This fits with cultural geography's perspective of humans making landscapes (Whatmore, 2006) and the need for a relational approach to understand landscapes (Milgin et al., 2020). Resurrecting the use of the word *cosmos* from the ancient Greek, Humboldt expressed this view in 'Kosmos', a publication in which nature, art, feeling, poetry, and science were combined (Wulf, 2015). Within it, data were presented lyrically and intricately in tables, maps, drawings, narratives, and poetic formats inspired by the greatest litterateurs of the time. Humboldt did not believe in the separation of feeling from nature (Mohan & Tamma, 2021) and as such his works are interwoven with descriptive passages written with deep feeling and romanticism.

Ernst Heinrich Philipp August Haeckel was a German zoologist, naturalist, eugenicist, philosopher, physician, professor, marine biologist and artist, famed for his work on the unicellular organisms radiolarians. Haeckel was greatly influenced by Humboldt. He too merged the worlds of science and art to better understand nature. As exploration and colonisation increasingly mapped and divided the world into smaller and smaller parts, Haeckel sought to reunify its fragments. He did not believe in differentiating the organic and inorganic but instead advocated for Monism—the synergy between mind and matter, art and science and beauty and understanding (Wulf, 2015). He described the radiolarians as 'hidden treasures' (Wulf, 2015) and represented them meticulously and with reverence. To Haeckel, the natural world was something to be marvelled at—surely something so attractive should be cared for? The images presented in Haeckel's 'Art Forms in Nature' publication became the foundations for the artistic style art nouveau (Halpern & Rogers, 2013). Haeckel's art and obvious wonderment for the natural world cultivated both his own relationship with nature and those who interacted with his work.

Concurrently with Haeckel, but situated in the United States, Scottish-born John Muir has been described as "the archetype of our oneness with the earth" (Anderson, 1998). Discontent with observing nature from afar, Muir spent as much time 'in nature' as possible. Although he initially set out to be a botanist, he found this disciplinary identity too restricting (Wulf, 2015) and instead termed himself an amateur nature lover (Worster, 2005). Muir turned to writing to convey the depth of his feelings for landscapes he found himself in, especially Yosemite. He wrote a 'love letter to sequoias' using the sap from their trunks as ink, describing an almost romantic relationship with trees. (The writing is described as still shining reddish purplish today (Wulf, 2013)). Writing in this way was his attempt to show the American public the wonder that nature had to offer (Wulf, 2015). His companions often described him as literally jumping for joy at the sight of natural scenery and singing to the landforms around him (Wulf, 2015). Muir's publications and campaigning as a conservationist led to the land comprising Yosemite to be set aside as a national park in 1890. His love letters evoked emotional responses from the wider public, convincing them that nature should be appreciated and conserved (Worster, 2005).

Humboldt, Haeckel and Muir shared a common trait of being in awe of nature and its complexity, representing its connected beauty through both science and other knowledge forms. By transcending disciplinary boundaries they acknowledge the infinite array of non-human voices and cherish the intricacies of it all (Santos & Wainwright, 2024). The disciplinary lens of Geography provides a space to explore such geoethical and affective relationship (Sharp et al., 2022). Riverscapes are living entities (RiverofLife et al., 2022; Poelina et al., 2023) that link time and space in complex emergent realities. They are the sites, facilitators, and mediators of multitudes of reciprocal relationships. Riverscapes hold memories, bringing together past and present to create the future (Poelina et al., 2022). Thus, they can provide ways to think through processes of understanding (Wilcock et al., 2013). Given that there is not a single voice to be heard, but infinite voices, differing processes of understanding or modes of encounter can foreground these, their overlaps and their differences (Lantto Klein et al., 2024). Prospectively, the use of polyvocal riverscape stories to explore these multifarious interactions can provide both an enactment and an embodiment of the more-than-human nature of Boelens et al.'s (2023) riverscape ontologies and geoethical futures (Sharp et al., 2022).

8.5 Conclusion

The river is an island, a refuge, a space for sharing. Riverscapes are integrators and spaces of common (and divisive) ground. Art, culture, and science come together in Indigenous knowledge systems, in eco-poetics, and as storied knowledge through the disciplinary vehicle of geography. Taking a relational and affective perspective, this chapter has considered how emotional connections and reciprocal relationship between humans and non-humans shape place, identity and ethical considerations. This chapter offers narratives and approaches towards a river story that might deepen understandings of the physical landscape and peoples' relationship with it. While both useful and effective individually, the juxtaposition and sharing of this chapter's narratives may help to weave together a more complete and affective story of the complex riverscape that is the Waimatā. Building on these understandings, we propose that a radical reorganisation of process is required to keep up with the radical reorganisation of land and riverscapes currently occurring across the globe, including in the Waimatā catchment.

Changing boundary conditions force a deeper and more nuanced understanding of riverscapes. For this reorganisation to occur, we contend that it is essential to foreground Indigenous ways of knowing and relatedness and—for Western scholars—return to being in awe of (and in) landscapes, as Humboldt, Haeckel, and Muir so often were. Deep emotional connections to riverscapes must be fostered to create the care required for symbiotic living within (as a part of) them. Engaging with affective geographies, often through differently-scientific means (e.g. ecopoetics) acknowledges that landscapes are not passive backdrops but are shaped through emotional entanglements. Thinking about rivers as living entities with voices of their own (Salmond et al., 2019) presents a generative perspective to reflect upon how the state of riverscapes depends on how people think about them (Wooltorton et al., 2022). Furthermore, we must take care to acknowledge the variety of voices within a landscape and that differing modes of interaction with these landscapes foreground different realities (Lantto Klein et al., 2024).

Therefore, river stories must encompass the multiple ways that riverscapes can be expressed. When shared through a more-than-human lens, from the perspective of the riverscape, and in line with previously proposed riverscape ontologies (Boelens et al., 2023), a reciprocal relationship that is key to living more generatively with rivers can begin to form. River stories allow for the re-imagining of possible futures—a profound reorganising of things—and can inform and give rise to them.

References

Abram, D. (1996). *The spell of the sensuous: Perception and language in a more-than-human world.* Vintage Books. https://doi.org/10.1093/isle/5.1.159

Ahmed, S. (2004). *The Cultural Politics of Emotion* (2). Edinburgh University Press. https://www--jstor-org.ezproxy.auckland.ac.nz/stable/10.3366/j.ctt1g09x4q

Albrecht, G., Sartore, G. M., Connor, L., Higginbotham, N., Freeman, S., Kelly, B., Stain, H., Tonna, A., & Pollard, G. (2007). Solastalgia: The distress caused by environmental change. *Australasian psychiatry: Bulletin of Royal Australian and New Zealand College of Psychiatrists, 15*(1), 95–98. https://doi.org/10.1080/10398560701701288

Albrecht, G. A. (2019). *Earth emotions: New words for a new world.* Cornell University Press. Retrieved October 8, 2024, from https://books.google.co.nz/books?id=2q94DwAAQBAJ&printsec=frontcover&source=gbs_ge_summary_r&cad=0#v=onepage&q&f=false

Anderson, E. P., Jackson, S., Tharme, R. E., Douglas, M., Flotemersch, J. E., Zwarteveen, M., Lokgariwar, C., Montoya, M., Wali, A., Tipa, G. T., Jardine, T. D., Olden, J. D., Cheng, L., Conallin, J., Cosens, B., Dickens, C., Garrick, D., Groenfeldt, D., Kabogo, J., et al. (2019). Understanding rivers and their social relations: A critical step to advance environmental water management. *WIREs Water, 6*(6), e1381. https://doi.org/10.1002/wat2.1381

Anderson, W. (1998). *Green Man: The archetype of our oneness with the earth. Compass Equestrian Limited.*. ISBN 978-0951703816

Ballance, P. (2017). New Zealand Geology: An illustrated guide. *Geoscience Society of New Zealand.* https://ndhadeliver.natlib.govt.nz/delivery/DeliveryManagerServlet?dps_pid=IE32307669

Boelens, R., Escobar, A., Bakker, K., Hommes, L., Swyngedouw, E., Hogenboom, B., Huijbens, E. H., Jackson, S., Vos, J., Harris, L. M., Joy, K. J., de Castro, F., Duarte-Abadía, B., Tubino de Souza, D., Lotz-Sisitka, H., Hernández-Mora, N., Martínez-Alier, J., Roca-Servat, D., Perreault, T., et al. (2023). Riverhood: Political ecologies of socionature commoning and translocal struggles for water justice. *The Journal of Peasant Studies, 50*(3), 1125–1156. https://doi.org/10.1080/03066150.2022.2120810

Brierley, G., & Fryirs, K. (2022). Truths of the Riverscape: Moving beyond command-and-control to geomorphologically informed nature-based river management. *Geoscience Letters, 9*(1), 1–26.

Brierley, G. J. (2010). Landscape memory: The imprint of the past on contemporary landscape forms and processes. *Area, 42*(1), 76–85. https://doi.org/10.1111/j.1475-4762.2009.00900.x

Brierley, G., Fryirs, K., Cullum, C., Tadaki, M., Huang, H. Q., & Blue, B. (2013). Reading the landscape. *Progress in Physical Geography: Earth and Environment, 37(5), 601–621.* https://doi.org/10.1177/0309133313490007

Coates, G., & Cox, G. J. (2002). *The rise and fall of the Southern Alps.* Canterbury University Press.

Cohen, J. B., Dannreuther, C., Fraundorfer, M., Mackie, C., Martin-Ortega, J., Mdee, A., & Sutil, N. S. (2023). Riverkin: Seizing the moment to remake vital relations in the United Kingdom and beyond. *People and Nature, 5*, 1877–1892. https://doi.org/10.1002/pan3.10534

Cosgrove, D. (2000). Cultural geography. In R. J. Johnston, D. Gregory, G. Pratt, & M. Watts (Eds.), *The dictionary of human geography* (Vol. 4, pp. 134–138). Blackwell Publishers. Retrieved October 8, 2024, from https://research-information.bris.ac.uk/en/publications/the-dictionary-of-human-geography-fourth-edition

Cunsolo Willox, A. (2012). Climate change as the work of mourning. *Ethics & the Environment, 17*(2), 137–164. https://doi.org/10.2979/ethicsenviro.17.2.137

Dell, K. (2021). Rongomātau – 'sensing the knowing': An indigenous methodology utilising sensed knowledge from the researcher. *International Journal of Qualitative Methods, 20.* https://doi-org.ezproxy.auckland.ac.nz/10.1177/16094069211062411

Fletcher, M.-S., Romano, A., Connor, S., Mariani, M., & Maezumi, S. Y. (2021). Catastrophic bushfires, indigenous fire knowledge and reframing science in Southeast Australia. *Fire, 4*(3), Article 3. https://doi.org/10.3390/fire4030061

Fuller, I. C., Brierley, G. J., Tunnicliffe, J., Marden, M., McCord, J., Rosser, B., Hikuroa, D., Harvey, K., Stevens, E., & Thomas, M. (2023). Managing at source and at scale: The use of geomorphic river stories to support rehabilitation of Anthropocene riverscapes in the East Coast Region of Aotearoa New Zealand. *Frontiers in Environmental Science, 11.* https://doi.org/10.3389/fenvs.2023.1162099

Green, K. (2022, March 31). *Flood-ravaged Gisborne hit by all the ingredients for the perfect storm.* Stuff. https://www.stuff.co.nz/national/explained/128148299/floodravaged-gisborne-hit-by-all-the-ingredients-for-the-perfect-storm

Grey, G. (1971). In H. W. Williams (Ed.), *Ngā mahi ā ngā tūpuna* (4th ed.). A. H. & A. W. Reed.

Halpern, M. K., & Rogers, H. S. (2013). Inseparable impulses: The science and aesthetics of Ernst Haeckel and Charley Harper. *Leonardo, 46*(5), 465–470. https://www.jstor.org/stable/43834078

Harvey, K., Brierley, G., Easton, L., Watson, L., & Tunnicliffe, J. (2021). Steps towards incorporating geomorphic considerations in farm and catchment management plans in the Waimatā Catchment, Aotearoa New Zealand. *10th Australian Stream Management Conference, Kingscliff, NSW*. https://www.waikereru.org/assets/documents/HarveyEtAlAugust2021.pdf

Head, L., & Gibson, C. (2012). Becoming differently modern: Geographic contributions to a generative climate politics. *Progress in Human Geography, 36*(6), 699–714. https://doi.org/10.1177/0309132512438162

Hikuroa, D. (2017). Mātauranga Māori: The ūkaipō of knowledge in New Zealand. *Journal of the Royal Society of New Zealand, 47*(1), 5–10. https://doi.org/10.1080/03036758.2016.1252407

Hikuroa, D., Brierley, G., Blue, B., Tadaki, M., & Salmond, A. (2022). *Restoring socio-cultural relationships with rivers: Experiments in fluvial pluralism from Aotearoa New Zealand*. https://doi.org/10.25455/wgtn.19664115

Hikuroa, D., Salmond, A., Brierley, G., & Lythberg, B. (2025). Listening to Rivers: Sharing river stories from Aotearoa New Zealand in a European context. *Shima, 19*(1), 21–44.

Hunt, T. (2023, January 11). *Gisborne and East Cape deserved a New Year break. It Wasn't to be*. Stuff. https://www.stuff.co.nz/national/130945955/gisborne-and-east-cape-deserved-a-new-year-break-it-wasnt-to-be

Huria, J. (1993). Review of *Deep River Talk: Collected Poems* by Hone Tuwhare. *Landfall, 186*.

Ingold, T. (2000). *The perception of the environment: Essays on livelihood, dwelling and skill* (1st ed.). Routledge. https://doi-org.ezproxy.auckland.ac.nz/10.4324/9780203466025

Intergovernmental Panel on Climate Change (IPCC). (2023). *Climate Change 2023 Synthesis Report. Contribution of Working Groups I, II and III to the Sixth Assessment Report of the Intergovernmental Panel on climate Change*. https://www.ipcc.ch/report/ar6/syr/downloads/report/IPCC_AR6_SYR_FullVolume.pdf

Jackson, S., Anderson, E. P., Piland, N. C., Carriere, S., Java, L., & Jardine, T. D. (2022). River rhythmicity: A conceptual means of understanding and leveraging the relational values of rivers. *People and Nature, 4*(4), 949–962. https://doi.org/10.1002/pan3.10335

Kant, I. (2001). *Prolegomena to any future metaphysics* (J. W. Ellington, Trans.; 2nd ed.). Hackett Publishing. (Original work published 1783). https://books.google.co.nz/books?id=h-GM2-3WWbAC&printsec=frontcover&source=gbs_ge_summary_r&cad=0#v=onepage&q&f=false

Kitson, J., Bucknum, M., & Meenar, M. (2023). Staging hydrophilic encounters – Experiential methods for creating dialogic listening space. *Social & Cultural Geography, 24*(5), 778–795. https://doi.org/10.1080/14649365.2021.1965196

Koppes, M. N. (2022). Braiding knowledges of braided rivers – The need for place-based perspectives and lived experience in the science of landscapes. *Earth Surface Processes and Landforms, 47*(7), 1680–1685. https://doi.org/10.1002/esp.5380

Krause, F., & Strang, V. (2016). Thinking relationships through water. *Society & Natural Resources, 29*(6), 633–638. https://doi.org/10.1080/08941920.2016.1151714

Lantto Klein, M., Dixon, D. P., & Williams, R. (2024). Fluvial encounters: Experimenting with a 'River's voice' amidst light-based datafication. *Progress in Physical Geography, 48*(5–6), 738–753. https://doi.org/10.1177/03091333241280543

Lave, R., Wilson, M. W., Barron, E. S., Biermann, C., Carey, M. A., Duvall, C. S., Johnson, L., Lane, K. M., McClintock, N., Munroe, D., Pain, R., Proctor, J., Rhoads, B. L., Robertson, M. M., Rossi, J., Sayre, N. F., Simon, G., Tadaki, M., & Van Dyke, C. (2014). Intervention: Critical physical geography. *The Canadian Geographer / Le Géographe Canadien, 58*(1), 1–10. https://doi.org/10.1111/cag.12061

Lidström, S., & Garrard, G. (2014). "Images Adequate to Our Predicament": Ecology, environment and ecopoetics. *Environmental Humanities, 5*(1), 35–53. https://doi.org/10.1215/22011919-3615406

Linton, J. (2010). *What is water?: The history of a modern abstraction*. University of British Columbia Press. http://ebookcentral.proquest.com/lib/auckland/detail.action?docID=3412905

Litchfield, N. J., & Berryman, K. R. (2005). Correlation of fluvial terraces within the Hikurangi Margin, New Zealand: Implications for climate and baselevel controls. *Geomorphology, 68*, 291–313. https://doi.org/10.1016/j.geomorph.2004.12.001

Lubinski, K. S., & Thoms, M. (2018). Rivers, scholars, and society: A situation analysis. In J. M. Kelly, P. V. Scarpino, H. Berry, J. Syvitski, & M. Meybeck (Eds.), *Rivers of the anthropocene* (1st ed., pp. 63–72). University of California Press. https://www.jstor.org/stable/10.1525/j.ctt2204r6w.12

Lyver, P. O., Akins, A., Phipps, H., Kahui, V., Towns, D. R., & Moller, H. (2016). Key biocultural values to guide restoration action and planning in New Zealand. *Restoration Ecology, 24*(3), 314–323. https://doi.org/10.1111/rec.12318

Marden, M., Mazengarb, C., Palmer, A., Berryman, K., & Rowan, D. (2008). Last glacial aggradation and postglacial sediment production from the non-glacial Waipaoa and Waimata catchments, Hikurangi Margin, North Island, New Zealand. *Geomorphology, 99*(1), 404–419. https://doi.org/10.1016/j.geomorph.2007.12.003

Marden, M., & Rowan, D. (1993). Protective value of vegetation on tertiary terrain before and during Cyclone Bola, East Coast, North Island, New Zealand. *New Zealand Journal of Forestry Science, 23*(3), 255–263. https://www.scionresearch.com/__data/assets/pdf_file/0011/59753/NZJFS2331993MARDON255_263.pdf

Marden, M., & Seymour, A. (2022). Effectiveness of vegetative mitigation strategies in the restoration of fluvial and fluvio-mass movement gully complexes over 60 years, East Coast region, North Island, New Zealand. *New Zealand Journal of Forestry Science, 52*. https://doi.org/10.33494/nzjfs522022x226x

Massey, D. (2001). Talking of space-time. *Transactions of the Institute of British Geographers, 26*(2), 257–261. https://doi.org/10.1111/1475-5661.00019

Mathews, F. (2009). The world hidden within the world: A conversation on ontopoetics. *Philosophy Activism Nature, 6*, 97–109. https://search-informit-org.ezproxy.auckland.ac.nz/doi/epdf/10.3316/informit.590634590746578

Mathews, F. (2022). Conservation needs to include a story about feeling. *Biological Conservation, 272*. https://doi.org/10.1016/j.biocon.2022.109668

Mazengarb, C., & Marden, M. (2024). [Dataset of Terrace Heights in the Waimatā Catchment Gisborne] [Unpublished data]. Colin Mazengarb, personal communication.

McCormack, D. P. (2003). An event of geographical ethics in spaces of affect. *Transactions of the Institute of British Geographers, 28*, 488–507. https://doi.org/10.1111/j.0020-2754.2003.00106.x

Milgin, A., Nardea, L., Grey, H., Laborde, S., & Jackson, S. (2020). Sustainability crises are crises of relationship: Learning from Nyikina ecology and ethics. *People and Nature, 2*(4), 1210–1222. https://doi.org/10.1002/pan3.10149

Mohan, A. V., & Tamma, K. (2021). The lasting contribution of Alexander von Humboldt to our understanding of the natural world. *Resonance, 26*(8), 1041–1050. https://doi.org/10.1007/s12045-021-1207-z

Muru-Lanning, M. (2016). *Tupuna Awa: People and politics of the Waikato River*. Auckland University Press.

Newman, J., & Sullivan, R. (2024). *koe: An Aotearoa ecopoetry anthology*. Otago University Press.

O'Grady, N. (2018). *Geographies of affect*. Oxford Bibliographies. https://doi.org/10.1093/obo/9780199874002-0186

Parsons, M., & Fisher, K. (2020). Indigenous peoples and transformations in freshwater governance and management. *Current Opinion in Environmental Sustainability, 44*, 124–139. https://doi.org/10.1016/j.cosust.2020.03.006

Phillips, J.D. (2007). The perfect landscape. *Geomorphology, 84*(3–4), 159–169. https://doi.org/10.1016/j.geomorph.2006.01.039

Pile, S. (2010). Emotions and affect in recent human geography. *Transactions of the Institute of British Geographers, 35*, 5–20. https://doi.org/10.1111/j.1475-5661.2009.00368.x

Poelina, A., Perdrisat, M., Wooltorton, S., & Mulligan, E. L. (2023). Feeling and hearing country as research method. *Environmental Education Research, 29*(10), 1486–1501. https://doi.org/10.1080/13504622.2023.2239531

Poelina, A., Wooltorton, S., Blaise, M., Aniere, C. L., Horwitz, P., White, P. J., & Muecke, S. (2022). Regeneration time: Ancient wisdom for planetary well-being. *Australian Journal of Environmental Education, 38*(3–4), 397–414. https://doi.org/10.1017/aee.2021.34

Poverty Bay Herald. (1895). The installation of Mayor. December 18, 2. https://paperspast.natlib.govt.nz/newspapers/PBH18951218.2.9

Reardon-Smith, M. (2023, April 12). *Aboriginal burning practices meet colonial legacies in Australia*. Edge Effects. https://edgeeffects.net/aboriginal-burning-australia/

RiverOfLife, M., River, U. S. R., Poelina, A., Wooltorton, S., Guimond, L., & Sioui Durand, G. (2022). Hearing, voicing and healing: Rivers as culturally located and connected. *River Research and Applications, 38*(3), 422–434. https://doi.org/10.1002/rra.3843

Rogers, A., Castree, N., & Kitchin, R. (2013). *A dictionary of human geography*. Oxford University Press. https://www.oxfordreference.com/display/10.1093/acref/9780199599868.001.0001/acref-9780199599868-e-1216#:~:text=1.,….%20

Royal, A. T. C. (2006). *Whenua – How the land was shaped – The North and South islands*. Te Ara – The Encyclopaedia of New Zealand. https://teara.govt.nz/en/whenua-how-the-land-was-shaped/page-2

Salmond, A. (2014). Tears of Rangi: People, water and power in New Zealand. *Hau: Journal of Ethnographic Theory, 4*(3), 285–309. https://doi.org/10.14318/hau4.3.017

Salmond, A. (2020). Afterword: "I Am the River, and the River Is Me". *Contemporary Pacific, 32*(1), 164–171. 307.

Salmond, A., Brierley, G., & Hikuroa, D. (2019). Let the Rivers Speak: Thinking about waterways in Aotearoa New Zealand. *Policy Quarterly, 15*(3), Article 3. https://doi.org/10.26686/pq.v15i3.5687

Salmond, A., Brierley, G., Hikuroa, D., & Lythberg, B. (2022). Tai Timu, Tai Pari, the ebb and flow of the tides: Working with the Waimatā from the Mountains to the Sea. *New Zealand Journal of Marine and Freshwater Research, 1–17*. https://doi.org/10.1080/00288330.2022.2096084

Santos, M., & Wainwright, J. (2024). River(s) wear: Water in the expanded field. *Cultural Geographies, 31*(4), 447–471. https://doi.org/10.1177/14744740241233699

Sauer, C. O. (1925). The morphology of landscape. In J. Leighly (Ed.), *Land and life: A Selection from the writings of Carl Ortwin Sauer* (pp. 315–350). University of California Press (1963). https://doi.org/10.2307/jj.13167997.20

Sharp, E. L., Brierley, G. J., Salmond, J., & Lewis, N. (2022). Geoethical futures: A call for more-than-human physical geography. *Environment and Planning F, 1*(1), 66–81. https://doi.org/10.1177/26349825221082168

Syvitski, J. (2012). Anthropocene: An Epoch of our making. *Global Change, 78*, 12–15. http://igbp.net/download/18.1081640c135c7c04eb480001178/1376383107963/NL78-anthropocene.pdf

Tapsell, P. (1997). The flight of Pareraututu. *Journal of the Polynesian Society, 106*(4), 323–374.

Thomas, M., Lythberg, B., Hikuroa, D., & Brierley, G. (2024). Problematising the Anthropocene: Geographic perspectives upon the riverscapes of Waimatā Catchment, Aotearoa New Zealand. *The Geographical Journal, e12598*. https://doi.org/10.1111/geoj.12598

Tuwhare, H. (1970). The River is an Island. In V. O'Sullivan (Ed.), *An anthology of twentieth century New Zealand poetry* (pp. 181–182). Oxford University Press.

Tuwhare, H. (1972). Wind Song and Rain. In *Sap-wood & Milk*. Caveman.

Twance, M. (2019). Learning from land and water: Exploring mazinaabikiniganan as indigenous epistemology. *Environmental Education Research, 25*(9), 1319–1333. https://doi.org/10.1080/13504622.2019.1630802

Whatmore, S. (2006). Materialist returns: Practising cultural geography in and for a more-than-human world. *Cultural Geographies, 13*(4), 600–609. https://doi.org/10.1191/1474474006cgj377oa

Wilcock, D., Brierley, G., & Howitt, R. (2013). Ethnogeomorphology. *Progress in Physical Geography, 37*(5), 573–600. https://doi.org/10.1177/0309133313483164
Wooltorton, S., Guimond, L., Reason, P., Poelina, A., & Horwitz, P. (2022). Voicing Rivers. *River Research and Applications, 38*(3), 371–375. https://doi.org/10.1002/rra.3939
Worster, D. (2005). John Muir and the modern passion for nature. *Environmental History, 10*(1), 8–19. https://www.jstor.org/stable/3985830.
Wulf, A. (2013). *Just a moment...* https://blogs.bl.uk/americas/2013/03/john-muir-is-going-sequoical-in-the-yosemite.html
Wulf, A. (2015). *The invention of nature*. John Murray.

Chapter 9
Whānau Care: A Model for Equitable Support for kaitiaki whānau

Jo Voice

Abstract Tamariki Māori (Māori children) represent 60 percent of children in state care in Aotearoa (New Zealand), despite Māori only being 16.5 percent of the population. In response to public outcry, policy changes in 2019 have reduced the number of tamariki Māori entering state care; however, they have resulted in under-resourced kaitiaki whānau (Māori family caregivers) providing care for traumatised tamariki. My analysis of the literature makes clear that the inequities experienced by kaitiaki whānau are the direct result of state policies and practises that continue to fail to recognise the importance and inter-connectedness of tamariki Māori well-being with that of their whānau (families) and their lands. This situation underscores ongoing issues in the state's approach to child welfare, particularly for Māori whānau, reflecting a disconnect between policy intentions, practical support, and a failure to fully embrace Māori perspectives in the care and protection of tamariki. Examination of these issues provides the rationale for the Whānau Care project, a model of practice co-designed with kaitiaki whānau, hāpu (family groups) and iwi (tribe), and relevant agencies to provide the proper support, training, and access to the funding needed for the care of at-risk tamariki Māori. My research demonstrates the urgent need for a transformative shift in child protection policies and practises in Aotearoa so that they become grounded in Māori principles of collective well-being and self-determination.

Keywords Child protection · Policy · Kaitiaki whānau · Tamariki Māori · Relationality

J. Voice (✉)
Whanau Care, Whanganui, New Zealand
e-mail: jovoice@whanaucare.com

M. Kennedy, E. O'Donnell (eds.), *People, Place and Nature in Indigenous-Settler Relations*, Indigenous-Settler Relations in Australia and the World 10, https://doi.org/10.1007/978-3-032-03600-1_9

9.1 Introduction

The issues within Aotearoa's child protection system are well documented (Māori Perspective Advisory Committee, 1988; Kaiwai et al., 2020; Savage et al., 2021; Waitangi Tribunal, 2021). In 2019, a widely publicised incident involving the contentious removal of a tamariki Māori into state care ignited a wave of criticism from the public and professionals. This incident led to a review and subsequent modification of the policies of Oranga Tamariki – Ministry for Children (OT), Aotearoa's statutory child protection agency, regarding removing children, particularly tamariki Māori. While some tamariki Māori still come into state care, the responsibility for caring for the majority of at-risk tamariki shifted to kaitiaki whānau (Oranga Tamariki, 2019a). For this chapter, kaitiaki whānau refers to the Māori caregiver's who make up part of the cohort of non-statutory family caregivers who provide permanent care for tamariki Māori identified by the state as at risk, so these tamariki do not need to enter into state care. Kaitiaki whānau receive limited support from OT because their legislation and policies do not identify them or provide for their needs. Four years after these policy changes, research shows that kaitiaki whānau and the tamariki they are caring for are experiencing more mental and emotional stress, financial hardship, and less equitable access to health and education supports than statutory whānau caregivers who provide care for tamariki that are, or who have been, in the custody of the state (Oranga Tamariki, 2019b, 2020, 2023b). Evidence shows that the combined financial cost to the State for a single child in State foster care is more than double the amount of financial support received by kaitiaki whānau when they provide the same support (Ministry of Social Development, 2023; Oranga Tamariki, 2021a).

As a Pākehā (non-Māori New Zealander) social worker, my interest in the issue of equitable outcomes for kaitiaki whānau stems from the 21 years I have worked in or adjacent to the field of child protection, supporting whānau through Aotearoa's child protection system. From 2017 to 2021, I managed a team of external contractors to OT. Our role was to exit children from the state's custody – often into the permanent care of their statutory whānau caregivers, who received an extensive package of ongoing support from OT, guaranteed to them in legislation and policy, in recognition of the high level of care that traumatised children require. By 2021, I had witnessed a growing number of social workers circumventing the need to bring at-risk tamariki Māori into state care by placing them into the care of kaitiaki whānau, often without providing them with adequate advice and support. This change in practice reduced the number of tamariki Māori entering care and removed a significant financial and staffing resource burden from the state. However, the kaitiaki whānau community has not benefited from these savings. From 2021 until 2022, as a Kaiārahi (navigator) in the Family Court, I witnessed the impact that these policy changes had on numerous kaitiaki whānau who sought assistance to navigate complex legal processes while also struggling to provide for the needs of the tamariki in their care.

Whānau Care is my response to the events and inquiries that led to the 2019 policy changes and the experience of caregivers, particularly kaitiāki whānau, in Aotearoa who are caring for at-risk tamariki without financial and social support from the statutory child protection system. The Whānau Care project seeks to create a model of practice that is co-designed with kaitiaki whānau, hāpu and iwi, and relevant agencies. The collaborative design process is embedded with the principles of Indigenous research methodologies and leadership theories whilst acknowledging the well-being derived from connection to place (Chilisa, 2020; Cram & Adcock, 2022; Moreton – Robertson & Walter, 2009; Smith, 2018; Thom & Grimes, 2022). The Whānau Care project, delivered on Indigenous land, acknowledges the impact of historical land dispossession on Māori wellbeing and aims to restore connection and belonging by engaging traditional landowners and kaitiaki whānau in culturally grounded design. Collective well-being and self-determination are the central themes woven into Whānau Care, as these are essential to enhance equitable outcomes for kaitiaki whānau. Whānau Care will describe an intervention from the point where a Statutory Care Provider first asks the new kaitiaki whānau to provide permanent care of their tamariki until the court grants final legal orders to the kaitiaki whānau. Supporting financial documents that evidence the financial viability of the intervention to funders will also be created as a part of the Whānau Care project.

9.1.1 Policy Failures

In 2019, national broadcasters aired the contentious removal of a tamaiti Māori (Māori child) into state care. With tamariki Māori representing 60 percent of children in state care, despite Māori only being 16.5 percent of the population, this incident reignited the concerns and discontent of whānau, hapū, and iwi towards OT's policies and processes, particularly regarding the removal tamariki Māori into state care (Hollingworth, 2019; Kaiwai et al., 2020; Oranga Tamariki, 2023c; Statistics New Zealand, 2018; Waitangi Tribunal, 2021). Inquiries and reviews into the actions of OT resulted in changes to policies that included extensive legal and managerial oversight of social worker's applications to remove children from the care of their parents (Oranga Tamariki, 2019a, 2023b). Under the new policy, 'Without Notice' applications (urgent Family Court applications where orders can be made without informing the respondent) were discouraged, with new policy focussing on placing tamariki with whānau without OT holding any court order regarding the placement arrangements, a requirement that was already included in legislation, but which had been forgone in practice (Hyslop, 2022; Oranga Tamariki, 2023b; Oranga Tamariki Act 1989. s.5. ss.(1)(c)). These changes made it time-consuming and technically challenging for social workers to secure a child's safety by bringing them into state care. It facilitated the shift in practise towards social workers focussing on encouraging kaitiaki whānau to provide legal day-to-day care of their at-risk tamariki under the *Care of Children Act 2004* or by agreement with birth whānau, financially supported by the Unsupported Child's Benefit (UCB) (a weekly welfare payment for people who are caring for children that are not their own) (*Care of Children Act,* 2004, s48; Keddell et al., 2021, 2022; Oranga Tamariki, 2021c, 2021d; Ministry of Social Development, 2023). The state began to view family-based care as the primary setting where the safety and emotional, psychological, and behavioural needs of at-risk children should be addressed (Oranga Tamariki, 2021c).

9.1.2 Inequitable Support for Carers

Overwhelmingly, OT's research and literature published by community agencies conclude that these changes to policy and practice have created inequity in support for the kaitiaki whānau (Bundle, 2021; Caring Families Aotearoa, 2023; Keddell et al., 2021, 2022; Oranga Tamariki, 2019b, 2020, 2021c). Non-statutory whānau caregivers, kaitiaki whānau included, receive less training, information, financial assistance, and access to well-being support than their statutory counterparts (Oranga Tamariki, 2021b, 2023a). In their 2021 report, *UCB as an Alternate Pathway* (Oranga Tamariki 2021c), OT recognised a link between their revised practices and the decrease in tamariki Māori entering state care since 2019. While these arrangements prevent tamariki Māori having unnecessary state involvement in their lives, placements with kaitiaki whānau can become unstable in the future because there is no established procedure for kaitiaki whānau to request help from OT when unforeseen challenges arise (Oranga Tamariki, 2021c). In a 2020 OT survey, stakeholder feedback suggested that the unequal financial aid and support given to non-statutory caregivers, kaitiaki whānau included, felt biased and unjust. Stakeholders argued that this policy penalised families for proactively trying to prevent children from entering state care (Oranga Tamariki, 2020). The survey showed that kaitiaki whānau were more stressed, felt less supported, and struggled to provide the care their tamariki needed on the UCB than non-Māori caregivers. They also carry extra expenses for connecting tamariki to their culture and whakapapa (Oranga Tamariki, 2020).

A follow-up survey in 2023 showed only slight improvement after an increase in financial entitlement (Oranga Tamariki, 2023a). This situation underscores a longstanding contradiction in the state's approach to the well-being of tamariki Māori. The state emphasises the importance of whānau care by changing policies and practices but inadequately supports those providing that care (Bradley, 1994; Hyslop, 2022; McRae & Nikora, 2006; Walker, 2001). While the state has recently devoted significant focus to determining whether the UCB adequately supports caregivers to meet the needs of the children in their care, there is presently no indication of plans to elevate the entitlements of kaitiaki whānau to be on par with those of statutory whānau caregivers (Minister for Children, 2019; Oranga Tamariki, 2019b, 2020, 2021c, 2023b).

Oranga Tamariki contracting information shows that it costs the state approximately $43,435 annually to keep a child in foster care (Oranga Tamariki, 2021a, p.34). In contrast, the state spends approximately $NZD17,000 annually to support a child living in the care of a kaitiaki whānau (Ministry of Social Development, 2023). Agencies such as Woven Whānau and Grandparents Raising Grandchildren offer peer support groups and advocacy to grandparent caregivers (Dallas, 2023; Grandparents Raising Grandchildren Trust, 2015). No literature outlined any government or non-government programmes in Aotearoa specifically funded to and focussed on supporting, providing additional financial assistance, advocating for, and training kaitiaki whānau – grandparents, great aunties and uncles, extended whānau, older siblings, cousins, whāngai whānau (adoptive family).

A leading child welfare research organisation in the United States references a recent shift in practice to placing at-risk children with kin, arguing for a "change in practice" driven by "a culture shift at the front line" that empowers and supports the extended family of vulnerable children who are providing for their care (Child Trends, 2016, as cited in El Hage, 2016. para.13). An emerging field of research, including anecdotal evidence collected by the Whanganui Non-Government Organisation, Woven Whānau, and led by Lynette Archer, is bringing attention to the kaitiaki whānau experience in Aotearoa, highlighting the impact of the 2019 policy change to the government (Caring Families, 2023; Lynette Archer, personal communication, July 19, 2023).

9.2 The Conflict: Policy Misaligned with Māori Worldview

The literature makes clear that the inequities experienced by kaitiaki whānau are the result of state policies and practises that continue to fail to recognise the importance and inter-connectedness of tamariki Māori well-being with that of their whānau (Kaiwai et al., 2020; Māori Perspective Advisory Committee, 1988; Savage et al., 2021; The Waitangi Tribunal, 2021). The policy changes post-2019 have reduced the number of tamariki Māori entering state care but have resulted in under-resourced kaitiaki whānau providing care for traumatised tamariki. These policies are in direct conflict with the value placed on the well-being of the collective (the kaitiaki whānau) as a source of well-being for the individual (tamariki Māori) as found in a Māori worldview (Rickard, 2014; Williams et al., 2019).

Scholarly research has identified that the individualistic Western ideas of family well-being that have dominated Aotearoa's child protection policy do not provide for better outcomes for tamariki and whānau Māori (Boulton & Gifford, 2014; Hyslop, 2022; Love, 2002). The Child Paramountcy Principle, held within child protection legislation, determines that the interests of the child should be paramount in guiding decision-making in child protection, but in practice, it often translates into disregarding the impact of decisions on whānau (Hyslop, 2022; Oranga Tamariki Act 1989, s.13. ss.1). Choate's (2019) research challenges the widespread use of Western theories and models when working with Indigenous families, pointing out the minimal evidence of their efficacy in enhancing well-being. From a Te Ao Māori perspective, well-being is found in "shared duties and collective rights", with whānau playing a crucial role in cultural understanding, personal identity, and health (Rickard, 2014, p. 27). International research supports this position. An Indigenous worldview is a space "where all things are interrelated and significant": the individual and the collective (Kwaymullina, 2005 as cited in Ryan & Evans, 2020. p. 54; Spiller et al., 2020. pp. 518 and 531).

9.2.1 Māori Resistance and Advocacy

Whānau, hāpu and iwi have long been pushing for a Māori-driven approach to tackle the disparate numbers of tamariki Māori in state care. In the 1970s and 80s, resistance movements such as Ngā Tamatoa, Arohanui Inc and the Polynesian Panthers advocated on behalf of whānau Māori for culturally aligned practice within child protection services (Savage et al., 2021). Similarly, the resurgence of the traditional act of whāngai in the Mātua Whāngai Programme of the 1980s sought to provide care for tamariki Māori within their iwi and devolve the provision of child protection services for tamariki Māori back to Māori (Walker, 2001). Successive inquiries into state care have recognised self-determination, or a by Māori for Māori approach, as a transformative approach to addressing the inequity experienced by whānau Māori in their interactions with the state (The Waitangi Tribunal, 2021; Kaiwai et al., 2020; Māori Advisory Committee, 1988; Savage et al. 2021).

In 1988, the *Puao-Te-Ata-Tu* report into racism in the Department of Social Welfare (now OT) summarised the issues within the state's approach to child protection as a "profound misunderstanding or ignorance of the place of the child in Māori society and its relationship with whānau, hāpu and iwi structures" (Māori Perspective Advisory Committee, 1988. p. 7). Subsequent reports have echoed this sentiment and have raised questions regarding the state's failure to implement the recommendations of *Puao-Te-Ata-Tu* by changing policy to reflect better the needs of the collective whānau Māori (Kaiwai et al., 2020; Savage et al., 2021; The Waitangi Tribunal, 2021).

Puao-Te-Ata-Tu described a vision of "shared decision making with appropriate Māori people" to "eliminate deprivation and alienation, equitable distribution of resources shared power and authority" (Māori Perspective Advisory Committee, 1988. p. 19 & 26). Te Tiriti o Waitangi (The Treaty of Waitangi) holds a guarantee from the state that Māori will have tino rangatiratanga (self-determination or sovereignty) over their kāinga (home) (The Waitangi Tribunal, 2021). The literature considered in this chapter is clear that the state has not kept this guarantee. In the case of the WAI Report 2915 (2021), regarding the Waitangi Tribunal claim brought against OT following the 2019 incident, the failure to keep this guarantee is called a "breach born of hostility to the promise itself" (p. 12).

Thirty-five years after *Puao-Te-Ata-Tu*, considerable research has focused on the devolution of decision-making in child protection to Māori. Hyslop's 2022 historical account of child protection in Aotearoa recommends Māori self-determination through devolution as a pathway to better outcomes for tamariki Māori. International research supports self-determination processes and holistic well-being approaches to decolonising systems, policies, and methods to reduce adverse outcomes for Indigenous children (MacKean et al., 2022). Central to this strategy is restoring Indigenous sovereignty and challenging systems of oppression, "a process of recovery, reclamation, reproduction and resurgence" (Williamson, 2021. p. 339).

Cornell's examination of poverty and self-determination in Australia, Aotearoa, Canada and the United States advocates for self-determination in nationhood and sovereignty (Cornell, 2006). The research shows how self-determination has provided the opportunity to enhance conditions for Indigenous populations. (Carroll et al., 2021; Cornell, 2006). Iwi and Kaupapa Māori organisations, informed by Māori ways of being and doing, represent self-determination in action (Boulton et al., 2013; Chant, 2013; Smith et al., 2019). The approach is by Māori for Māori and "advances Māori aspirations from a context in which Māori thinking, values, knowledge, language, cultural protocols and views of the world provide the basis of action" (Mane, 2009, p. 1).

There remains a significant gap between policy and practise regarding the state's support for kaitiaki whānau and the well-being of tamariki Māori. While the policy shifts post-2019 aimed to reduce the number of tamariki Māori entering state care, they inadvertently created a system where kaitiaki whānau receive insufficient support despite playing a crucial role in providing for these children's safety. This gap reflects a more profound dissonance between Western models of child welfare and Te Ao Māori perspectives on collective well-being. The ongoing challenge is ensuring that the support for kaitiaki whānau and tamariki Māori is adequate in terms of resources and culturally aligned. A more effective and equitable approach to child welfare in Aotearoa must recognise and resource the critical role of kaitiaki whānau in nurturing the well-being of tamariki Māori in a manner that allows them to be self-determining.

9.3 Towards a Better Outcome

Whānau Care seeks to address the inequities created by the 2019 policy changes by creating a model of practice via a co-design process with kaitiaki whānau that describes how community organisations can support the well-being of the tamariki and their kaitiaki whānau. In developing a practice model in response to the inequities kaitiaki whānau face when caring for traumatised tamariki, I have drawn from my experience supporting kaitiaki whānau, emergent research on the impact of the 2019 policy changes, and the evident lack of government and community support for kaitiaki whānau. The model of practice will include supporting financial documents, such as a budget and cost-benefit analysis, that can be used to train kaimahi (staff/workers) and can be presented to funders to outline the intervention process, the rationale behind its design and the intended outcomes. Shared decision-making, focusing on collective well-being and sustainable development, as discussed in Indigenous leadership theories, will also guide the project process (Maaka et al., 2018; Ryan & Evans, 2020; Spiller et al., 2020).

The Whānau Care project will be delivered on Indigenous land. For Māori, the belief that land is a source of wellbeing is embedded in their cultural values and practices. Land serves as a symbol of cultural identity and connection to ancestors and each other, forming a vital link between past, present, and future and providing a sense of belonging and connection to culture. Traditionally, land was shared, strengthening collective wellbeing and supportive networks. Contemporary challenges that whānau Māori face can be linked to historical land dispossession and the subsequent cascading economic and wellbeing ramifications (Thom & Grimes, 2022). If Whānau Care is to address issues of inequity, it must also acknowledge the whakapapa (origins/link to) of that inequity. Whānau Care aims to recognise the traditional landowners and the kaitiaki whānau that live on their whenua (land) by including them and their voice in the project design via consultation and participation in hui and wānanga in traditional spaces informed by indigenous knowledge (Ministry for the Environment, 2021; Ministry for Health, 2023; Moewaka Barnes & McCreanor, 2019). In this way, there is an opportunity for connection and belonging on traditional lands in traditional ways that acknowledge the restorative nature of these connections.

The project relies heavily on the transformative nature of self-determination. However, there is not one unifying definition of self-determination, and the ability to be self-determining can depend on the differing degrees of oppression in a person's life (Kuokkanen, 2019. pp. 22 & 25). Kuokkanen (2019) suggests that it is helpful to consider what is indispensable for individual and collective well-being when seeking to facilitate self-determination. While Whānau Care holds funding in its budget to remove barriers to participation, these barriers may not be immediately apparent. Assumptions must not be made about what the project facilitator can do to encourage and support the participation of individual kaitiaki whānau.

9.3.1 Relationality is Key

Kaupapa Māori (Māori ways of being and doing) and Indigenous methodologies centre relationality and relationships. The Whānau Care project will practice collaboration through its relationships with kaitiaki whānau and relevant support services in their communities (Bell et al., 2022; Chilisa, 2020; Cram, 2018; Moreton-Robertson & Walter, 2009; Smith, 1999, 2018). Whānau Care will be designed, kanohi ki te kanohi, in wānanga (collaborative learning spaces) with kaitiaki whānau. Whānau Care will be embedded with whānau goals and ideas, using knowledge, data, and evidence from whānau (Rewiti, 2022). This approach, integral to knowledge production within a Māori paradigm and Indigenous thinking more generally, has proven to lead to more effective services, significant social impact, and practical solutions (Burkett, 2012. p. 4; Moreton-Robertson & Walter, 2009; Smith, 2018). This process recognises that the dynamic and diverse nature of whānau ora (family health and well-being) for whānau Māori is grounded in Māori cultural values.

Kaupapa Māori methodologies require the design process to take shape consistent with the principles of Te Tiriti o Waitangi. This allows Whānau Care to become a tool for decolonisation that promotes the self-determination of Indigenous voices, an approach that "focuses on restoring mana to survivors and not further perpetuating harm" (Moreton-Robertson & Walter, 2009; Simpson & Ake, 2010; Savage et al., 2021, p.28). The Waitangi Tribunal defines the principles, or the "fundamental truths" of Te Tiriti, as participation, active protection, reciprocity, and redress (Te Puni Kōkiri, 2001. p.76). Along with collaborative hui (meetings) and wānanga, steps such as shared decision-making shared problem definition and shared problem-solving, whakawhanaungatanga (relationship building), discussing group tikanga (protocols that guide behaviour) and shared language, establishing the purpose of the co-design process, providing activities to encourage the flow of ideas, and inviting feedback are ways participation will be encouraged (Burkett, 2012; Farr, 2018; Lee, 2008; Mahuika & Mahuika, 2020).

9.4 Conclusion

The public airing of the state's removal of a tamaiti Māori in 2019 sparked significant controversy and again exposed longstanding issues in the state's approach to child welfare. Subsequent policy reforms sought to reduce state interventions by securing the care and protection of at-risk tamariki by placing them with kaitiaki whānau rather than bringing them into state care. However, these kaitiaki whānau receive less support than state-appointed caregivers, leading to stress, difficulty in accessing welfare support for the tamaiti and their caregiver, and potential instability in placements.

The literature shows that the state's actions were inconsistent with the needs of tamariki Māori and their whānau. It calls for genuine engagement with Māori perspectives on collective well-being and a commitment to self-determination to rectify the inequities in child welfare services. Research indicates that this differential support between kaitiaki whānau and statutory whānau caregivers is biased and unfair and occurs in the context of Western systems and policies that do not regard Māori ways of being and doing. Financial constraints are particularly acute for kaitiaki whānau, who also face the cultural responsibilities of connecting children to their heritage.

International research echoes the challenges faced by kaitiaki caregivers (Dean et al., 2023; Gupta-Kagan, 2019; Harding et al., 2020; McHugh, 2003; Silver Century Foundation, 2018; Smart et al., 2022). The Māori-centric worldview emphasises the collective over the individual, which clashes with the Western-centric approach that dominates the child welfare system. Advocacy for a Māori-led approach to child protection has a long history, with movements and initiatives highlighting the need for culturally aligned practices.

The Whānau Care project represents a step towards correcting inequity and improving the support for kaitiaki whānau and tamariki Māori. The project's plan for an inclusive approach to participant engagement should ensure that kaitiaki whānau are not merely participants but co-creators of the Whānau Care model. This inclusivity ensures that the project's direction is guided by those it intends to benefit, facilitating a collective journey towards well-being. Whānau Care underscores the importance of place by acknowledging the traditional landowners and the restorative potential of reconnecting with traditional lands and practices. The commitment to ethical practice within this project will be primarily managed through the adherence to professional codes of conduct and legislation and through a continuous evaluative process that holds the vision and goals of kaitiaki whānau at its core.

The Whānau Care project is expected to impact various domains - from the immediate community to the policy level, where advocacy for kaitiaki whānau well-being may lead to systemic changes. While the project holds promise, it also faces potential challenges, including securing sustainable funding, maintaining cultural appropriateness, and achieving collective agreement within diverse kaitiaki whānau communities. The project's adaptability, commitment to reflective practises, and willingness to be guided by the needs and wisdom of the community will be essential to its success.

Whānau Care seeks to become a culturally responsive, ethical, and inclusive approach to addressing the needs of vulnerable tamariki and their kaitiaki whānau. With its implementation, Whānau Care aspires to be more than a model of practice; it seeks to create hope and be a mechanism for transformational change.

References

Bell, A., Yukich, R., Lythberg, B., & Woods, C. (2022). Enacting settler responsibilities towards decolonisation. *Ethnicities, 22*(5), 605–618. https://doi.org/10.1177/14687968211062675

Boulton, A., & Gifford, H. (2014). Whānau Ora; He Whakaaro Ā Whānau: Māori family views of family well-being. *The International Indigenous Policy Journal, 5*(1), 1–18. https://doi.org/10.18584/iipj.2014.5.1.1

Boulton, A., Tamehana, J., & Brannelly, T. (2013). Whānau-centred health and social service delivery in New Zealand: The challenges to, and opportunities for, innovation. *Mai Journal, 2*(1), 18–32. https://journal.mai.ac.nz/sites/default/files/Vol%202%20%281%29%20024%20Boulton.pdf

Bradley, J. (1994). Iwi and the Maatua Whāngai Programme. In R. Munford & M. Nash (Eds.), *Social work in action* (pp. 178–199). The Dunmore Press Ltd.

Bundle, K. (2021). *Submission No SSA 18 to the Social Services Select Committee on the Social Security (Financial Assistance for Caregivers) Amendment Bill.* Grandparents Raising Grandchildren Trust. New Zealand. https://www.grg.org.nz/site/grg/files/Submissions/GRG%20Sub%20SSA18-Financial%20Assistance%20for%20Caregivers%20Bill.pdf

Burkett, I. (2012). An introduction to co-design. *Sydney: Knode, 12*. https://www.yacwa.org.au/wp-content/uploads/2016/09/An-Introduction-to-Co-Design-by-Ingrid-Burkett.pdf

Care of Children Act 2004 (NZ). (2004). Retrieved October 20, 2023, from https://www.legislation.govt.nz/act/public/2004/0090/latest/DLM317233.html.

Caring Families Aotearoa. (2023). *Care and protection green paper: Te pepa kakariki atawhai me te whakamaru*. Caring Families Aotearoa. https://www.caringfamilies.org.nz/3d-flip-book/care-and-protection-green-paper-2023/

Carroll, S., Cornell, S., & Jorgensen, M. (2021). Can a self-determination strategy improve Indigenous health care? In D. Smith, A. Wighton, S. Cornell, & A. Vai Delaney (Eds.), *Developing governance and governing development: International case studies of Indigenous futures* (pp. 201–221). Roman & Littlefield.

Chant, L. (2013). *Hauora Kotahitanga: Māori health experiences as models for co-operative co-existence between indigenous and non-indigenous peoples*. Doctor of Philosophy in Community Health thesis, University of Auckland, Auckland. https://researchspace.auckland.ac.nz/handle/2292/20440

Chilisa, B. (2020). *Indigenous research methodologies*. Sage Publications.

Choate, P. (2019). The call to decolonise: Social work's challenge for working with Indigenous peoples. *The British Journal of Social Work, 49*(4), 1081–1099. https://doi.org/10.1093/bjsw/bcz011

Cornell, S. E. (2006). *Indigenous peoples, poverty and self-determination in Australia, New Zealand, Canada and the United States*. Native Nations Institute for Leadership. Management, and Policy. https://indigenousgov.hks.harvard.edu/publications/indigenous-peoples-poverty-and-self-determination-australia-new-zealand-canada

Cram, F. (2018). Conclusion: Lessons about Indigenous evaluation. In F. Cram, K. A. Tibbetts, & J. LaFrance (Eds.), *Indigenous evaluation. New directions for evaluation, 159*, 121–133. https://doi.org/10.1002/ev.20326

Cram, F., & Adcock, A. (2022). Kaupapa Māori research. In *Handbook of qualitative cross-cultural research methods* (pp. 56–84). Edward Elgar Publishing. https://doi.org/10.4337/9781800376625.00013

Dallas, M. (2023, May 05). Grandparents needs fall through cracks. *Feilding-Rangitikeī Herald*. https://www.pressreader.com/new-zealand/feilding-rangitikei-herald/20230504/page/11

Dean, L., Urbanová, M., & Akauola, J. (2023). *Making sense of being in care, adopted, or whāngai – Literature review*. Wellington, New Zealand: Oranga Tamariki—Ministry for Children. https://ot.govt.nz/assets/Uploads/About-us/Research/Latest-research/Making-sense-of-being-in-care/Makingsense_Literature-review.pdf

ElHage, A. (2016). Keeping children in the family instead of foster care. *Institute for Family Studies*. https://ifstudies.org/blog/keeping-children-in-the-family-instead-of-foster-care#:~:text=According%20to%20the%20AECF%2C%20compared%20to%20kids%20in,children%20in%20state%20custody%2C%20according%20to%20Child%20Trends.

Farr, M. (2018). Power dynamics and collaborative mechanisms in co-production and co-design processes. *Critical Social Policy, 38*(4), 623–644. https://doi.org/10.1177/0261018317747444

Grandparents Raising Grandchildren Trust. (2015). *About GRG*. Grandparents Raising Grandchildren Trust. Retrieved March 2, 2023, from https://www.grg.org.nz/About+GRG.html

Gupta-Kagan, J. (2019). America's Hidden Foster Care System. *Stanford Law Review, 72*, 841. https://www.stanfordlawreview.org/print/article/americas-hidden-foster-care-system/

Harding, L., Murray, K., Shakespeare-Finch, J., & Frey, R. (2020). The wellbeing of foster and kin carers: A comparative study. *Children and Youth Services Review, 108*, 104566. https://www.sciencedirect.com/science/article/pii/S019074091930636X

Hollingworth, A. (2019). Oranga Tamariki: Māori leaders issue strong call to action. *NewsHub*. https://www.hub.co.nz/home/new-zealand/2019/07/oranga-tamariki-m-ori-leaders-issue-strong-call-to-action-over-uplifts.html

Hyslop, I. (2022). *A political history of child protection: Lessons for reform from Aotearoa New Zealand*. Policy Press.

Kaiwai, H., Allport, T., Herd, R., Mane, J., Ford, K. Leahy, H., Varona, G., & Kipa, M. (2020). *Ko te wā whakawhiti: It's time for change. A Māori Inquiry into Oranga Tamariki*. Whānau Ora Commissioning Agency. https://whanauora.nz/maori-inquiry/

Keddell, E., Fitzmaurice, L., & Cleaver, K. (2021). *The Prevention Project: Supporting whānau and reducing baby removal, Research Report*. University of Otago. https://preventionproject-whanau.squarespace.com/research-report

Keddell, E., Fitzmaurice, L., Cleaver, K., & Exeter, D. (2022). A fight for legitimacy: Reflections on child protection reform, the reduction of baby removals, and child protection decision-making in Aotearoa New Zealand. *Kōtuitui: New Zealand Journal of Social Sciences Online, 17*(3), 378–404. https://doi.org/10.1080/1177083X.2021.2012490

Kuokkanen, R. (2019). Self-determination. In *Restructuring relations: Indigenous self-determination, governance, and gender* (pp. 23–59). Oxford University Press. https://doi.org/10.1093/oso/9780190913281.003.0002

Kwaymullina, A. (2005). Seeing the light: Aboriginal law, learning and sustainable living in country. *Indigenous Law Bulletin, 6*(11), 12–15.

Lee, Y. (2008). Design participation tactics: The challenges and new roles for designers in the co-design process. *Co-design, 4*(1), 31–50. https://doi.org/10.1080/15710880701875613

Love, C. (2002). Māori perspectives on collaboration and colonisation in contemporary Aotearoa/New Zealand child and family welfare policies and practices [Conference presentation]. In *Positive Systems of Child Welfare Conference*. Waterloo, ON. https://scholars.wlu.ca/cgi/viewcontent.cgi?article=1017&context=pcfp

Maaka, M. J., Wong, K. L., Perry, W. K., & Johnston, P. M. (2018). Indigenous leadership: A complex consideration. In *Handbook of indigenous education* (pp. 229–248). https://doi.org/10.1007/978-981-10-1839-8_71-1

MacKean, T., Shakespeare, M., & Fisher, M. (2022). Indigenous and non-Indigenous theories of well-being and their suitability for well-being policy. *International Journal of Environmental Research and Public Health, 19*(11693), 1–16. https://doi.org/10.3390/ijerph191811693

Mahuika, N., & Mahuika, R. (2020). Wānanga as a research methodology. *ALTERNATIVE: An International Journal of Indigenous Peoples, 16*(4), 369–377. https://doi.org/10.1177/1177180120968580

Mane, J. (2009). Kaupapa Māori: A community approach. *MAI: A New Zealand Journal of Indigenous Scholarship, 3*(1) https://www.researchgate.net/publication/40426881_Kaupapa_Maori_A_community_approach

Māori Perspective Advisory Committee. (1988). *Puao-Te-Ata-Tu (Day break): The Report of the Ministerial Advisory Committee on a Māori Perspective for the Department of Social Welfare*. New Zealand Government.

McHugh, M. (2003). A further perspective on kinship care: Indigenous foster care. *Developing Practice: The Child. Youth and Family Work Journal, 8*, 14–24. https://search.informit.org/doi/epdf/10.3316/informit.869507010511407

McRae, K., & Nikora, L. (2006). Whāngai: Remembering, understanding and experiencing. *MAI Review, 2006, 1* (Intern Research Report 7), 1–18. http://www.review.mai.ac.nz

Minister for Children. (2019). Reforming financial assistance for caregivers: Proposed response to the review of the foster care allowance, orphan's benefit, unsupported child's benefit and related payments. *Caregiving*. https://www.orangatamariki.govt.nz/caregiving/financial-help/review-of-financial-assistance-for-caregivers/

Ministry for Health. (2023). *Pae Tū: Hauora Māori strategy. Publications*. https://www.health.govt.nz/publication/pae-tu-hauora-maori-strategy

Ministry for the Environment. (2021). *Our land 2021*. https://environment.govt.nz/publications/our-land-2021

Ministry of Social Development. (2023). *Unsupported child's benefit*. Work and Income. Ministry of Social Development, New Zealand Government. Retrieved March 7, 2023 from https://www.workandincome.govt.nz/products/a-z-benefits/unsupported-childs-benefit.html.

Moewaka Barnes, H., & McCreanor, T. (2019). Colonisation, hauora and whenua in Aotearoa. *Journal of the Royal Society of New Zealand, 49*(sup1), 19–33. https://doi.org/10.1080/03036758.2019.1668439

Moreton-Robinson, A., & Walter, M. (2009). Indigenous methodologies in social research. In M. Walter (Ed.), *Social research methods* (2nd ed., pp. 1–18). Oxford University Press.

Oranga Tamariki – Ministry for Children. (2019a). *Practice review: Professional Practice Group Practice Review into the Hastings Case.* https://www.orangatamariki.govt.nz/about-us/news/hawkes-bay-practice-review/

Oranga Tamariki – Ministry for Children. (2019b). *Caregivers raising children with the orphans benefit and the unsupported child's benefit: A survey of caregivers.* Oranga Tamariki – Ministry for Children. Wellington, New Zealand. https://www.orangatamariki.govt.nz/about-us/research/our-research/

Oranga Tamariki – Ministry for Children. (2020). *Review of financial assistance for caregivers: Summary of engagement Findings.* Oranga Tamariki – Ministry for Children. Wellington, New Zealand. https://www.orangatamariki.govt.nz/about-us/research/our-research/

Oranga Tamariki – Ministry for Children. (2021a). *Funding policy and guidance for shared care.* https://www.orangatamariki.govt.nz/working-with-children/information-for-providers/partnered-care/funding/.

Oranga Tamariki – Ministry for Children. (2021b). *Support for permanent caregivers.* Oranga Tamariki – Ministry for Children Practice Centre. Retrieved March 2, 2023, from https://practice.orangatamariki.govt.nz/our-work/care/transitioning-between-placements-or-out-of-care/support-for-permanent-caregivers/

Oranga Tamariki – Ministry for Children. (2021c). *The UCB as an alternate pathway: Exploration of UCB supported care which followed a report of concern to Oranga Tamariki in 2019.* Wellington, New Zealand: Oranga Tamariki—Ministry for Children. https://www.orangatamariki.govt.nz/about-us/research/our-research/

Oranga Tamariki – Ministry for Children. (2021d). *Use of legal orders to support a safe stable and loving home for tamariki in permanent care.* Oranga Tamariki – Ministry for Children Practice Centre. Retrieved March 2, 2023, from https://practice.orangatamariki.govt.nz/our-work/care/transitioning-between-placements-or-out-of-care/use-of-legal-orders-to-support-a-safe-stable-and-loving-home-for-tamariki-in-permanent-care/

Oranga Tamariki – Ministry for Children. (2023a). *Orphan's benefit and unsupported child's benefit: Caregiver engagement report.* Oranga Tamariki Ministry for Children. https://www.orangatamariki.govt.nz/about-us/research/our-research/orphans-benefit-and-unsupported-childs-benefit-caregiver-engagement-report/

Oranga Tamariki – Ministry for Children. (2023b). Without notice application for interim custody of a tamaiti or rangatahi. *Practice Centre.* Oranga Tamariki. Retrieved October 02, 2023, from https://practice.orangatamariki.govt.nz/policy/without-notice-application-for-interim-custody-of-a-tamaiti-or-rangatahi/#when-to-file-an-application-without-notice. https://www.orangatamariki.govt.nz/caregiving/becoming-a-caregiver/

Oranga Tamariki – Ministry for Children. (2023c). Statistics about how we work with children: Key data tables. *Oranga Tamariki.* Retrieved on October 10, 2023 from https://www.orangatamariki.govt.nz/about-us/information-releases/statistics-about-how-we-work-with-children/

Oranga Tamariki Act 1989 (NZ). (1989). Retrieved October 28, 2023, from https://www.legislation.govt.nz/act/public/1989/0024/latest/whole.html

Rewiti, A. (2022). Developing a kaupapa whānau framework to explore social, cultural and health benefits of a whānau-inspired initiative. *MAI Journal 2022, 11*(2), 128–139. https://doi.org/10.20507/MAIJournal.2022.11.2.4

Rickard, T. (2014). *He iwi moke, he whanokē: iwi social services, policy and practice* [Master's Thesis, Massey University]. Massey University Thesis and Dissertations. https://mro.massey.ac.nz/handle/10179/5768

Ryan, T., & Evans, M. (2020). The wisdom of differentiating between Indigenous leader and Indigenous leadership. In A. Intezari, C. Spiller, & S. Yang (Eds.), *Practical wisdom, leadership and culture: Indigenous, Asian and Middle-Eastern perspectives* (pp. 46–63). Routledge.

Savage, C., Moyle, P., Kus-Harbord, L., Ahuriri-Driscoll, A., Hynds, A., Paipa, K., Leonard, G., Maraki, J., & Leonard, J. (2021). *Hāhā-uri, hāhā-tea - Māori Involvement in State Care 1950–1999.* Report prepared for the Crown Secretariat. Ihi Research.

Silver Century Foundation. (2018). *They're falling through the cracks: Grandmothers raising grandchildren.* Silver Century Foundation. https://www.silvercentury.org/2018/01/theyre-falling-through-the-cracks-grandmothers-raising-grandchildren/

Simpson, M., & Ake, T. (2010). Whitiwhiti Korero: Exploring the researcher's relationship in cross-cultural research. *Journal of Intercultural Communication Research, 39*(3), 185–205. https://doi.org/10.1080/17475759.2010.534861

Smart, J., Muir, S., Hughes, J., Goldsworthy, K., Jones, S., Cuevas-Hewitt, L., & Vale, C. (2022). *Identifying strategies to better support foster, kinship and permanent carers.* Australian Institute of Family Studies. https://aifs.gov.au/research/research-reports/ldentifying-strategies-better-support-foster-kinship-and-permanent-carers

Smith, L. T. (1999). *Decolonising methodologies: Research and indigenous peoples.* Bloomsbury Academic.

Smith, L. T. (2018). Indigenous insight on valuing complexity, sustaining relationships, being accountable. In R. Hopson & F. Cram (Eds.), *Tackling wicked problems in complex ecologies: The role of evaluation* (pp. 45–66). Stanford University Press.

Smith, V., Moore, C., Cumming, J., & Boulton, A. (2019). Whānau Ora: An indigenous success story. In J. Leutjens, M. Mintrom, & P. 't Hart (Eds.), *Successful public policy: Lessons from Australia and New Zealand* (pp. 505–526). Australian National University Press.

Spiller, C., Maunganui Wolfgramm, R., Henry, E., & Pouwhare, R. (2020). Paradigm warriors: Advancing a radical ecosystems view of collective leadership from an Indigenous Māori perspective. *Human Relations, 73*(4), 516–543.

Statistics New Zealand. (2018). *2018 Census: Place summaries.* Statistics New Zealand, New Zealand Government. Retrieved October 20, 2023, from https://www.stats.govt.nz/tools/2018-census-place-summaries/new-zealand

Te Puni Kōkiri. (2001). *The principles of The Treaty of Waitangi as expressed by the courts and The Waitangi Tribunal.* https://www.tpk.govt.nz/en/o-matou-mohiotanga/crownmaori-relations/he-tirohanga-o-kawa-ki-te-tiriti-o-waitangi

Thom, R. R. M., & Grimes, A. (2022). Land loss and the intergenerational transmission of wellbeing: The experience of iwi in Aotearoa New Zealand. *Social Science & Medicine, 296*, 114804. https://doi.org/10.1016/j.socscimed.2022.114804

Waitangi Tribunal. (2021). *He Pāharakeke, He Rito, Whakakīkīnga Whāruarua: Oranga Tamariki Urgent Inquiry* (Report No. WAI 2915). Ministry of Justice, New Zealand Government. https://www.waitangitribunal.govt.nz

Walker, S. (2001). *He Maatua Whāngai Programme o Otepoti from a caregiver perspective* [Master's Thesis, University of Otago]. https://core.ac.uk/download/360556519.pdf

Williams, T., Ruru, J., Irwin-Easthope, H., Quince, K., & Gifford, H. (2019). Care and protection of tamariki Māori in the family court system. *Ngā Pae o te Māramatanga.* https://www.maramatanga.ac.nz/sites/default/files/teArotahi_19-0501%20Ruru.pdf

Williamson, B. (2021). Healing, decolonisation and governance. In *Developing governance and governing development: International case studies of indigenous futures.* Rowman and Littlefield International.

Part III
Restoring Country, Restoring Justice: Pathways to Transformation

Chapter 10
First Nations Water Holder: The Future for Cultural Water in the Murray Darling Basin

Melissa Kennedy

Abstract This chapter examines the potential for an Indigenous-led governance model in the Murray-Darling Basin (MDB) through the concept of the First Nations Water Holder (FNWH). It argues that current water management frameworks, which prioritize economic and environmental considerations, fail to address the cultural and spiritual needs of Indigenous communities whose deep connections to the land and water have spanned millennia. The FNWH represents a transformative approach, enabling Indigenous peoples to reclaim control over water entitlements to restore cultural economies, practices, and ecological health. Central to this vision is the concept of cultural water, a distinct category that prioritizes Indigenous relationships with water, including ceremonial use and biodiversity maintenance, alongside environmental flows. This chapter explores the historical context of water governance in Australia, highlighting the exclusion of Indigenous perspectives and the need for systemic change. It discusses how the FNWH can act as a critical tool for securing water rights, promoting reconciliation, and empowering First Nations communities to exercise sovereignty over their water resources. Ultimately, the chapter calls for a reimagining of water governance that weaves together Indigenous knowledge, environmental sustainability, and the revitalization of Indigenous cultural practices in the MDB.

Keywords Water justice · Murray Darling Basin · First Nations · cultural flows · water governance

M. Kennedy (✉)
School of Social Science, Monash University, Clayton, VIC, Australia
e-mail: melissa.kennedy1@monash.edu

M. Kennedy, E. O'Donnell (eds.), *People, Place and Nature in Indigenous-Settler Relations*, Indigenous-Settler Relations in Australia and the World 10, https://doi.org/10.1007/978-3-032-03600-1_10

10.1 Introduction

Water is a vital resource for life and a central to the cultural identities and practices of Indigenous peoples. In the Murray-Darling Basin (Fig. 10.1), which extends across a vast stretch of southeastern Australia, the river systems hold profound spiritual, cultural, and ecological significance for Indigenous peoples who have lived and cared for this land for millennia (Moggridge & Thompson, 2021; Hemming et al., 2019). However, the governance of water in this region has been dominated by Western legal, economic, and environmental frameworks that prioritise water as a commodity and resource for industrial, agricultural, and urban development (Marshall, 2017). This exclusion of Indigenous worldviews and knowledge from water management has created a deepening crisis for both the environment and the First Nations communities that depend on the river systems for their cultural, economic, and spiritual well-being.

As a Tati Tati Traditional Owner from Murray River Country, Australia, with kinship relations to Wadi Wadi, Mutti Mutti, and Latji Latji, I belong to the waters of the Murrumbidgee and Milloo (Murray River). A descendant of river people, my inheritance is an enduring responsibility to Country, and to speak for and with Country. I also walk in the world of the West, as a researcher and water policy consultant under the dark umbrella of colonial institutions.

The history of water governance in Australia is a story of dispossession and marginalisation (Hartwig et al., 2021). With the arrival of European settlers, the traditional knowledge systems of Indigenous peoples were ignored, and water became subject to Western legal definitions, which saw it as a tradable commodity under property rights laws. These systems not only led to the degradation of water ecosystems but also stripped Indigenous peoples of their ability to manage and care for the rivers, lakes, and streams that were, and continue to be, central to their cultural practices (Marshall, 2017; Leonard, et al., 2023). In the MDB, this conflict between Western water governance and Indigenous perspectives on water management has resulted in an ongoing struggle for water justice (O'Donnell et al., 2023). Despite some efforts to incorporate Indigenous interests into water policy, the prevailing approach remains insufficient, largely ignoring the vital role of cultural water in restoring both ecological and community health.

In response to this ongoing inequity, the concept of the First Nations Water Holder (FNWH) has emerged as a radical yet necessary solution for transforming water governance in the Basin. The FNWH would be an Indigenous-led institution designed to secure and manage water entitlements for cultural purposes, allowing Indigenous communities to reclaim their right to govern water in a way that aligns with their traditional knowledge and values. This vision challenges the current water management systems that view water solely through the lens of economic utility and environmental conservation. Instead, it presents a holistic approach that seeks to restore water justice by prioritizing the spiritual, cultural, and environmental needs of Indigenous communities alongside the ecological health of the region's river systems.

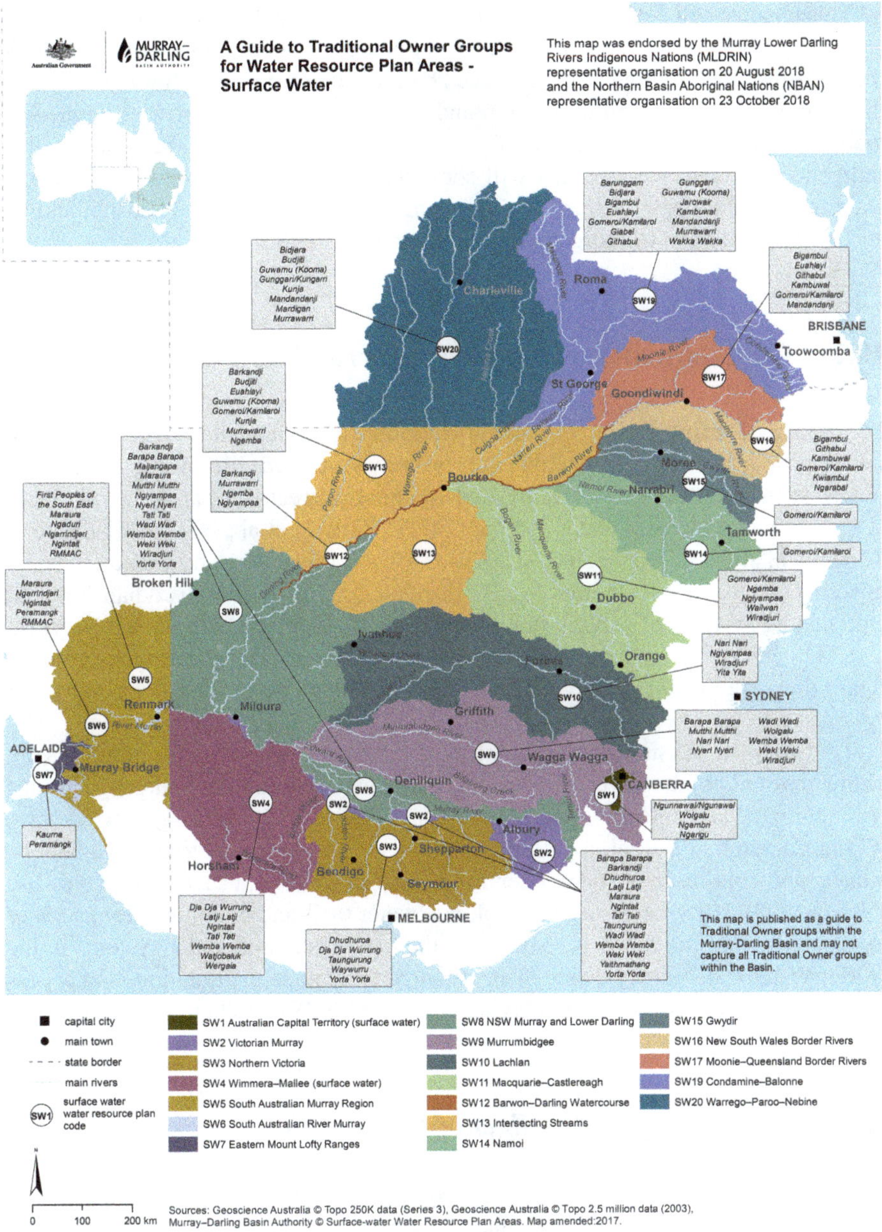

Fig. 10.1 Murray Darling Basin. (Source: Geoscience Australia & Murray Darling Basin Authority (License: Creative Commons Attribution 4.0 International). Available on https://data.gov.au/data/dataset/murray-darling-basin-boundary)

This chapter will explore the need for an Indigenous-led approach to water governance, with a particular focus on the First Nations Water Holder. It will address the historical and ongoing struggles faced by Indigenous communities in the MDB, the vital importance of cultural water, and how the FNWH can serve as a transformative model for water justice. By examining the potential benefits of this Indigenous governance model, I will show that the FNWH is a tool for securing water rights and a crucial step toward reconciliation, environmental sustainability, and the revitalization of Indigenous communities in the Basin.

10.2 The Fundamental Conflict: Murray-Darling Basin

The Murray-Darling Basin (Fig. 10.1) embodies a fundamental conflict between two opposing worldviews: one that sees water as a sacred, life-giving force intertwined with Indigenous sovereignty, and another that treats it as an economic commodity to be extracted, traded, and controlled (Hartwig et al., 2021; Moggridge & Thompson, 2021; O'Donnell et al., 2023). This section explores how colonial water governance has reshaped the MDB, displacing Indigenous custodianship and entrenching systems of exploitation that continue to degrade both the environment and Indigenous rights. Examining the tensions between these perspectives highlights the struggle for water justice in one of the world's most contested river systems.

The MDB, in Australia's southeast, is a catchment of over one million square kilometres and home to over 55 sovereign Indigenous Nations and language groups with intimate and interconnected relationships with their rivers and waterways (Costanza-van den Belt et al., 2022). While diverse in customs, traditions, beliefs, values, and culture, they share a central principle: water is a sacred source of life (Nelson et al., 2018). As a result of colonial water theft and ongoing colonial water mismanagement, the MDB is also one of the most distressed water systems in the world. Water is constructed as a mere resource for human exploitation, and access to water is fiercely contested and highly politicised (O'Donnell et al., 2023).

10.2.1 Geography and Indigenous Connection

Across tens of thousands of years, First Nations have demonstrated a sustainable coexistence with water and land, a harmony achieved through the continuation of traditional customs and the integration of water science (Costanza-van den Belt et al., 2022; Moggridge & Thompson, 2021). However, this symbiotic relationship has been disrupted by the theft of land and water during recent colonial periods. The consequence has been the systematic exclusion of Indigenous peoples' knowledge, interests, and rights at all levels of water planning processes (Ayre & Mackenzie, 2013; Marshall, 2017). The MDB is now a transboundary river system supporting the majority of Australia's irrigated agriculture (Connell, 2015) in a fragmented realm of water legislation and policy created based on the priorities and values of an introduced culture, with introduced concepts of property rights and ownership (Burdon et al., 2015). Indigenous customs, science, and holistic values for Country that are rooted in cultural principles are rarely incorporated into water planning processes (Ayre & Mackenzie, 2013). This exclusionary legacy underscores the urgent need for a paradigm shift in water governance to honour and integrate the wealth of Indigenous wisdom that has sustained the land and rivers for millennia.

10.2.2 Impact of Colonial Water Governance

The modern history of water policy in Australia, shaped by Anglo-European socio-political values, prioritises economic advancement and property rights at the expense of water's inherent agency and cultural significance (Burdon et al., 2015). The arrival of Europeans marked a pivotal moment, initiating large-scale alterations to cultural and environmental landscapes, resulting in the depletion of floodplain-dependent wetlands and a concerning deterioration in the health of rivers and surrounding woodlands (Taylor et al., 2016; Woods et al., 2022). The language of property rights became deeply embedded in legislation, shaping a discourse of water management centred around legal ownership, treating water as a commodity (Burdon et al., 2015).

Basin First Nations' reciprocal relationship with water is in what Arsenault et al. (2018) describe as a state of fundamental moral conflict with the resource-driven commodification of water by the West. The illegitimacy of British occupation of Indigenous lands was enabled by the flawed legal doctrine of terra nullius, or land belonging to no one. What followed was an unbundling of land and water rights, a Western construct of property rights where separate bundles of rights were created in law for land and water, directly contrasting with Indigenous constructs of Country, which are a web of connections (Hartwig et al., 2021; O'Donnell & Kennedy, 2023). The separation of land and water into property units is also enabled by aqua nullius, or water belonging to no one, a term coined by Wiradjuri researcher Dr. Virginia Marshall. Marshall (2017) argues water rights are an illegitimate legal fiction; the Crown unlawfully acquired water in the same way it unlawfully acquired land.

Furthermore, the dispossession of water from First Nations peoples and their ongoing exclusion from contemporary water markets have limited economic growth and opportunities for Indigenous communities and organisations (Marshall, 2019; Mooney, 2014; O'Donnell et al., 2021). In 2020, Indigenous organisations held less than 0.2% of water access entitlements in the MDB (Hartwig & Jackson, 2020; Jackson & Head, 2020). Mooney (2017) argues Indigenous communities want to build enterprises for employment and economic stimulation; however, it is challenging for them to acquire water on the open market in competition with irrigators and agri-businesses. The exclusion of Indigenous water science from settler water management also exacerbates the environmental crisis whilst undermining the legitimacy and sustainability of both water science and governance (O'Donnell et al., 2023). While legislative obligations require environmental water managers to incorporate First Nations values and priorities into watering programs, there is currently little opportunity for this to occur (Moggridge et al., 2019).

10.2.3 Current State of Water Management

In the MDB, two centuries of agricultural expansion, water extraction, floodplain harvesting, and the manipulation of rivers have resulted in extensive declines in native fish abundance, water quality, and waterway health (Williams, 2017). In response, there were wholesale federal policy changes in the 1990s in an attempt to protect the conditions of waterways; however, First Nations values and priorities were largely ignored (Hemming et al., 2019). Scholars and environmentalists have also criticised the changes as the primary driver of change was increasing water efficiency while maintaining a highly productive, resource-driven agricultural economy (Godden et al., 2020). In the early 2000s, increasing levels of environmental degradation sparked several new water policy initiatives implemented to address the sustainability issue. Most significant was the *National Water Initiative* (NWI), the 2004 federal policy that created 'environmental flows' that are separate and permanent water entitlements for the environment (Taylor et al., 2016). The NWI was the first national agreement to acknowledge 'Indigenous water needs', although it did not require these to receive the same standard of protection as other users' rights to water. Following the NWI, the *Water Act 2007* was similarly created to improve policies to guarantee enough water would remain in river systems to protect and restore the conditions of the Basin, including the establishment of the Commonwealth Environmental Water Holder (CEWH) to acquire and manage environmental water entitlements. By 2010, both NSW and Victoria had created programs to acquire and/or manage environmental water entitlements. The Victorian Environmental Water Holder (VEWH) is similar to the CEWH, but with greater institutional independence as a statutory corporation (O'Donnell, 2012). The CEWH and the VEWH hold permanent rights to water (water access entitlements) that provide them with annual water allocations, which they use to achieve environmental outcomes, which can include selling water on the water market to other users, provided that the proceeds of the sale are invested in environmental outcomes.

Both policies paved the legislative pathway for the *Murray Darling Basin Plan 2012* (Basin Plan), a $AUD13 billion initiative to set limits on how much water can be extracted from the basin system (McConnel, 2019). These limits are called Sustainable Diversion Limits (SDLs) and didn't come into effect until 2019, seven years after the Basin Plan was established. The SDLs require 3200 gigalitres of water to be returned to the environment, some 600 gigalitres short of the minimum volume recommended by scientific research conducted by the Murray Darling Basin Authority (MDBA) in 2010 (Williams, 2017). There are unanswered questions about the science supporting the SDL volumes that have troubled many, as numerous scientific reports indicate the targets are far too low and cannot achieve the intended outcomes of basin sustainability (CSIRO, 2011), with some calculations of the volume required as high as 7600 gigalitres (Williams, 2017).

In July 2022, the federal Government officially declared the Basin Plan would fail to achieve the water recovery targets by the intended end date of 2025 (Kingsford, 2022). The Basin Plan is also contentious among environmental activists as while it aims to reduce annual surface water extraction, it increases the amount of permissible groundwater extraction by over 1700GL per year, even though surface and groundwater are proven to be connected (Pittock et al., 2015). Policy commitment failures and suspect environmental science aside, all three significant pieces of water policy reform fail to hit the mark regarding First Nations' inherent rights to water or their inclusion in water management decisions.

In 2022, despite the evident and impending failure of the Basin Plan, the newly elected Federal Labor government responded to the previous Government's failures to follow through on water commitments by promising to 'find' more water for the environment (Jackson et al., 2021). The pledge, in part, is to reinstate the water buyback scheme as the most cost-effective way of returning water to much-needed systems – a scheme that the previous Liberal Government abolished (Carpenter, 2022). While a buyback program could provide optimistic results for the environment, First Nations are still without power or agency within these decision-making processes. Ayre & Mackenzie (2013) argue authorities have failed to account for Indigenous interests and rights in water, including Indigenous knowledge in water planning processes. Taylor et al. (2016) agree and suggest the lack of advancement in Indigenous water policy is due to the absence of sustained commitment by governments, the result of non-legislative policy initiatives and a four-year political cycle of budgets and elections. In 2018, the federal government pledged $AUD40 million to purchase water rights for First Nations, yet by 2022, not a single dollar of that commitment had been spent.

10.3 The Need for a Paradigm Shift: Cultural Water

For First Nations peoples in Australia, water is often understood as a living relative for whom everyone has a sacred and inherited responsibility to care for (Marshall, 2019; O'Bryan, 2017; Poelina et al., 2019; Wilson & Inkster, 2018). Water holds great significance for Indigenous communities as it connects people to place, the natural world, and their relationships with all things in their environment (Leonard et al., 2023; Moggridge & Thompson, 2021; Poirier & Schartmueller, 2012). North American Indigenist research approaches and frameworks centre tribal methodologies using Indigenous ways of knowing (Arsenault et al., 2018). Cribb et al. (2022) detail Indigenous frameworks implemented in Aotearoa from treaty settlements using alternate ways of knowing, being, and doing in managerial discourses of land and water.

In Australia, the concept leading the way in challenging the status quo of water management emerged from the southern MDB mid-2000s – cultural flows. This philosophy of water was first reported in the Echuca Declaration, released in 2007 by the Murray and Lower Darling Rivers Indigenous Nations (MLDRIN), a coalition of over 20 sovereign First Nations. The Declaration defined cultural flows as water entitlements owned by First Nations groups to benefit the cultural, environmental, social, and economic conditions of each Nation (MLDRIN, 2007; O'Donnell et al., 2021). The Echuca Declaration firmly established cultural flows as a legitimate water management process, detailing how water management should adhere to community values in acquiring, managing, and using water to advance Basin Nations' agendas. It further asserts Southern Basin Nations' priorities to reinforce Indigenous economic objectives, sustain and enhance cultural knowledge(s), enable access to water resources for First Nations peoples, and build capacity within community organisations (MLDRIN, 2007).

Following the release of the Echuca Declaration, a seven-year research project was designed and implemented by a cross-disciplinary research team supported by MLDRIN and the MDBA. The National Cultural Flows Research Project significantly altered the discourse of how Indigenous interests in water could be integrated into water processes, with several other related studies that have developed knowledge of water dependant cultural values (Moggridge & Thompson, 2021; Woods et al., 2017). The research aims were to develop a methodology to assist First Nations and water managers in creating and implementing a cultural flows framework (Woods et al., 2017).

The need for the framework came about from several key drivers: the need to acknowledge inherent, unceded Indigenous water rights; the exclusion of Indigenous Nations from holding water rights that could be used for economic outcomes; and environmental watering regimes that were and are still failing to meet cultural watering objectives. As Moggridge and Thompson (2021) explain, there is an ongoing disregard for relationships between Traditional Owners and their Country.

Despite cultural flows swiftly becoming a widely used and understood concept amongst First Nations and academics, there has been little translation to tangible implementation within environmental management, government-led programs, or legislative change, even with ongoing advocacy and research. Such research includes the *Cultural Water for Cultural Economies* report, a joint project from MLDRIN and the University of Melbourne, funded by the Victorian Government to establish the pathways to independent water entitlements for First Nations peoples (O'Donnell et al., 2021). The MDBA now also recognises cultural flows and environmental flows as distinct concepts that must be managed as such (MDBA, 2022). However, as Hemming et al. (2019) argue, Indigenous interests and contributions to water policy continue to be pigeonholed as intangible cultural contributions with no real governing authority.

10.3.1 The Cultural Water Paradigm

As an extension of the National Cultural Flows Research Project methodology and framework, the Cultural Water Paradigm (Fig. 10.2) has been developed to establish the groundwork to transform current water-management policy and apply values of caring for Country, cultural economy, and environmental health to contemporary water use and management (O'Donnell & Kennedy, 2023; O'Donnell et al., 2023). Settler colonial water management still separates water into different services to be traded against one another; in contrast, Indigenous knowledge shows that cultural water economies are holistic and built on healthy Country to support healthy people (O'Donnell et al., 2023). Although yet to be applied in full, elements of the paradigm are already indicating the power of Indigenous holistic water management, highlighting how while science identifies the importance of integrated water management, it still fails to recognise the elements of cultural and spiritual wellbeing that are essential for First Nations water management.

10.3.2 Cultural Economies

Over the past decade, the concept of cultural economy has emerged from the Southern Murray-Darling Basin as a way for Traditional Owners to articulate the need for water beyond environmental health – to support economic growth and

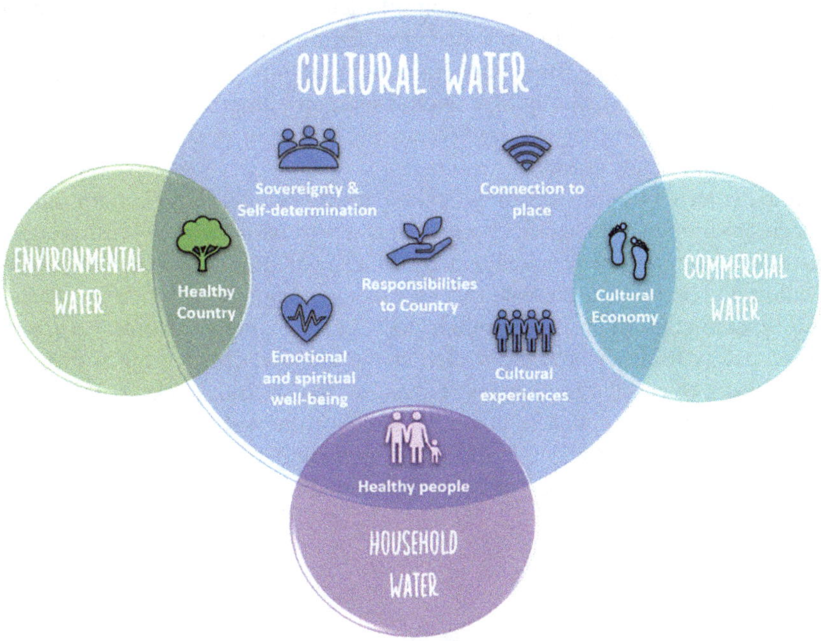

Fig. 10.2 Cultural water paradigm. (Image credit: M. Kennedy (O'Donnell & Kennedy, 2023))

self-determination (O'Donnell et al., 2021). Unlike Western definitions that reduce economy to financial value, cultural economy encompasses a broader, more holistic understanding of economic activity that is deeply intertwined with culture, Country, and community. It recognizes that economic sustainability for First Nations people is not separate from cultural and environmental well-being but is instead embedded within them. This perspective challenges mainstream water management frameworks that often exclude or restrict economic uses of water under the guise of environmental protection.

Cultural economy manifests in diverse ways and is inherently different for each Nation from diverse landscapes. It may include the intergenerational transfer of knowledge, the trade of goods and wisdom, and the practice of cultural traditions such as smoking ceremonies for healing and providing safe passage for guests on Country. It may also include critical activities like growing medicinal plants, maintaining native grasses for weaving, and supporting cultural businesses that sustain both people and Country. While cultural economy can involve financial income – such as creating jobs for people to work on Country – it is ultimately about survival and prosperity in a way that aligns with Indigenous law and responsibilities to place. In the context of capitalism and imposed colonial governance structures, currency and financial sustainability have become necessary components of cultural economy, but they do not define it. Instead, cultural economy remains a living expression of how First Nations communities continue to adapt, resist, and thrive within imposed systems while maintaining their deep connection to Country.

Cultural economies are a crucial component of the Cultural Water Paradigm, as they recognize that water is essential for environmental and cultural well-being and for sustaining Indigenous livelihoods and self-determination. This means that while environmental water could, in theory, support some of these outcomes, it would still be constrained by existing regulatory frameworks that prohibit commercial or economic use. By embedding cultural economies within water policy, First Nations can exercise greater control over water resources, fostering economic independence alongside cultural and ecological restoration.

While the term cultural economy is not a new idea emerging from the Basin, as a conceptual framework specific to place, it is a novel approach to Western notions of economy, and waterway management. Mazzucato and Niemeijer (2000) argue that cultural economies are created when market economies are filtered through a society's socio-cultural and economic history. In other literature, cultural economy has been a way to describe sectors such as the arts and creative industries, or the urban impacts of economic changes (Gibson & Kong, 2005). There is little to no inclusion of Indigenous concepts in either study or the wider literature. In the context of water policy reform in the MDB, understanding Traditional Owner priorities for cultural economies will be critical to understanding water justice agendas.

10.4 Towards a First Nations Water Holder

For Indigenous peoples, sustainable water management is embedded in cultural lore to ensure healthy people and healthy waterways for future generations (Moggridge & Thompson, 2021). The well-being of people is fundamentally connected to the well-being of their lands, and maintaining a connection to one's Country is paramount to personal well-being (Garvey et al., 2021; Wilson & Inkster, 2018). Owning water rights alone will be insufficient in achieving water justice when there exists power imbalances and inequities that First Nations must face. A critical goal for Indigenous water justice will be empowering Indigenous people with the capability to implement their rights by applying principles of self-governance (Poirier & Schartmueller, 2012).

10.4.1 A Revolutionary Approach

An FNWH represents a revolutionary shift by actualizing Indigenous sovereignty and recognizing the inherent rights of First Nations to access, manage, and govern water according to their own laws and traditions. Unlike current water governance structures that allocate water through colonial legal and economic frameworks, an FNWH would empower First Nations to exercise authority over water in alignment with cultural governance principles. This means that First Nations would not only have a legal stake in water management but would also be able to implement holistic, place-based water stewardship practices that reflect the interconnected relationships between water, land, people, and spirit. By embedding First Nations' governance within water policy, an FNWH directly challenges the legacy of aqua nullius, ensuring that water rights are recognized as inseparable from Indigenous sovereignty and self-determination.

This innovative approach has the potential to overturn the entrenched Western dichotomy between environment and economy, demonstrating that water can and must support both cultural economies and ecological health. Under a cultural water paradigm, economic activity – such as traditional trade, cultural businesses, and employment on Country – is not in opposition to environmental sustainability but is instead an integral part of a thriving, interconnected system. A First Nations-led water management framework would foster collaborative governance between upstream and downstream Nations, ensuring that water is shared and cared for in ways that sustain the entire Basin rather than being fragmented into competing interests.

By embedding cultural flows within an FNWH, this model acknowledges that water is more than just a resource to be extracted – it is a living entity that supports life, culture, and economic wellbeing in ways that Western water management has long failed to recognize. The National Cultural Flows Research Project details how a watering regime can succeed by following First Nations' values and processes to achieve cultural and environmental objectives (Woods et al., 2017). The expected outcomes of a cultural flow include improved overall ecosystem health through enhanced water quality, totemic species, and native vegetation, as well as strengthened physical, mental, and spiritual health of Indigenous peoples (Moggridge & Thompson, 2021). Integrating the National Cultural Flows Research Program as a foundational element that underscores a commitment to robust methodology and values in cultural water programs.

Self-determination requires Indigenous participation and power in the federal system with binding authority over a given jurisdiction (Tilley, 2011). This is where a cultural water paradigm and an FNWH must be in place for successful water reform to occur. A cultural water paradigm provides the philosophical and practical framework for managing water holistically, ensuring that cultural, spiritual, environmental, and economic needs are all considered together rather than in isolation. However, without an FNWH to hold and manage water rights on behalf of First Nations, the principles of the cultural water paradigm risk being sidelined in a system that prioritizes market-based allocation. The FNWH is necessary to actualize the paradigm by securing and managing water entitlements in ways that align with Indigenous governance, enabling communities to restore cultural flows, support cultural economies, and strengthen their inherent rights to water. Together, these structures challenge the colonial legacy of water management and create a pathway for meaningful, Indigenous-led water reform.

10.4.2 Practical Considerations: Water Acquisition

Diverging from the existing paradigm, where government entities like the CEWH and the VEWH manage environmental water, the FNWH, as a non-governmental organisation, would take charge of cultural water entitlements. The National Cultural Flows Research Program will be central to this visionary framework, providing essential tools and methodology to underpin cultural water programs (MLDRIN, 2017; Nelson et al., 2018; Woods et al., 2017). As a non-government organisation, an FNWH would hold and manage cultural water entitlements, including the permanent water access entitlements and the annual water allocations, similar to how the CEWH and VEWH hold environmental water. The FNWH will develop cultural watering plans to deliver water on Country using placed-based priorities and programs in partnership with Indigenous Nations and organisations.

However, the critical question arises: where would these cultural water entitlements originate? Indigenous organisations hold less than 0.2% of water access entitlements in the MDB, highlighting a stark inequity in water distribution that severely limits First Nations' ability to exercise self-determination in water governance (Hartwig & Jackson, 2020). Recognising the need for a paradigm shift, water entitlements must also be reallocated to support the cultural water paradigm. This shift is not just about securing water for Indigenous use—it represents a fundamental challenge to the settler-colonial framework that artificially separates water into distinct economic, environmental, and cultural categories. The cultural water paradigm rejects this division by affirming that water management must be holistic, integrating healthy people, cultural economies, and Indigenous sovereignty. By securing meaningful water allocations, First Nations can uphold their responsibilities to Country, sustain cultural livelihoods, and reinforce self-determination in ways that are currently denied under existing water regimes.

For the FNWH model to succeed, there must be new and separate water entitlements to the current environmental water regime. The 2004 legislation that led to the creation of water solely for the environment was agreed to by the Council of Australian Governments (COAG) (Taylor et al., 2016). Building on this foundation, there is an opportunity to advocate for embedding a new type of water entitlement in legislation: Cultural Water. Grounded in the premise that current environmental water targets in the Basin Plan fall short of sustainability requirements, the proposal contends that new cultural water entitlements, championed by the Commonwealth, could bridge this gap, offering a means for the Basin to survive and thrive. As previously stated, the science already shows that the environmental water targets in the current (and failing) Basin Plan are insufficient to meet sustainability requirements (Kingsford, 2022; Williams, 2017).

New cultural water entitlements created by the Commonwealth can compensate for the shortfall in what the Basin needs, but their significance extends beyond mere allocation. As part of the cultural water paradigm, these entitlements challenge the colonial separation of environment and economy by demonstrating that the two are deeply interconnected. By recognizing water as both a life-giving force and an economic asset, this approach redefines water management in a way that upholds Indigenous sovereignty and self-determination.

10.4.3 Relationship to Existing Water Policies

The *Restoring Our Rivers Amendment Act*, passed by the Senate in late 2023, may provide a promising pathway toward establishing a FNWH body. With the passing of this legislation, the Labor government reaffirmed its dedication to increasing water rights for First Nations in the MDB, increasing the initial allocation for water acquisition from $AUD40 million to $AUD100 million (Gain et al., 2024).

Efforts are now underway to deploy these funds, with an interim governance body, made up of Basin First Nations people, currently directing the purchasing framework and policy; the interim arrangements would see the Commonwealth be the temporary holder of any entitlements purchased until the enduring governance mechanism is developed and established. This arrangement would ensure that the water entitlements are transferred to and managed by a body led by First Nations, for the benefit of First Nations (Kennedy, 2024). The passage of the *Restoring Our Rivers Amendment Act 2023* and the accompanying commitments holds profound significance for the establishment and operation of an FNWH. These developments highlight several key implications:

(1) **Dedicated water entitlements for First Nations in the basin**

For the first time, there is a tangible commitment to ensuring that water entitlements within the basin are explicitly allocated to First Nations. Although the initial commitment is still small, it represents a critical step toward addressing historical inequities and affirming First Nations' inherent rights to manage water resources on our own terms. Such entitlements can empower communities to prioritize cultural, environmental, and economic outcomes aligned with their needs and values.

(2) **Emerging political support**

The allocation of funding and the acknowledgment of the FNWH concept demonstrates a level of political recognition and support for water justice. This political backing is essential for the success of the FNWH, as it underscores a growing awareness of the importance of equitable water governance. While this momentum is promising, ongoing advocacy will be necessary to ensure that this support translates into concrete action.

(3) **Validation of the FNWH concept as a timely and necessary initiative**

The renewed focus on water reform in the basin affirms that the FNWH proposal resonates with the basin's current needs. It positions the FNWH as a critical mechanism to address water scarcity, restore environmental flows, and uphold First Nations' stewardship of their lands and waters. This alignment with the broader agenda strengthens the FNWH's legitimacy and potential to deliver meaningful change.

(4) **Urgency to move beyond theory into practice**

To capitalise on this momentum, the FNWH concept must evolve from an idea into an operational reality. This requires clear frameworks, robust governance structures, and mechanisms for engaging First Nations communities in decision-making. Current opportunities for advancing water justice will not wait; delays could risk diminishing the gains achieved so far. Translating the concept into action is crucial to meeting the moment and securing long-term benefits for First Nations and the basin.

In essence, these developments offer a rare and valuable opportunity to bring water justice into sharper focus. By building on this foundation, the FNWH can become a transformative force in the basin, ensuring that First Nations have access to water entitlements and are empowered to lead in managing and safeguarding this vital resource for future generations.

10.5 Conclusion

Water is essential for all life. However, for Australia's First Nations, its significance extends beyond the practical management and manipulation of water for agricultural, economic, and consumption purposes. The enduring political consequences of colonial society have effectively excluded First Nations people and their unique ways of understanding Country from pivotal water policy and legislative decision-making processes (Ayre & Mackenzie, 2013). Critically assessing the shortcomings of the existing water governance model, particularly exemplified by the Basin Plan's failure to fulfil the cultural needs of First Nations communities, underscores the urgency for a paradigm shift. Despite the growing collaboration among Indigenous leaders, scholars, and knowledge custodians, efforts to overcome these barriers have encountered significant challenges (Moggridge & Thompson, 2021).

Notably, this shift becomes all the more imperative when the Basin Plan falls short of delivering on ecological targets despite its objectives grounded in Western science and sustainability (Kingsford, 2022). After the federal policy changes in the 1990s aimed at safeguarding waterways, First Nations values and priorities for water were primarily overlooked (Hemming et al., 2019). Critics have even argued that the primary driver of these changes was increasing water efficiency while maintaining a resource-driven agricultural economy (Godden et al., 2020). This scepticism has been borne out as the subsequent decades witnessed heightened environmental degradation, prompting the implementation of various water policy initiatives.

Following the changing political agendas, urgent calls from First Nations communities in the MDB for radical transformation to water management highlight the critical need for a paradigm shift. While there is a demand for Indigenous leadership in organisations like the CEWH and VEWH, the argument presented here advocates for a more comprehensive approach. The inadequacy of the current water management paradigm becomes even more apparent when ecological targets, based on both Western science and sustainability requirements, remain unmet (Kingsford, 2022). This realisation emphasises the necessity of a new management approach, moving beyond the constraints of the existing framework.

An FNWH is a new approach deeply rooted in cultural values and place-based principles essential for the sustainability of the Basin. This proposal not only responds to the shortcomings of the present system but also presents an opportunity for a more inclusive and practical approach to water governance in Australia. The crucial question of sourcing cultural water entitlements prompts the suggestion of legislative action by the Commonwealth, akin to the recognition of the need for environmental water entitlements in 2007. There will be a need for new and distinct water entitlements specifically dedicated to cultural purposes. In drawing parallels with the historic agreement by the COAG that led to the creation of environmental water, an FNWH future is envisioned with a similar collaborative approach to embed Cultural Water entitlements in legislation.

In conclusion, establishing an FNWH represents a progressive step towards acknowledging and preserving the cultural significance of water. The proposal builds upon past initiatives for environmental water and underscores the need for a dedicated legislative framework to secure and manage cultural water entitlements. This innovative model holds the potential to contribute significantly to the holistic and sustainable management of water resources in Australia while also fostering greater collaboration and partnership with Indigenous communities in caring for waterways.

Furthermore, the inherent conflict between modern water management paradigms and Indigenous ontologies is irreconcilable within the current system. Rather than persistently attempting to amend legislation that has historically excluded First Nations Peoples, the FNWH is posited as a bold departure—an opportunity to break free from this cycle of exclusion and create a new paradigm for water management. With current political support, the FNWH's innovative and radical approach could pave the way for a harmonious coexistence of sustainability objectives and the cultural needs intrinsic to the MDB. In essence, the FNWH stands as a visionary proposal and a beacon of hope for a future where water management is deeply rooted in environmental stewardship and cultural respect.

Acknowledgements Melissa Kennedy is an Advisory Board member for the Aboriginal Water Entitlements Program. She has not written this chapter in her capacity as a member of the Advisory Board, but rather in her personal capacity as a Traditional Owner and researcher.

References

Arsenault, R., Diver, S., McGregor, D., Witham, A., & Bourassa, C. (2018). Shifting the framework of Canadian water governance through indigenous research methods: Acknowledging the past with an eye on the future. *Water, 10*(1), 49. https://doi.org/10.3390/w10010049

Ayre, M., & Mackenzie, J. (2013). "Unwritten, unsaid, just known": The role of Indigenous knowledge(s) in water planning in Australia. *Local Environment, 18*(7), 753–768. https://doi.org/10.1080/13549839.2012.665864

Burdon, P., Drew, G., Stubbs, M., Webster, A., & Barber, M. (2015). Decolonising Indigenous water 'rights' in Australia: Flow, difference, and the limits of law. *Settler Colonial Studies, 5*(4), 334–349. https://doi.org/10.1080/2201473x.2014.1000907

Carpenter, T. (2022). New hope for the Murray-Darling basin. *Green Left Weekly* (1348), 10. https://search.informit.org/doi/10.3316/informit.497271069375159

Commonwealth Scientific and Industrial Research Organisation (CSIRO). (2011). *A Science review of the implications for South Australia of the guide to the proposed Basin plan: Synthesis.* Goyder Institute for Water Research. Technical Report Series No. 11, Adelaide.

Cribb, M., Mika, J. P., & Leberman, S. (2022). Te Pa Auroa na Te Awa Tupua: The new (but old) consciousness needed to implement Indigenous frameworks in non-Indigenous organisations. *AlterNative, 18*(4), 566–575. https://search.informit.org/doi/10.3316/informit.785597666158885

Gain, A. K., Milne, C., & Fritsch, O. (2024). Water justice for first nations. *Australasian Journal of Environmental Management, 31*(3), 261–268. https://doi.org/10.1080/14486563.2024.2409525

Garvey, G., Anderson, K., Gall, A., Butler, T. L., Cunningham, J., Whop, L. J., Dickson, M., Ratcliffe, J., Cass, A., Tong, A., Arley, B., & Howard, K. (2021). What Matters 2 Adults (WM2Adults): Understanding the foundations of aboriginal and Torres Strait Islander wellbeing. *International Journal of Environmental Research and Public Health, 18*(12), 6193. https://doi.org/10.3390/ijerph18126193

Gibson, C., & Kong, L. (2005). Cultural economy: A critical review. *Progress in Human Geography, 29*(5), 541–561.

Godden, L., Jackson, S., & O'Bryan, K. (2020). Indigenous water rights and water law reforms in Australia. *Environmental Planning and Law Journal, 37*(6), 655. www.researchgate.net/publication/359711574

Hartwig, L., & Jackson, S. (2020). *The status of Aboriginal water holdings in the Murray-Darling Basin: ARI Report No. 2020/004*. Australian Rivers Institute, Griffith University. http://hdl.handle.net/10072/400302

Hartwig, L., Jackson, S., Markham, F., & Osborne, N. (2021). Water colonialism and Indigenous water justice in south-eastern Australia. *International Journal of Water Resources Development, 38*(1), 30–63. https://doi.org/10.1080/07900627.2020.1868980

Hemming, S., Rigney, D., Bignall, S., Berg, S., & Rigney, G. (2019). Indigenous nation building for environmental futures: Murrundi flows through Ngarrindjeri country. *Australasian Journal of Environmental Management, 26*(3), 216–235. https://doi.org/10.1080/14486563.2019.1651227

Jackson, S., Hartwig, L., & Carmody, E. (2021). Treading water on Indigenous water rights: The serious deficiencies of water allocation planning and management in NSW under the Murray Darling Basin plan. *Pandora's Box*, 72–97. https://search.informit.org/doi/10.3316/agispt.20220516066984

Jackson, S., & Head, L. (2020). Australia's mass fish kills as a crisis of modern water: Understanding hydrosocial change in the Murray-Darling Basin. *Geoforum, 109*, 44–56. https://doi.org/10.1016/j.geoforum.2019.12.020

Kennedy, E. (2024). $100m Indigenous Murray-Darling Basin water fund faces devaluation amid rising prices. *ABC News*. Retrieved November 21, from https://www.abc.net.au/news/rural/2024-08-04/murray-darling-basin-aboriginal-water-concerns-government/104155452

Kingsford, R. (2022). It's official: The Murray-Darling Basin Plan hasn't met its promise to our precious rivers. So where to now? Retrieved September 20, 2023, from https://theconversation.com/its-official-the-murray-darling-basin-plan-hasnt-met-its-promise-to-our-precious-rivers-so-where-to-now-188074

Leonard, K., David-Chavez, D., Smiles, D., Jennings, L., 'Anolani Alegado, R., Tsinnajinnie, L., Manitowabi, J., Arsenault, R., & Begay, R. L. (2023). Water Back: A review centering rematriation and indigenous water research sovereignty. *Water Alternatives, 16*(2), 374–428.

Marshall, V. (2017). *Overturning aqua nullius*. Aboriginal Studies Press.

Marshall, V. (2019). Removing the Veil from the 'Rights of Nature': The dichotomy between first nations customary rights and environmental legal personhood. *Australian Feminist Law Journal, 45*(2), 233–248. https://doi.org/10.1080/13200968.2019.1802154

Mazzucato, V., & Niemeijer, D. (2000). The cultural economy of soil and water conservation: Market principles and social networks in eastern Burkina Faso. *Development and Change, 31*(4), 831–855.

McConnel, C. (2019). Critical human water needs: Failing to comply with the objects of the water act and human rights obligations. *Environmental and Planning Law Journal, 36*(3), 212–228. https://search.informit.org/doi/10.3316/agispt.20190608011804

Moggridge, B. J., Betterridge, L., & Thompson, R. M. (2019). Integrating Aboriginal cultural values into water planning: A case study from New South Wales, Australia. *Australasian Journal of Environmental Management, 26*(3), 273–286. https://doi.org/10.1080/14486563.2019.1650837

Moggridge, B. J., & Thompson, R. M. (2021). Cultural value of water and western water management: An Australian Indigenous perspective. *Australasian Journal of Water Resources, 25*(1), 4–14. https://doi.org/10.1080/13241583.2021.1897926

Mooney, W. (2014). Watershed on Indigenous rights needed in Victoria. *Chain Reaction, 120*, 17. https://search.informit.org/doi/10.3316/informit.170631934417203

Murray Lower Darling Rivers Indigenous Nations. (2007). *Echuca Declaration*. MLDRIN. https://www.mldrin.org.au/wp-content/uploads/2018/07/Echuca-Declaration-Final-PDF.pdf

Murray Lower Darling Rivers Indigenous Nations, Northern Basin Indigenous Nations, & North Australian Indigenous Land and Sea Management Alliance. (2017). *Cultural flows. A guide for first nations*. https://www.mdba.gov.au/sites/default/files/pubs/cultural-flows-a-guide-for-first-nations.PDF

Murray-Darling Basin Authority. (2022). *Water for first nations*. Murray-Darling Basin Authority: Australian Government. https://www.mdba.gov.au/about-basin/water-for-first-nations

Nelson, D. R., Godden, L., & Lindsay, B. (2018). *A pathway to cultural flows in Australia*. National Cultural Flows Research Project. https://www.mdba.gov.au/sites/default/files/publications/a--pathway-to-cultural-flows-in-australia_1.pdf

O'Bryan, K. (2017). Giving a voice to the river and the role of Indigenous people: The Whanganui river settlement and river management in Victoria. *Australian Indigenous Law Review, 20*, 48–77. https://search.informit.org/doi/10.3316/informit.912095333629949

O'Donnell, E. (2012). Institutional reform in environmental water management: The New Victorian environmental water holder. *Journal of Water Law, 22*, 73–84.

O'Donnell, E., Godden, L., & O'Bryan, K. (2021). *Cultural water for cultural economies*. MLDRIN & University of Melbourne. https://law.unimelb.edu.au/__data/assets/pdf_file/0008/3628637/Final-Water-REPORT-spreads.pdf

O'Donnell, E., & Kennedy, M. (2023). *The legacy of Aqua Nullius is causing a sustainability disaster*. University of Melbourne.

O'Donnell, E., Kennedy, M., Garrick, D., Horne, A., & Woods, R. (2023). Cultural water and Indigenous water science. *Science, 381*(6658), 619–621. https://doi.org/10.1126/science.adi0658

Pittock, J., Williams, J., & Grafton, R. (2015). The Murray-Darling Basin plan fails to deal adequately with climate change. *Water: Journal of the Australian Water Association, 42*(6), 28–32.

Poelina, A., Taylor, K. S., & Perdrisat, I. (2019). Martuwarra Fitzroy River Council: An Indigenous cultural approach to collaborative water governance. *Australasian Journal of Environmental Management, 26*(3), 236–254. https://doi.org/10.1080/14486563.2019.1651226

Poirier, R., & Schartmueller, D. (2012). Indigenous water rights in Australia. *The Social Science Journal, 49*(3), 317–324. https://doi.org/10.1016/j.soscij.2011.11.002

Taylor, K. S., Moggridge, B. J., & Poelina, A. (2016). Australian indigenous water policy and the impacts of the ever-changing political cycle. *Australasian Journal of Water Resources, 20*(2), 132–147. https://doi.org/10.1080/13241583.2017.1348887

Tilley, J. (2011). Governing 'on country': Using institutions to close the gap in Indigenous disadvantage. *Alternative Law Journal, 36*(1), 36–41. https://doi.org/10.1177/1037969x1103600108

Williams, J. (2017). Water reform in the Murray–Darling Basin: A challenge in complexity in balancing social, economic and environmental perspectives. *Journal & Proceedings of the Royal Society of New South Wales, 150*, 68–92.

Wilson, N. J., & Inkster, J. (2018). Respecting water: Indigenous water governance, ontologies, and the politics of kinship on the ground. *Environment and Planning E: Nature and Space, 1*(4), 516–538. https://doi.org/10.1177/2514848618789378

Woods, C., Dell, K., & Carroll, B. (2022). Decolonizing the business school: Reconstructing the entrepreneurship classroom through Indigenizing pedagogy and learning. *Academy of Management Learning & Education, 21*(1), 82–100. https://doi.org/10.5465/amle.2020.0366

Woods, R., Rigney, G., & Smyth, L. (2017). The national cultural flows research project. *Native Title Newsletter, 2*, 22–23. https://search.informit.org/doi/10.3316/agispt.20173874

Chapter 11
Walking Together: Djaara Women's Journey to Restore Food Plants and Heal Country

Nina Roberts, Aunty Marilyne Nicholls, the Djaara Women's Knowledge Group, Julie Andrews, and Susan Lawrence

Abstract For many generations, Djaara women have been unable to collectively practice caring for Country due to the immense and multi-layered impacts of colonisation. In response, a Djaara Women's Knowledge Group has been formed. Together with a post-graduate ecology student, they have begun exploring pathways for an unfolding of knowledge to return traditional food plants to Country. In this chapter, we explore our cross-cultural knowledge-sharing journey. It brings together the community of Djaara women—often geographically dislocated and disparate—who are walking towards understanding and Healing Country, and a non-Indigenous researcher walking alongside. This researcher helps uncover and share new information about the plants in their current context, facilitating a cross-cultural dialogue. The journey has highlighted the importance of several key principles for walking together: (1) Putting cultural safety first—acknowledging power dynamics between knowledge systems and the grief and loss felt when reconnecting with damaged Country; (2) Earning trust through long-term relationship building; (3) Communicating research regularly and accessibly; and (4) Being adaptable in maintaining engagement—especially during challenges like the pandemic and floods. Finally, we reflect on the challenges and successes we've encountered in learning to walk together, with the goal of bringing back the food plants and healing Country.

N. Roberts (✉)
Department of Ecological, Plant and Animal Sciences, La Trobe University, Bundoora, VIC, Australia
e-mail: Nina.Roberts@latrobe.edu.au

A. M. Nicholls · the Djaara Women's Knowledge Group
DJAARA (Dja Dja Wurrung Clans Aboriginal Corporation), Bendigo, VIC, Australia

J. Andrews
Department of Social Inquiry and Gabra Biik Wurruwila Wutja Indigenous Research Centre, La Trobe University, Melbourne, VIC, Australia

S. Lawrence
Department of Archaeology and History, La Trobe University, Bundoora, VIC, Australia

M. Kennedy, E. O'Donnell (eds.), *People, Place and Nature in Indigenous-Settler Relations*, Indigenous-Settler Relations in Australia and the World 10, https://doi.org/10.1007/978-3-032-03600-1_11

Keywords Cross-cultural ecology · Cultural revival · Embodied knowledge · Healing Country

11.1 Introduction to Country and Custodians

The practices and research we are sharing are situated on Dja Dja Wurrung Country, in central Victoria (Fig. 11.1). The Traditional Owners of this Country are known as *Djaara*, their language name for their Country is *Djandak*. Djaara have lived on their traditional lands for millennia. The practices that sustain people, plants, and animals who share Country[1] are embedded in all aspects of Djaara culture and spiritual connections to place. Through Djaara eyes, this has been the case since creation. In the Djaara creation story, Bunjil (the creator spirit taking the form of a Wedgetail Eagle) and spirit helpers created Djandak, including people (RSA, 2013). Djandak encompasses diverse landscapes and vegetation, including grasslands and grassy woodland on the Northern Plains, and open forests and woodlands elsewhere, which vary with the landforms, soils, climate and geology.

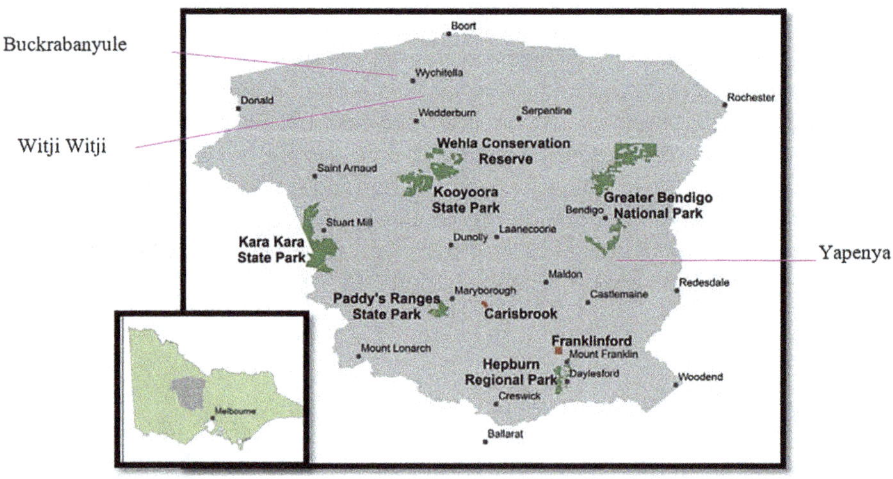

Fig. 11.1 Map showing the location of Dja Dja Wurrung Country in Victoria (south-east Australia) and the three sites chosen by the Djaara Women's Knowledge Group for women's work on Country during this project for healing and returning cultural plants. Six Joint Management Parks are indicated by green fill. (Adapted from DDWCAC, 2017)

[1] The term Country is widely used by Australian First Nations people, including Djaara people, to refer to land and all living and non-living entities who are present within the landscape. In this cultural usage it has qualities of a proper noun (Rose, 1996) and is capitalized.

British colonisation of Djandak in the early nineteenth century had devastating and wide-reaching impacts on both people and the land and waters. First, the Djaara population was decimated by a smallpox epidemic, then later, colonising pastoralists took over much of Djandak, preventing access to hunting grounds, springs and plant foods (Attwood, 2017; RSA, 2013). Murnong (Yam Daisy, *Microseris walteri*), a staple food plant cared for and harvested by Djaara women using fire and digging sticks, was almost completely eradicated from the landscape by sheep in the early years of colonisation (Attwood, 2017; Gott, 1983).

Less than a decade after this devastating impact of pastoralism, the mostly hilly and less fertile parts of Djandak (that had been less desirable to pastoralists) became the epicentre of the Australian gold rush of the 1850s and 1860s (Attwood, 2017), driving massive deforestation and soil disturbance and resulting in what Djaara people today refer to as 'Upside-down Country' (DDWCAC, 2017). Today, despite the tragic history of starvation, disease, and violence that pushed Djaara people from their Country, the descendants of the survivors are reviving culture and the dialogue that their ancestors had for thousands of generations with Djandak (DDWCAC, 2017). There is much healing to be done.

We respectfully acknowledge the important role of Djaara Elders, past and present, in caring for Culture and healing Country, and the contribution of Djaara youth in bringing their energy to carrying this forward.

Djandak holds the memory of Djaara ancestors and hears her people's presence. Despite the massive disruption to people, culture and Country experienced from colonisation, the connection of Djaara people to Country remains. This is a cultural landscape, and cultural knowledge remains embedded in the landscape everywhere.

11.2 Our Project

11.2.1 Who Is Involved

The *Yam Paddock Project*, is a collaboration between university researchers (principally represented by Nina Roberts during her doctoral research at La Trobe University) and a group of Djaara women, supported by a Research Agreement made by the University, Central Victorian Biolinks Alliance (a not-for-profit ecological group) and Dja Dja Wurrung Clans Aboriginal Corporation (trading as DJAARA). The project grew out of a common understanding that many important Dja Dja Wurrung food plants, including the once staple root-vegetable Murnong, need care or restoration and that cultural practices such as fire and digging are likely an essential part of restoration.

The collaboration aimed to initiate cross-cultural dialogue and knowledge-building about restoration ecology. Once the project began, it became apparent that a significant first step towards meeting that goal would be creating opportunities for Djaara women to strengthen their relationships with ancestorial Country; that is, for the Djaara women to spend time on Country together and have time to yarn about Country together first and foremost, and secondarily to yarn about Country with the non-Indigenous research team members.[2] It is from a bank of such relationship strengthening experiences that possibilities for cross-cultural dialogue have grown. This project is inherently relational and personal, and we will therefore begin with introductions:

Aunty Marilyne Nicholls is a multi-clan woman and respected Djaara cultural elder. She is a master weaver with knowledge of traditional methods passed down through her family line. Noticing that men dominated the on-Country activities and decision making, Aunty Marilyne advocated for creating a Djaara Women's Knowledge Group. She felt it was culturally important to increase opportunity and capacity among Djaara women to claim an active role on Country, especially caring for women's food and fibre plants. Project funding supported Aunty Marilyne's employment as the project officer through DJANDAK, the enterprise arm of DJAARA.

The Djaara Women's Knowledge Group (WKG) is a group of Djaara women of different ages and family groups. The Djaara WKG has existed as an evolving email list over the four years of the research collaboration. The Djaara WKG members are invited to join online and on-Country events and activities and receive regular research updates. The vision of having a group of Djaara women leading the re-activation of women's business on Djandak predates this research project; however, the financial support that accompanied the research project has supported activities and facilitated knowledge sharing (both between group members and between the research student and the members). The women in the group have varying levels of experience on Country: some have worked in natural resource management (NRM) jobs for years, and others were relatively new to on-Country practices like planting or burning.

> We need to keep the focus on Djaara women's cultural practice and reconnecting with Country – Aunty Marilyne, Djaara WKG member, 2021

Nina Roberts is a non-Indigenous woman of British immigrant/coloniser parents, with a deep interest in the way people relate to and value the non-human. When Nina graduated from high school in the early 1990s, further study in ecology was the clearest pathway she could see to valuing and caring for the non-human world in her future work. She came to this project curious about what a dialogue between ecology and Djaara women re-establishing cultural practice might look like when First Nations empowerment is a driving research principle.

---

[2]Yarning, a cultural form of conversation that may include the telling of stories and sharing of information, has been written about and adopted as a culturally safe research method for research involving First Nations people; it was introduced to the academy by Indigenous scholars and is now used globally by both Indigenous and non-Indigenous scholars, who recognize yarning as an Indigenous methodology (Bessarab & Ng'andu, 2010; Barlo et al., 2020; Kennedy et al., 2022). Storying is a related Indigenous methodological approach (Phillips & Bunda, 2018).

Nina's supervisory team at La Trobe University included two ecologists, Associate Professor John Morgan and Associate Professor Jim Radford, archaeologist Professor Susan Lawrence, and Aboriginal Studies Professor Julie Andrews. Susan and Julie had realised that women's perspectives were missing from practices on Country and academic scholarship about traditional food plants. Their conversations with Aunty Marilyne allowed the Yam Paddock Project to emerge. This cross-disciplinary team of academics created a research agreement with Dja Dja Wurrung Clans Aboriginal Corporation to pave the way for this research, facilitating an interdisciplinary research approach. Prof Julie Andrews is the only Indigenous member of the University supervisory team. Aunty Marilyne and colleague Dr. Nathan Wong (who works for Dja Dja Wurrung Clans Enterprises—DJANDAK) were external supervisors.

The story of the Djaara WKG will be the focus of this chapter, with some explorations of the perspective of the research student. We will, therefore, include passages of text that are the words of Djaara WKG members shared with and recorded by Nina.

11.2.2 Finding Common Purpose Through Plants

From the early days of this collaboration, the newly formed Djaara WKG, under the guidance of Aunty Marilyne, wanted to spend time on Country building their collective knowledge and confidence in participating in (and leading) women's caring for Country practices. Some expressed a long-term vision and goal: that their actions and developing knowledge would allow them to visit the chosen sites in years to come with children and/or grandchildren and be able to teach and experience cultural harvesting of women's plants. This aspiration aligns with goals around access to bush tucker articulated in the Dja Dja Wurrung Country Plan (DDWCAC, 2017). The opportunity of engaging in collaborative research about food plants also aligns with aspirations to increase the knowledge, capacity and presence on Djandak of Djaara members, as expressed in DJAARA strategy documents (e.g. Djandak Wi (Country Fire) Strategy 2024–2034, DJAARA, 2024).

There was initial caution about working together with a non-Indigenous ecology research student. Expression of this distrust included sentiments such as:

> We don't want just another ecology report. – Djaara WKG member 2021

> We don't want cultural practice tested by science; we know it from our own way of knowing – Djaara WKG member 2021

Ecological research was perceived as privileging Western ways of knowing over the Djaara women's cultural ways of knowing, in this case, about plants and Country. The Djaara WKG members wanted the group and the collaboration to be a space where they had the authority to re-establish cultural ways of knowing Country. They wanted to practice using traditional tools and methods of caring for soil and plants (e.g. digging sticks and cultural fire). Furthermore, the group culture strongly emphasised the spiritual and ceremonial dimensions of women reconnecting with Country. Given the group's caution about scientific methods of knowledge generation marginalising other ways of knowing, the potential for Nina to play a role in documentation became the way forward for the still-emerging collaboration:

> We're re-starting up practices that haven't been around, and I think documenting them is a really good way of learning on the journey and reflecting and growing that in a way. – Bec Phillips, Djaara WKG member, 31 August 2023

To facilitate the on-Country activities of the Djaara WKG (and, in turn, to document them), Nina propagated plants in the shade houses at the University. After sourcing local seeds of key tuberous food plants, Murnong and Bulbine Lily, she grew several hundred of these as tube stock ready for planting by the Djaara WKG. She also prepared seed bags (Murnong, Bulbine, Chocolate Lily and Wallaby Grass) for group members to directly seed.

11.2.3 Sites and Activities Chosen

A property called Yapenya (Fig. 11.1) was the primary focus of group activities in the first two years of the project. This former sheep-grazing property encompasses sites of high cultural significance, including Djaara stone arrangements (registered under the Aboriginal Heritage Act 2006). It came under freehold ownership of DJAARA after the Recognition and Settlement Agreement was signed between DJAARA and the State Government in 2013 (DDWCAC, 2017; RSA, 2013). In choosing this site as a place for focus early in the collaboration, the Djaara WKG ensured they had autonomy from land management partner agencies (that would have some level of involvement or oversight if the site was on public land). Additionally, they would have the privacy and space to engage in their process of caring for Country without having to address the curiosity or scrutiny of landowners, land managers or the broader public before they were ready.

Throughout the project, the activities of the Djaara WKG have included meeting on Country, yarning about Country (weaving in historical, ecological and cultural knowledge sources to some yarns); conducting women's cultural fire and using digging sticks for plant reintroduction trials and weeding. The experiences built confidence among group members, empowering some to share cultural practices and knowledge with the broader community, both Indigenous (e.g. Firesticks Alliance National Women's Fire Workshop in 2024, Right Plant Right Way workshops in 2022 and 2024) and non-Indigenous (e.g. working with a local conservation group to educate landowners about Djaara cultural ways of caring for Country, including demonstrating a cultural burn).

The challenges to bringing cultural food plants back to sites where they had disappeared since colonisation are many and complex. Bringing plants back requires more than reintroducing the plant and tending to it; there is first healing of Country needed. Additional considerations to sourcing seeds or plants included planting in an appropriate season and manner, returning to the site regularly to carry out culturally informed caring-for-Country practices, and understanding what steps are needed to heal Country from the land uses since colonisation. The current challenges include changes to soils from compaction and/or application of fertilisers, the introduction of many highly competitive weedy plant species and pests (such as rabbits, Red Legged Earth Mites and slugs), which predate on the palatable leaves and roots (especially of tender young plants).

Returning tuberous plants such as Murnong and Bulbine Lilies to sites where Djaara women could care for them in cultural ways proved to be beyond the timeframe of this project: many steps to healing are needed. The importance of maintaining focus on the cultural plants, regardless of timeframe or outcomes, was that it opened an opportunity for the Djaara women to come together and explore their processes of knowledge creation and connection to Country. Secondarily, Nina's presence in the collaboration as an outsider with a background in ecological research allowed her to provide tangible assistance and, later, two-way cross-cultural ecological learning within a relational framework. The longer timeframe needed for the plants to be returned further highlights that for Djaara women, healing Country and rebuilding connections are long-term, ongoing processes that extend far beyond the limited dimensions of an academic research project.

11.2.4 An Historic Cultural Burn

> Cultural burning is healing and connects Djaara women to Country in a good way. – Aunty Marilyne Nicholls, Djaara WKG member, 2021

Djaara reviving a cultural fire practice (Djandak Wi) has been a significant part of re-establishing their presence on Country, strengthening connections and healing both people and Country (DJAARA, 2024). Djaara-led culturally informed burning has been facilitated by the Recognition and Settlement Agreement between the DJAARA Aboriginal Corporation and State Government of Victoria reached in 2013 and the 2019 Victorian Traditional Owner Cultural Fire Strategy (The Victorian Traditional Owner Cultural Fire Knowledge Group, 2019). Some of the Djaara WKG members had participated in Djandak Wi when this project started; however, the group recognised the importance of all members having this opportunity and of women undertaking cultural fire independently of men in ways that allowed them time and space for ceremony.

In 2021, supported by this project, members of the Djaara WKG carried out their first burn: the first all-female traditional owner burn on their Country since colonisation (see Fig. 11.2). It was the first action in working towards trialling the return of cultural food plants to their chosen women's site on the Yapenya property, which has a history of sheep grazing and no trees within the study site (some fringing the site). The burn was carried out on a fine autumn day, ideal for applying gentle

Fig. 11.2 Photos documenting the first Djaara Women's Knowledge Group burn in 2021, showing the culturally informed practice: (**a**) green branchlets selected as fire dampeners; (**b**) painting up with ochre and charcoal on hands and faces; (**c**) and (**d**) use of traditional fire lighting tools; (**e**), (**f**), (**g**) (overleaf) directing and controlling fire with grass firesticks, garni (digging sticks) and green branchlets

Fig. 11.2 (continued)

cultural fire. Pasture grasses (primarily annuals) and weeds such as Paterson's Curse and Flat Weed were abundant, and there was a moderate layer of thatched dry grass on top of the ground at this time of year. It is on a gentle slope close to a broad valley, with granite outcrops sheltering the site on one side. Birds and kangaroos are abundant in this landscape.

Djaara women's preparation for the cultural burn included taking safety measures to ensure fire containment, as required by DJANDAK (Dja Dja Wurrung Enterprises). Once these preparations were done, the women attended to cultural preparations such as painting faces and hands with ochre, singing in Dja Dja Wurrung language of their intentions to Country and ancestors, and practising with traditional fire-making tools.

Six Djaara women and one child were present for this important event (Fig. 11.2). Yarning and knowledge sharing were integrated at all stages, including practising skills and teaching cultural protocols. There was an intense focus in the group; everyone was fully present, and there was clarity about the significance and purpose of the women's knowledge group on this day. Nina, Susan (Nina's supervisor), and Catalina (a former post-graduate researcher who was also collaborating with the Djaara WKG at the time) were also present as observers and documenters only. The photos and video footage were taken at the request of the Djaara women and are the property of the Djaara women to share as they wish (Fig. 11.2).

In yarns with the Djaara WKG members later in the collaboration, the sense of spiritual and cultural connectedness with Country that the women experienced from this burn were emphasised, as were feelings of pride and empowerment:

> We used grass matches to light it. As soon as we lit it up, the two eagles flew over, to give us that safe passage to burn country. [This burn event] was women returning spirit back on Country through cultural burning.
>
> – Sharnie Hamilton, Djaara WKG member, 6 Dec 2024

> To have all those women out there, from our elders, our Aunties, right down to our young kids, and then right down to the next generation of younger kids, so three generations of females out there burning Country … it started up their own song and storyline. It sparked a fire within their sprit, within their Murrup.
>
> – Sharnie Hamilton, Djaara WKG member, 19 Dec 2024

Some Djaara WKG members who were not present on the day of this significant women's burn also expressed feeling uplifted by it when shown photos and video footage:

> Country needs the smoke, it needs the language, it needs the people, so seeing this … just makes my spirit happy, to know the women are back out there caring for Country in this way.
>
> – Aunty Vicki Murray, Djaara WKG member, 6 December 2024

11.2.5 Learning with Bodies and Garni

For the Djaara WKG members, throughout this project, opportunities to build familiarity with using digging sticks came to be strongly associated with building a memory bank of sensory knowledge and feelings of connectedness with ancestors and Country. Before this project was initiated, the significance of women's diggings sticks (garni) was already known to many group members as an essential part of their knowledge story. During an earlier project, also trialling some reintroduction of food and fibre plants, an opportunity to make and use garni had arisen (Fig. 11.3), setting in motion a cultural knowledge activation process for Djaara women:

> In making our garni we really got to understand the wood we were working with We shared the story of the seven sisters [in which women were the first fire keepers], and talked about how a garni was the main implement of women, you know: it's our power, and the extension of our arm, we always had it with us; it was our main tool. And putting it into practice gives you a different kind of knowing.
>
> And from [making our garni] a garni song and dance came out. And so, it was a way of reviving that ceremony around the garni with Djaara women.
>
> – Bec Phillips, Djaara WKG member, April 2024

The activation of women's ceremony, as described by Bec, and the ways this was bound together with caring for the plants in traditional cultural ways set the focus for many of the activities of the Djaara WKG members during this project.

After the first burn at the Yapenya study site, the group trialled planting Murnong and lilies into patches dug with digging sticks. Both direct seeding and transplanting of young tube stock grown by Nina were trialled.

Fig. 11.3 Bec Phillips and Timani Nicholls (Djaara WKG members) use garni on Country for planting Kangaroo Grass as part of a demonstration for a landowner/land manager group. May 2023

> Digging sticks give you a great connection to Country, [especially] to soil. To bring such an historical tool that was used over thousands of years ago, you know to bring it into this century … reliving that digging stick farming experience … is very empowering.. You see all the women we had out on Country using it, doing it. Really freeing spirit and really really empowering those women, it makes my heart sing.
> – Sharnie Hamilton, Djaara WKG member, 19 December 2024

In a yarn early in 2024, Djaara WKG members Aunty Marilyne Nicholls and Bec Phillips reflect on the vital role of embodied experiences on Country, which often have resulted in deep feelings of sensory and spiritual attunement to Country.

> Bec: It was like a spirit memory woke up inside when you seen all the Djaara women in the fields, with their garni, digging away in this beautiful peace.
> Aunty Marilyne: Yes, definitely.
> Bec: They're just there, in their relationship with Country. And you could just see how good it was for our own mental health and wellbeing when you have practices like that as a part of your everyday.
> Aunty Marilyne: yes. … it's kind of like [the practice] has been asleep and suddenly it's awoken. Cos the spirit of it's still there, see.
> [It is helping us] to understand more of our culture. In the sense of not just understanding how to work with our garni, but also how to remember ceremony and how to remember our connection with star constellations and how to work with different soil types and what we're working with now.

Bec further elaborates on the importance of caring intention and being awake to all the information available through the senses, including those facilitated by using a Garni while being present on Country:

> There's like a switch that happens, you know, with intention. … it happens through that whole journey of "where might be a good place to put them" …. Like, "oh, there's a big kangaroo track through here, well, we're not going to plant them right there, as the kangaroos will just eat them", or there's a whole lot of rabbit holes through here. You know, so we had to think … where is probably a good place where they're going to last. … I think to be a good carer for country you shift your thinking to the thing you're caring for. So, what do I need if I were a Murnong? … you feel like you're more wholistically with Country in all of its elements and layers when your thinking shifts to what you're caring about and how you're doing that … in the time that it needs, not in modern day 'let's get this done we've got other things to do'. Building the relationship with that land and that plant, and your Garni.
> – Bec Phillips, Djaara WKG member, April 2024

This yarn between Bec, Aunty Marilyne and Nina touched on the multiple layers of benefit to Djaara women by making space for embodied knowledge practices: knowledge that is developed through the senses and held in the body, not just in the head. Further to feelings of connectedness and renewal of stories and cultural practices, Djaara women were feeling what their interactions with County mean for other entities, including the earth itself. Bec's description of building relationships with land and plants is an example of the ethics of reciprocity, respect and obligations between humans and the non-human world that is integral to Traditional

Ecological Knowledge systems. The description of stories and songs being awakened shows that caring for Country is part of a reciprocal restoration process with culture (Kimmerer, 2011).

Within our project, a focus on caring for the plants on Country was a fertile ground on which multi-directional relationships (between both human and non-human aspects of Country) could grow. Taking action related to healing Country (for example, trying out the reintroduction of food plants using digging sticks and direct seeding and transplanting) and became a mode of adaptive doing (Rawluk et al., 2023) that increased experiences of Country (for example feeling the way the digging stick breaks the soil differently after rain, or seeing/hearing fire spread through the dry top layer of grass). These are the ways knowledge is created and shared in the moment. The multiple benefits to the Djaara women are shown using the visual metaphor of the radiating petals of an open Murnong flower in Fig. 11.4.

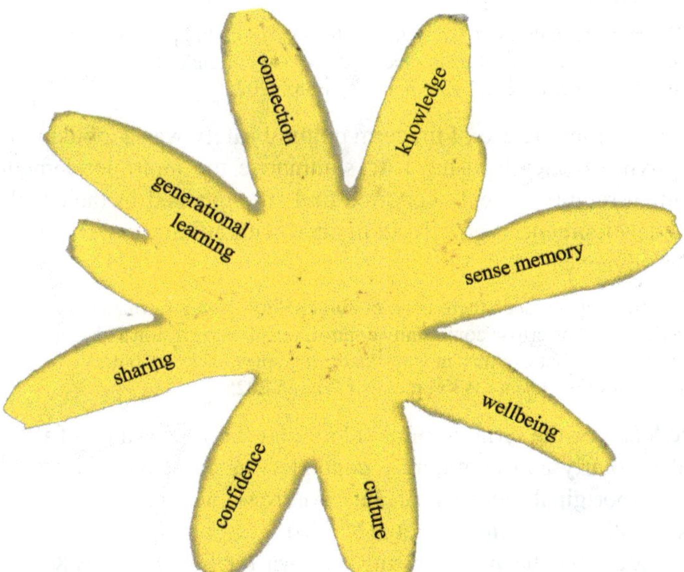

Fig. 11.4 The multiple benefits of experiences on Country to Djaara WKG members, represented as petals on an open Murnong flower

11.3 Unpacking the Challenges

11.3.1 Establishing Cultural Safety on Country

Cultural safety is a term that has originated in the health services sector (Gerlach, 2012)[3] for the purpose of welcoming First Nations clients, framing cultural awareness for staff and inclusion of Aboriginal and Torres Strait Islander health practitioner expertise and knowledge. In this paper, we apply cultural safety in the context of Aboriginal women working with and on Country. The aim of this project was to increase and document Djaara women's cultural practice knowledge and create partnerships within a research context with non-Djaara researchers; it is important to work within the framework of cultural safety for this approach. From its inception in the context of nursing in Aotearoa New Zealand in the 1980s (Ramsden, 1993), cultural safety has been used to raise critical awareness of the power structures of colonisation, encouraging those working with and for Indigenous people and communities to be self-aware and challenge the status quo to actively validate and show respect for Indigenous experiences, protocols and knowledge (Commonwealth of Australia, 2021; Gerlach, 2012; Kennedy et al., 2022).

> For us Djaara women, cultural safety means having a suitable place where your spirit as a Djaara woman feels comfortable and feels at ease and not under threat
> – Aunty Marilyne, Djaara WKG member, March 2025

In this collaboration, the use of the term cultural safety was adopted in response to Aunty Marilyne frequently using it to summarise the multiple elements of what Djaara women require to feel supported and strengthened in their wellbeing and knowledge development. It goes hand in hand with cultural governance, according to Aunty Marilyne:

> Without cultural governance there is no cultural safety. This governance includes policies and procedures that prioritise community control, self-determination and support for Djaara members to learn about Country in ways that bring spirit back to Country
> – Aunty Marilyne, Djaara WKG member, March 2025

The antecedence of the term cultural safety and the importance of the concept in Aboriginal scholarly and community contexts was emphasised by research team member and Aboriginal scholar Prof. Julie Andrews, who has drawn on related concepts in her writing (e.g. Andrews et al., 2023).

Cultural safety has been a recurring concern for the Djaara WKG members in various contexts. One element is ensuring that cultural ways of knowing and being are not marginalised or usurped by the Western scientific knowledge model (already

[3] The principles associated with the term 'cultural safety' have been widely applied in Aboriginal community settings in Australia, even when the term is not employed: e.g. Victorian Aboriginal Health Services Yarning program: https://www.vahs.org.au/yarning-safenstrong/ and https://www.vacca.org. Bin-Sallik (2003) has argued for the adoption of both language and principles of 'cultural safety' in the tertiary education sector. We have encountered a single use of it in conservation science, in a coral conservation project on sea Country in Australia (Daly et al., 2024).

mentioned in Sect. 11.2.2; our path to creating safety around this issue is further discussed in Sect. 11.4.1). A second element of cultural safety relates to the cultural understanding of being able to exercise authority and make group decisions when working on their own Country, drawing on their own (existing and emerging) cultural knowledge.

Another context in which cultural safety arose was selecting places on Country for the group to practice culture and caring for Country. At the start of this project, the Djaara WKG members visited several potential study sites. Aunty Marilyne strongly expressed to Nina the importance of the women having confidence that they would have perpetual access to the chosen sites and not come back (potentially years later) to find a locked gate, as this would be repeating past traumas of dispossession. Furthermore, the preference for privacy and autonomy meant Djaara women did not select a site in one of the Joint Management Parks on Dja Dja Wurrung Country that are also publicly accessible as National or State Parks.

In addition to the (DJAARA-owned) Yapenya property, two other sites that the Djaara WKG members felt offered adequate cultural safety were both on private property owned by Bush Heritage Australia (BHA) (Fig. 11.1). BHA had been working closely with Djaara at these locations around the time of this project and had an agreement in place to ensure ongoing access for the Djaara women to the sites they selected.[4]

Consideration of gender also contributes to establishing cultural safety for the Djaara women. Historically, the opportunities for being on Country have been through employment in NRM. The NRM sector is male-dominated, and this is true within Traditional Owner organisations, government agencies, and the private sector (James et al., 2021; Daniels et al., 2022). Aunty Marilyne wanted to see the Djaara women (including those with no previous experience) feel safe to get involved on Country and see themselves as having an active role, especially at women's sites. She emphasised that women have different ways of doing things and need autonomy and space to explore that together. In her role as project officer for the Yam Paddock Project at DJANDAK, she was typically under the management line of non-Indigenous men with NRM and/or ecology expertise. Consequently, she felt the need to advocate for autonomy and privacy around the project to better establish cultural safety for herself and the other group members.

11.3.2 Systemic Barriers

Caring for Country requires access to Country and is most feasible when there are legal rights in place (Williamson, 2021). For Djaara people, dispossession and subsequent imposition of a colonial property rights system led to a multi-generational

[4] The access agreement was in the form of a Memorandum of Understanding signed in 2021. Later, a shared Strategic Landscape Plan was developed by Bush Heritage and DJAARA (Bush Heritage Australia and DJAARA, 2024).

disruption to practices such as cultural burning (DDWCAC, 2017). In 2012, Djaara Traditional Owners achieved a Recognition and Settlement Agreement (RSA) with the Victorian Government. The RSA has facilitated Djaara access to agreed areas of public land (DDWCAC, 2017). However, with 87% of Dja Dja Wurrung Country being privately owned (DDWCAC, 2017), the freedom of Djaara people to access land for cultural practice remains heavily restricted.

Further to establishing legal access, the experiences of the Djaara women demonstrate that there are multi-layered systemic barriers to conducting cultural practices on Country. Djaara women need to navigate the administrative, safety and planning aspects of carrying out on-Country practices, especially cultural fire, which often creates a high dependency on the organisational structure (in this case DJANDAK), where staff are appointed who are familiar and confident in these systems. Often, these staff members are non-Djaara and/or men, although this has begun to shift even within the timeframe of this project. The systems themselves were also found to be disempowering to Djaara women:

> We need ways that are cultural ways, [ways that] feel right for all aspects of wellbeing. The structural and systematic approach has been limiting for Djaara women, however a shift to a cultural system would be empowering. This is an ongoing challenge in the project.
> – Aunty Marilyne, Djaara WKG member, 6 Oct 2023

The administrative layers that must be negotiated inhibit the ability of the Djaara WKG to make decisions that are responsive to listening to Country. This was seen in the second year of caring for the Yapenya women's site when waiting on permissions meant one opportunity to burn in 'just right' conditions (during a site visit made by Aunty Marilyne and Bec Phillips in early April) slipped past. By the time a return to the site with permissions in place was arranged, conditions had changed, with Autumn rain and new green growth making it hard to put fire in the site.

Further barriers raised by Aunty Marilyne have related to the lack of funding for the women's time and expenses, needing support for meeting requirements for fire safety training, fire uniforms and boots and many Djaara women living off Country with family caring and work commitments increasing the challenges around participation. There were times during this project that Djaara women's frustrations with systemic barriers were acutely felt as holding back the activities and autonomy of Djaara WKG activities on Country:

> The admin is managed by men still. There is still the gap [in accommodating women]. I don't think they're ready for us.
> – Aunty Marilyne, Djaara WKG member, 29 Jan 2024

However, there were positive signs of these barriers diminishing during this project, largely due to increased confidence and experiences among many Djaara women. Further gains have been made by an increase in the number of Djaara women employed by the DJAARA Aboriginal Corporation.

> [Practising fire] has brought empowerment back to the women. And the women are still out there today doing cultural burns. The women's group have built up partnerships too …
> – Sharnie Hamilton, Djaara WKG member, 19 Dec 2024

11.3.3 Barriers Due to Events of the Time

The disruption that came with the COVID-19 pandemic in 2020 (the year this project commenced) and 2021–2022 profoundly influenced how this collaboration unfolded. In the years since the peak of the pandemic impact, it remains challenging to fathom the extent of this impact on the way we all were: in work, our communities, and our personal lives. The restrictions on movements and gatherings added to existing challenges of bringing people together, limiting the opportunities for connection and development of relationships between the Djaara WKG and the outsider research team, within the Djaara WKG, and between Djaara WKG members and Country. In the period of relative 'normality' of movement in early 2021, important progress was made, including the Djaara WKG hosting a cultural awareness day for all project team members and funders at the Yapenya site. The Djaara WKG members and research partners gathered to discuss plans for the study sites and holding both a women's cultural burn and a planting day using digging sticks.

As the second wave of lockdowns hit in mid-2021, we strove to adapt and continue to work within the restrictions. Some adaptions included follow-up planting and weeding being carried out by one or two group members rather than as a group; Nina continued to monitor the site alone regularly, under an approved permit allowing her to leave Melbourne; holding Zoom meetings with Djaara WKG members and the research partners (often including guest speakers who could share on topics related to what the Djaara WKG were interested in learning about on Country); and the commencement of a project newsletter distributed by email for sharing project updates and other information of potential interest to the group, including seasonal photos of plants on Country.

Coming out of lockdowns in 2021 and 2022 brought further challenges, including extensive damaging floods across central and northern Victoria in October 2022, in locations where many of the Djaara WKG members lived. Opportunities to come together on Country were curtailed by the direct impacts of flooding on group members and their families, with many roads flooded and flood-damaged, affecting the accessibility of sites. These events were substantial among the challenges we faced, and it is important to note that events of a similar nature are likely to be part of our collective future together on Country as we increasingly feel the ramifications of destabilised earth systems. Researchers have shown causal links between global land use, land cover and anthropogenic climate change and locally felt impacts such as more frequent extreme weather events and exposure to novel diseases including pandemics (Carlson et al., 2021; Seneviratne et al., 2021).

11.4 What We Learned from Walking Together

11.4.1 Finding a Relational and Experiential Research Framework

Past and present power dynamics between the dominant academic knowledge systems and knowledge systems of colonised peoples cast a problematic backdrop to research involving Aboriginal communities: the word 'research' itself 'stirring up silence' and 'bad memories' (Smith, 2022). Comments from Djaara WKG members that drew attention to this distrust of academic research helped Nina to deepen her reflexive awareness as an outsider researcher. The feeling of unease from the Djaara women set in motion the need to re-centre her conception of the research, shifting it into an intercultural and interdisciplinary space.

> Figuring out how (or if) I could conduct ecological research that would benefit and empower the Djaara WKG was the principle challenge I have grappled with and grown through.
>
> I found it was first necessary to loosen the grip of disciplinary norms on my thinking about what my work involved. I had expected that questions would flow from the WKG that I could use ecological methods to answer in a helpful way, but instead I had to let go of what I thought I could or should do as a researcher, and instead feel for a way forward through trying to understand the shape of the emerging collaboration, that centred the Djaara women's experiences whilst also learning about the plants. This was done while planning, facilitating and documenting healing Country activities.
>
> – Nina Roberts, student researcher, 2024

We found slowing down was a key learning of the collaboration; it's a lesson noted by other cross-cultural collaborative researchers (Castleden et al., 2012; Chen et al., 2023; Wright et al., 2023). Slowing down is a key part of allowing relationships and a relational research framework to develop, and we believe that in this project was part of establishing cultural safety. As simple as it sounds, it's not so easy when working within funding timeframes, reporting requirements and institutional deadlines. Slowing down this project allowed the development of trust between the Djaara WKG members and the student researcher. It acknowledged that Djaara women were sometimes processing grief and loss on reconnecting with a Country that has been damaged extensively since colonisation. The challenging circumstances of the time, the lockdowns and floods, provided the unexpected gift of time to listen and go slowly.

Another crucial factor facilitating slowing down has been the source of project funding. Financial support for this project has come from the philanthropic sector rather than more traditional sources of research funding, such as the Australian Research Council and the University. The specific nature of the philanthropic funding has opened the space for dialogue with the funders around timelines and outcomes. They have been willing to approve extended timeframes and have accommodated the evolving expectations about the results of the project. This has enabled far greater flexibility and a slower pace than would have ordinarily been the case. Instead of the three-year timeline usual for most grants, the project has been able to stretch to 8 years while results have been measured by the objectives of the women rather than the quantity of peer-reviewed publications.

Slowing down was necessary to grow trust that this research wasn't going to repeat the entrenched colonial power dynamics that partnerships between non-indigenous agencies and Traditional Custodians still often do (Weir, 2023). We had opportunities to develop and demonstrate reflexivity and understanding, in other words, nurture cultural safety within the collaboration. Some indicators that cultural safety was achieved are:

- Feelings of mutual respect between visitors/researchers and the participating Djaara women
- Inclusion and celebration of Aboriginal women's ways of knowing and cultural practices in research outputs such as conference presentations (which in turn contributed to the strength of the cultural practice, creating confidence for the women to develop the practice further)
- The women feel strong and proud of their collective knowledge and activities and develop their voices (a significant impact of the research).

Looking back over the research project to date, we can see the shifting emphasis within the relational process of undertaking this project together, from concerns with establishing cultural safety to deepening relationships and trust, and how both feed into the experiences that built knowledge for the Djaara women (Fig. 11.5).

Indigenous research methodologies, as explained by Indigenous scholars, are knowledge-building grounded in Indigenous ways of being and doing (Hughes & Barlo, 2020; Singh & Major, 2017; Smith, 2022; Tynan, 2021; Wilson, 2008). Relationality is a central aspect of these methodologies: the researcher is always explicitly present and 'in relationship with' all elements of the research—they are also in the process of 'becoming'. They are being shaped by all these elements, and vice versa (Bawaka Country et al., 2015). Responsibilities to relationships come first: "Research must not jeopardise responsibilities with kin; family, Country, and the futurities of those relationships" (Tynan, 2021). These responsibilities extend also to relationships and obligations to ancestors:

> As Djaara Women we have a responsibility to practice our culture that goes back thousands of generations, season to season; it is still current now and is still based on cultural practice.
> – Aunty Marilyne Nicholls, Djaara WKG member, 2024

These responsibilities to relationships, whether past, present or future, underpin the ethical guiding principle of relational accountability in Indigenous research methodologies (Wilson, 2008). The Indigenous, relational research method of *Yarning* or 'shared storying' (Bessarab & Ng'andu, 2010; Dickson, 2020; Fredericks et al., 2011; Leeson et al., 2016; Martin & Mirraboopa, 2003; Walker et al., 2014) was used in this project to encourage relational dialogue and connection. There were yarns between the Djaara women, between the Djaara women and Nina, and yarns with Country. Yarns with Country extend the connection, reciprocity and embodied listening and learning of Yarning (Hughes & Barlo, 2020). As Wright et al. (2023)

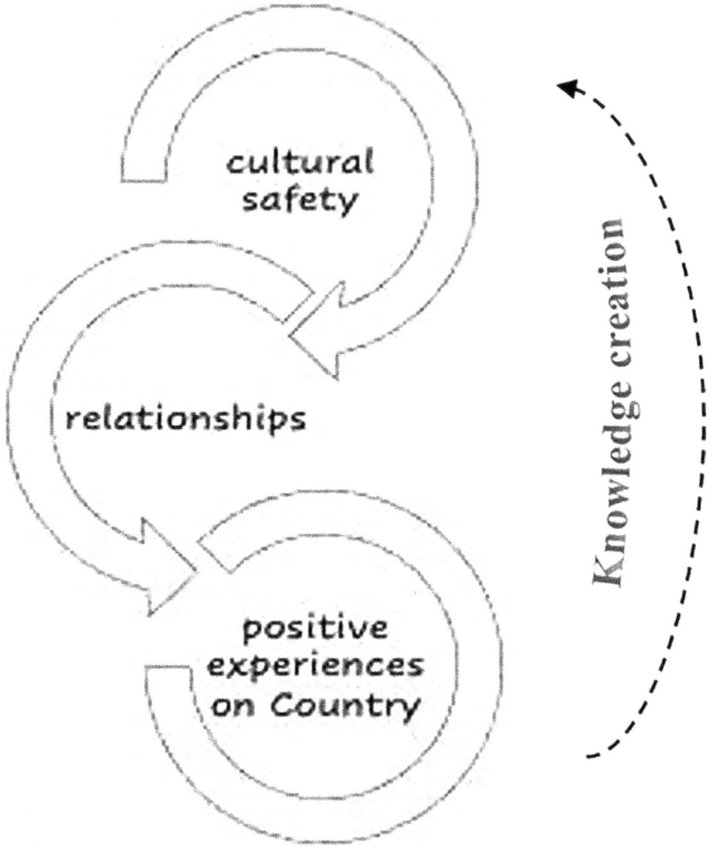

Fig. 11.5 The process of co-creating knowledge relevant and valuable to Djaara women healing Country under cultural governance

explain: Country cannot be detached from Storying because they are intertwined as a collective being:

> Country not only opens the meanings of the collective memory, but also the collective spirit of all that is living, past and present. Going On Country also recognises Aboriginal sovereignty and acknowledges the stories of what has been taken away and of what continues to thrive and grow. On Country is not just a cultural experience, but historical and political—one of resistance. It is only after the On Country and the Shared Storying that it is possible to set the foundations for working together. (Wright et al., 2023, p. 119)

Yarning with Country requires 'visceral receptivity' (Hughes & Barlo, 2020), engaging all the senses to create a 'sensory narrative' (Cameron, 2022). Djaara WKG member Bec Phillips expressed the need for this receptivity:

> Country is our biggest teacher; if you listen to it, Country will show you what you need to know. In talking to Country and letting it talk back to you or show you what you need to do, you are in the service of it.
>
> – Bec Phillips, Djaara WKG member, August 2023

Finding ways to acknowledge and centre the yarns with Country, the visceral receptivity, and the sensory narratives was challenging for Nina when working within the accepted language and methods of Western academic ecology. Bringing your bodily senses and sense-making into relationship with all that is in your research doesn't fit within a positivist (scientific) approach to knowing the ecology of places, whereby we collect information as if we could erase our presence from what we are studying. Knowing the plants better and understanding the current challenges for returning them to Country remains a strong motivation for finding ways to communicate across these different (seemingly incompatible) knowledge systems.

11.4.2 Attending to Project Information Flow

Over this multi-year collaborative project, we have found that the production of a bi-monthly newsletter (originally a Covid innovation for keeping group members informed and engaged) has been a valuable way of establishing an information flow from the research back into the Djaara WKG. The newsletters also provided other functions, such as a seasonal diary of Country and a way to share personal stories about their relationships to Country from different Djaara women within the group. The medium we chose was simple—an image-rich PDF document, typically about 6 pages long, distributed via email. Alternative media such as (limited access) video updates, podcasts, blogs or even spatial apps could serve this purpose even more effectively if accessible to the group. The newsletter has served a key role in ensuring that research documentation is presented to all Djaara WKG group members (who chose to engage) regularly and in an accessible format. We are now considering how to keep the information in those newsletters in a format that continues to be easily accessed by Djaara women as a resource.

Regular updates and information about the research are essential in the context of a long-term collaboration. A part-time PhD (as this has been) typically spans more than 7 years, and publication of the research in peer-reviewed journals may add further years to this. That's a long time for collaborators to wait to see the shape of what is coming from the research and to have access to it as a potential resource. The newsletters have been direct documentation of the unfolding research and Djaara WKG activities. They have been used as a teaching tool to bring new group members up to date with what has already happened in the project and as a way of including women who haven't been able to be present at group gatherings. During the long years of the doctoral research process, the newsletters have been a consistent gesture of reciprocity to the Djaara women from Nina and an archive of our time walking together.

11.5 Conclusions

In this chapter, we have sought to share the achievements of the Djaara WKG in engaging with cultural ways to deepen knowledge of Country. We have also shared the challenges and successes of the Women's Knowledge Group collaborating with a non-Indigenous ecology research student. The relationships built in this process have laid the groundwork for a dialogue about how research can support the Djaara WKG members as they reconnect with Country and explore culturally informed caring for Country activities. It has been a valuable time for building and attending to relationships with each other and Country.

It has been evident in this collaboration that Djaara women's ways of knowing and being needed to be given space and priority in the research. It was hard to see where an ecology study fit in with the cultural revival processes of the Djaara WKG at the beginning; collective experiences of Country and cultural practices needed to come first. The project gave WKG members opportunities for time on Country practising culture and to have their cultural knowledge revival documented.

> Cultural Knowledge is like medicine for the women. [since my involvement in this project] I'll never look at a digging stick the way I used to. Also, I have more experience with cultural ways of burning.
> – Aunty Marilyne, Djaara WKG member, 6 Oct 2023

In parallel with the revival of cultural practices on Country by Djaara women, Nina was able to provide support such as sourcing seeds and growing seedlings, gathering relevant information from a range of sources and monitoring and observing the plants, pests and weeds to supplement the women's expanding knowledge of, and relationship with, their Country in its current state.

> It's been really empowering … to have [Djaara] women and yourself out on Country … science and culture side by side makes my heart sing.
> – Sharnie Hamilton, Djaara WKG member, 19 Dec 2023

We are still walking together as we write this. Though this project will soon conclude, many ideas for future learning journeys the Djaara WKG could take are present in the group. Each step taken towards such learning and experience-building is a step towards reclaiming sovereign knowledge of Country. We feel a positive foundation for future collaborations has been built.

Acknowledgements We wish to acknowledge the project funders: the Sali family and the Helen McPherson-Smith Trust. We would also like to acknowledge all supervisors and project steering committee members for their time and input in supporting and shaping this work: Assoc. Prof. Jim Radford, Assoc. Prof. John Morgan, Dr. Nathan Wong, Dr. Sophie Bickford and Dr. Jack Pascoe. Thank you to John Morgan and Jim Radford for their review of the draft manuscript.

Photo credits: Photo in Fig. 11.2g taken by Catalina Labra Ode. All other photos were taken by Nina Roberts.

References

Andrews, J., Burns, E., James, C., & Rajčan, A. (2023). Bordieu's habitus clivé in voicing, feeling, being Aboriginal. *Journal of Sociology, 59(2), 290–299.* https://doi.org/10.1177/14407833221144103

Attwood, B. (2017). *The good Country: The Djadja Wurrung, the settlers and the protectors.* Monash University Publishing.

Barlo, S., Boyd, W., Pelizzon, A., & Wilson, S. (2020). Yarning as Protected Space: Principles and protocols. *AlterNative, 16(2), 90–98.* https://doi.org/10.1177/1177180120917480

Bessarab, D., & Ng'andu, B. (2010). Yarning about yarning as a legitimate method in indigenous research. *International Journal of Critical Indigenous Studies, 3*(1), 37–50.

Bin-Sallik, M. (2003). Cultural safety: Let's name it! *The Australian Journal of Indigenous Education, 32*(1), 21–28.

Bush Heritage, & DJAARA. (2024). *Wurreka Galangu shared strategic plan 2024–2034.* Bush Heritage Australia and DJAARA.

Cameron, L. (2022). Indigenous ecological knowledge systems—Exploring sensory narratives. *Ecological Management and Restoration, 23*(S1), 27–32. https://doi.org/10.1111/emr.12534

Carlson, C. J., Albery, G. F., & Phelan, A. (2021). Preparing international cooperation on pandemic prevention for the Anthropocene. *BMJ Global Health, 6, 10.1136/bmjgh-2020-004254*(3).

Castleden, H., Morgan, V. S., & Lamb, C. (2012). "I spent the first year drinking tea": Exploring Canadian university researchers' perspectives on community-based participatory research involving indigenous peoples. *Canadian Geographer, 56*(2), 160–179. https://doi.org/10.1111/j.1541-0064.2012.00432.x

Chen, S. L., Baumflek, M. J., Sampson, T., & Cabe, T. (2023). Doing research together: Wasdi (*Allium tricoccum*) plants guide dynamic research collaborations in Cherokee landscapes. *Ecology and Society, 28*(2). https://doi.org/10.5751/ES-14000-280213

Commonwealth of Australia. (2021). *Keeping our kids safe: Cultural safety and the national principles for child safe organisations.*

Country, B., Wright, S., Suchet-Pearson, S., Lloyd, K., Burarrwanga, L., Ganambarr, R., Ganambarr-Stubbs, M., Ganambarr, B., & Maymuru, D. (2015). Working with and learning from Country: Decentring human authority. *Cultural Geographies, 22*(2), 269–283. https://doi.org/10.1177/1474474014539248

Daly, J., Van Issum, H., Cummins, M., Evans-Illidge, E., Randall, C. J., Boyd, S., Cveticanin, L., Dundas-Hargraves, K., Liddell, B., Hobbs, R. J., & O'Brien, J. K. (2024). Cross-cultural approach to biobanking of living coral cells on Woppaburra Sea Country in the southern inshore great barrier reef. *Australasian Journal of Environmental Management*, 1–16.

Daniels, C. W., Rangers, N. Y., Russell, S., & Ens, E. J. (2022). Empowering young aboriginal women to care for Country: Case study of the Ngukurr Yangbala rangers, remote northern Australia. *Ecological Management and Restoration, 23*(S1), 53–63. https://doi.org/10.1111/emr.12538

Dickson, M. (2020). Learning ethics from an echidna: Embedding indigenous knowledges at the core of ethical research practice. *Methodological Innovations, 13*(3). https://doi.org/10.1177/2059799120976929

Dja Dja Wurrung Clans Aboriginal Corporation. (2017). *Dhelkunya Dja: Dja Dja Wurrung country plan 2014–2034.* http://www.djadjawurrung.com.au/fact-sheets

DJAARA. (2024). *Djandak Wi (country fire) strategy Dhelkunya Wi (healing fire) 2024–2034.*

Fredericks, B., Adams, K., Finlay, S., Fletcher, G., Andy, S., Briggs, L., & Hall, R. (2011). Engaging the practice of yarning in action research. *Action Learning and Action Research Journal, 17*(2), 7–19. http://eprints.qut.edu.au/48053/

Gerlach, A. J. (2012). A critical reflection on the concept of cultural safety. *Canadian Journal of Occupational Therapy, 79*(3), 151–158.

Gott, B. (1983). Murnong—*Microseris scapigera*: A study of a staple food of Victorian aborigines. *Australian Aboriginal Studies, 2*, 2–2.

Hughes, M., & Barlo, S. (2020). Yarning with Country: An Indigenist research methodology. *Qualitative Inquiry*. https://doi.org/10.1177/1077800420918889

James, R., Gibbs, B., Whitford, L., Leisher, C., Konia, R., & Butt, N. (2021). Conservation and natural resource management: Where are all the women? In *ORYX* (Vol. 55(6), pp. 860–867). Cambridge University Press. https://doi.org/10.1017/S0030605320001349

Kennedy, M., Maddox, R., Booth, K., Maidment, S., Chamberlain, C., & Bessarab, D. (2022). Decolonising qualitative research with respectful, reciprocal, and responsible research practice: A narrative review of the application of yarning method in qualitative aboriginal and Torres Strait islander health research. *International Journal of Equity Health, 21*(1), 134.

Kimmerer, R. W. (2011). Restoration and reciprocity: The contributions of traditional ecological knowledge. In D. Egan, E. E. Hjerpe, & J. Abrams (Eds.), *Human dimensions of ecological restoration: Integrating science, nature, and culture* (pp. 257–276). Island Press. https://doi.org/10.5822/978-1-61091-039-2_18

Leeson, S., Smith, C., & Rynne, J. (2016). Yarning and appreciative inquiry: The use of culturally appropriate and respectful research methods when working with aboriginal and Torres Strait islander women in Australian prisons. *Methodological Innovations, 9*. https://doi.org/10.1177/2059799116630660

Martin, K., & Mirraboopa, B. (2003). Ways of knowing, being and doing: A theoretical framework and methods for indigenous and Indigenist re-search. *Journal of Australian Studies, 27*(76), 203–214. https://doi.org/10.1080/14443050309387838

Phillips, L., & Bunda, T. (2018). *Research through with and as storying*. Routledge.

Ramsden, I. (1993). Kawa Whakaruruhau: Cultural safety in nursing education in Aotearoa (New Zealand). *Nursing Praxis in New Zealand, 8*(3), 4–10.

Rawluk, A., Ritchie, E., Neale, T., Smith, W., Pascoe, J., Murray, M., Doherty, T., Carter, R., Bourke, M., Falconer, S., Nimmo, D., Price, J., White, M., Bates, P., Wong, N., Nelson, T., & Atkinson, A. (2023, September). Tomorrow's Country: Practice-oriented principles for Indigenous cultural fire research in south-east Australia. 1–16. https://doi.org/10.1111/1745-5871.12596.

Recognition and Settlement Agreement between Dja Dja Wurrung Clans Aboriginal Corporation and the State of Victoria. (2013).

Rose, D. B. (1996). *Nourishing terrains : Australian aboriginal views of landscape and wilderness*. Australian Heritage Commission.

Seneviratne, S. I., Zhang, X., Adnan, M., Badi, W., Dereczynski, C., Di Luca, A., Ghosh, S., Iskandar, I., Kossin, J., Lewis, S., Otto, F., Pinto, I., Satoh, M., Vicente-Serrano, S. M., Wehner, M., & Zhou, B. (2021). Weather and climate extreme events in a changing climate. In V. Masson-Delmotte, P. Zhai, A. Pirani, S. L. Connors, C. Péan, S. Berger, N. Caud, Y. Chen, L. Goldfarb, M. I. Gomis, M. Huang, K. Leitzell, E. Lonnoy, J. B. R. Matthews, T. K. Maycock, T. Waterfield, O. Yelekçi, R. Yu, & B. Zhou (Eds.), *Climate change 2021: The physical science basis. Contribution of working group I to the sixth assessment report of the intergovernmental panel on climate change* (pp. 1513–1766). Cambridge University Press, Cambridge and New York, NY. https://doi.org/10.1017/9781009157896.013

Singh, M., & Major, J. (2017). Conducting indigenous research in Western knowledge spaces: Aligning theory and methodology. *Australian Educational Researcher, 44*(1), 5–19. https://doi.org/10.1007/s13384-017-0233-z

Smith, L. T. (2022). *Decolonising methodologies: Research and indigenous peoples* (3rd ed.). Bloomsbury Academic and Professional.

The Victorian Traditional Owner Cultural Fire Knowledge Group. (2019). *The victorian traditional owner cultural fire strategy*.

Tynan, L. (2021). What is relationality? Indigenous knowledges, practices and responsibilities with kin. *Cultural Geographies*. https://doi.org/10.1177/14744740211029287

Walker, M., Fredericks, B., Mills, K., & Anderson, D. (2014). "Yarning" as a method for community-based Health Research with indigenous women: The indigenous women's wellness research program. *Health Care for Women International, 35*(10), 1216–1226. https://doi.org/10.1080/07399332.2013.815754

Weir, J. K. (2023). Expert knowledge, collaborative concepts, and universal nature: Naming the place of indigenous knowledge within a public-sector cultural burning program. *Ecology and Society, 28*(1). https://doi.org/10.5751/ES-13822-280117

Williamson, B. (2021). *Cultural burning in NSW Challenges and opportunities for policy makers and Aboriginal peoples*. Working Paper No. 139/2021, A. N. U. Centre for Aboriginal Economic Policy Research.

Wilson, S. (2008). *Research is ceremony: Indigenous research methods*. Fernwood Publishing.

Wright, M., Culbong, T., Webb, M., Sibosado, A., Jones, T., Guima Chinen, T., & O'Connell, M. (2023). Debakarn Koorliny Wangkiny: Steady walking and talking using first nations-led participatory action research methodologies to build relationships. *Health Sociology Review*, 1–18. https://doi.org/10.1080/14461242.2023.2173017

Chapter 12
Securing Water Access: Economic Futures for Victoria's Traditional Owners

Cormac Mercer

Abstract The relationship between Aboriginal and/or Torres Strait Islander people and water represents a complex array of connections consolidated across generations and millennia through ancestral First Laws that emphasise custodianship of living waters, reciprocity, and intergenerational equity. Despite this, Aboriginal and/or Torres Strait Islander interests in water have been overlooked by the racialised settler water regime. Aqua nullius has enabled the dislocation and dispossession of water from Aboriginal and/or Torres Strait Islander communities. Within this context, I will seek to explore the opportunities and constraints for Traditional Owner groups seeking to access, use and manage water for economic development in Victoria, and argue that greater water access and use by Traditional Owner groups in Victoria is integral to wealth creation and economic benefits for present and future Aboriginal and/or Torres Strait Islander communities.

Keywords Aqua nullius · Traditional owners · Self-determination · Water policy

12.1 Introduction

Access to, use and management of water by Aboriginal and/or Torres Strait Islander communities cannot be 'decoupled from the relationship with the environment and water resources because Aboriginal water concepts are central to community and kinship relationships' (Marshall & Kirby, 2017).

These connections have been emphasised and reaffirmed through numerous declarations and statements, such as the Boomanulla Statement (2002), the Garma International Aboriginal Water Declaration (2008), the Mary River Statement (2009), the Echuca Declaration (2010), and the Fitzroy River Declaration (2016).

C. Mercer (✉)
Melbourne Law School, University of Melbourne, Parkville, VIC, Australia
e-mail: mercercormac0@gmail.com

M. Kennedy, E. O'Donnell (eds.), *People, Place and Nature in Indigenous-Settler Relations*, Indigenous-Settler Relations in Australia and the World 10, https://doi.org/10.1007/978-3-032-03600-1_12

Each instance strongly advocates for Indigenous rights in the face of the ongoing processes of colonisation, as well as central provisions of the *United Nations Declaration on the Rights of Indigenous Peoples* ('*UNDRIP*').

Aboriginal and/or Torres Strait Islander community water holdings and the water governance systems in place across Australia which govern such holdings, have been referred to as 'abysmally low' (Grafton et al., 2022); 'overlooked' (Whaler, 2023); 'inequitable' (Marshall & Kirby, 2017, p. 94); '[a] racialised regime' (O'Donnell et al., 2022, p. 60); and designed around the 'invisibility' of Aboriginal and/or Torres Strait Islander communities (May, 2022).

It is within this context that the dislocation and dispossession of water rights held by Aboriginal and/or Torres Strait Islander communities has occurred, because of aqua nullius, which 'render[s] invisible existing [Aboriginal and/or Torres Strait Islander communities'] relationships to water' (Marshall & Kirby, 2017, p. 133; Morgan et al., 2004).

In light of aqua nullius, I will seek to explore the opportunities and constraints for Traditional Owner groups seeking to access, use and manage water for economic development in Victoria as a result of the *Water is Life* policy (DELWP, 2022), and argue that greater water access and use by Traditional Owner groups in Victoria is integral to economic enterprise and economic certainty for Aboriginal and/or Torres Strait Islander communities (McAvoy, 2006; Jackson & Langton, 2011). This argument will also explore the inadequacy of water allocation for Traditional Owner groups based on 'cultural flows', given these allocations affirm 'discourses about individual and cultural identities and the maintenance or dissolution of social boundaries' (Strang, 2006, p. 155).

I will argue that the reallocation of water from existing users to Traditional Owner groups, in recognition of their pre-existing rights and the inequities water holdings, is the most straightforward opportunity for Traditional Owners in Victoria to acquire water for economic purposes (O'Donnell, 2021); as McAvoy correctly asserts, 'in a commodified world the only real power to control exploitation is to own the commodity' (McAvoy, 2006, p. 101).

This is an apt moment to both explore and demand the economic rights and interests of Aboriginal and/or Torres Strait Islander communities in water in the Victorian context. The *Advancing the Treaty Process with Aboriginal Victorians Act 2018* (Vic) heralds a unique form of legal relationship in Australia; the potential exists to design 'a changing environment … centred on the idea of [an] equal and ongoing relationship between nations' (McMillan et al., 2020, p. 629).

The Victorian Government's recent *Water is Life* policy will be explored, including the opportunities and constraints it provides for Traditional Owner groups. This exploration will be contextualised through an examination of the federal native title framework, the concept of aqua nullius, and the Victorian legislative framework, all of which provide the structure in which the *Water is Life* policy seeks to operate.

The extent and value of present water holdings held by Traditional Owner groups, in particular in Victoria, revealing the scale of present inequities, will be outlined. This analysis will reveal that entry into the water market by Traditional Owners can overcome the injustice of double dispossession of water for Aboriginal and/or

Torres Strait Islander communities and enable the stimulation and growth of intergenerational wealth (Mooney et al., 2020).

12.1.1 Positionality Statement

I am a proud Gadigal and Dharug man with strong community connections to the Bundjalung community on the north coast of New South Wales.

My deep connection to water stems from my Nan, who has taught me to have fundamental respect for its lifeways.

I am a graduate of Melbourne Law School, where I was fortunate to develop and research ideas around the use and ownership of water in this country, its colonial past and present-day iterations.

It is from this context that I will explore the opportunities and constraints for Traditional Owner groups in Victoria seeking to access, use and manage water for economic development in Victoria, given the recent *Water is Life* policy.

12.1.2 Water Use by Traditional Owner Groups for Cultural Purposes

The aqua nullius concept, or water belonging to no one, forms the core context which the *Water is Life* policy seeks to ameliorate. Aqua nullius has provided the misplaced understanding that First Nations people do not have a body of law which governs the care and management of water (Marshall & Kirby, 2017). As a result, Aboriginal and/or Torres Strait Islander rights to water are not necessarily valued under the common law, 'with statutory interpretation … often in conflict with an Aboriginal property rights paradigm' (Marshall & Kirby, 2017, p. 94). As Marshall & Kirby correctly assert, the depth of Aboriginal customary property rights in water resources requires a 'paradigm shift' to recognise that First Nations characterisation of water sits on shaky ground with the Western legal system (2017, p. 82).

There are clear tensions between existing policies and legislative frameworks which isolate the use of water by Traditional Owner groups to cultural purposes (or 'cultural flows'), which restricts the potential of such water for economic uses by Traditional Owner groups.

For example, s 223(1)(a) of the *Native Title Act 1993* (Cth) restricts the native title rights and interests of communities and/or individuals, where found, to a strict focus on 'traditional law and the traditional customs observed by Indigenous peoples'. Not only does this not provide adequate weight to the concept of dormant and/or revitalised laws, customs and practices, there is limited opportunity for the recognition of rights to utilise water for commercial purposes given that native title solely recognises 'traditional, or pre-sovereignty, interests' (Marshall & Kirby,

2017, p. 79; Macpherson, 2019). Similar impediments exist at a Victorian jurisdiction level contained in the *Traditional Owner Settlement Act 2010* (Vic) ('the *TOS Act*'), as will be explored in Sect. 12.3.2.

The restriction of the water entitlements of Traditional Owner groups to cultural entitlements as a distinct category of use by existing legislative and policy frameworks presents risks for Traditional Owner groups. As will be explored, a pertinent risk is the continued marginalisation of Traditional Owner groups from access to water entitlements for economic development.

12.2 The Extent of Water Holding Held by Aboriginal and/or Torres Strait Islander Communities

The Murray-Darling Basin ('MDB') has been referred to as the 'most sophisticated and longest-running water market in the world' (Ann Wheeler & Garrick, 2020, p. 132). Yet, the scale of inequity in water holdings held by Traditional Owner groups (Hartwig & Jackson, 2020) is a stark indication of the need to reallocate existing water holdings to Traditional Owner groups so as to enable an environment of autonomous and self-determined decision making to foster economic development (Marshall & Kirby, 2017; Jackson & Langton, 2011).

To understand how entry into the water market by Victorian Traditional Owner groups can enable economic certainty and enterprise for Indigenous communities, it is relevant to understand the extent of the disparity in surface and groundwater holdings. Such disparities have led academics to assert that a new form of Aboriginal dispossession is occurring (Sheehan & Small, 2007). A quantitative analysis of the extent of Aboriginal water rights and interests in Australia has 'generally been ignored in research … [enabling] historical marginalisation' (Marshall & Kirby, 2017, p. 152).

The area of the MDB incorporates the lands of more than 55 First Nations groups, who comprise 10% of the MDB population according to Costanza-van den Belt et al. (2022). Evidence indicates that the situation is likely worse when examining data from Victoria in isolation. For example, the volume of water access entitlement in Victoria totals 3,833,008 (ML), yet, the water access entitlement volume held by Traditional Owners sits at 6797 (ML)—less than 0.2 per cent of all water access entitlements in Victoria (DELWP, 2022). In 2021, 1% of the Victorian population identified as Aboriginal and/or Torres Strait Islander (ABS, 2022). Such inequities are further magnified relative to the Indigenous estate, which refers to all land that is held communally by Aboriginal and/or Torres Strait Islander people through diverse forms of land title, being 438 million hectares, or 57% of Australia's total landmass (Jacobsen et al., 2020).

The extent of the inequity with regard to water rights is fundamentally clear (O'Neill et al., 2021).The same analysis by Hartwig et al. (2021), estimated that in 2015–2016 market terms, Indigenous held surface water entitlements in the MDB were valued at approximately A$18.4 million, with groundwater entitlements valued at approximately A$772,800. When combined, these figures equate to 0.12% of the total $16.5 billion market value of all broader entitlements in water within the MDB. In 2018, the Australian Government committed $40 million to acquire access entitlements for Traditional Owner groups in the MDB. In 2022, the value of entitlements in the southern MDB was $AUD30 billion (Aither., 2022), so the funding commitment would only acquire 0.3% of the market value. As of February 2023, none of this funding had been deployed to facilitate acquisitions for Traditional Owner groups (Jasper, 2023). Subsequent research has revealed that Traditional Owner groups have forgone financial opportunities due to this lack of funding deployment, with approximately $11 million in water entitlement foregone (Aither, 2023). This demonstrates the capacity of water holdings to generate potential economic development for First Nations groups as well as underscoring the hurdles First Nations groups face in seeking governments to make good on policy pronouncements.

An exploration of the statutory frameworks at a federal and state level that govern Traditional Owner group's access to water is now necessary to examine the various opportunities and/or constraints to water for economic development by Traditional Owner groups.

12.3 Existing Legal Frameworks

12.3.1 The Native Title Framework

The native title framework is the only generalised legal mechanism at a federal level for the recognition of Indigenous rights to use water (Macpherson, 2019). Native title has been referred to as 'an incomplete response to ongoing Indigenous exclusion from water law frameworks' (Macpherson, 2019).

Though the acknowledgement of native title by the High Court of Australia in *Mabo v Queensland [No 2]* was heralded as a 'significant victory', amendments to native title legislation have 'narrow[ed] the window of opportunity to reclaim … lands and waters' (Marshall & Kirby, 2017, p. 94). For example, s 223(1)(a) of the *Native Title Act 1993* (Cth) restricts the native title rights and interests of communities and/or individuals, where found, to a strict focus on 'traditional law and the traditional customs observed by Indigenous peoples'. Not only does the function of this section not provide adequate weight to the concept of dormant and/or revitalised laws, customs and practices, there is extremely limited opportunity for the recognition of rights to utilise water for commercial purposes given that native title solely recognises 'traditional, or pre-sovereignty, interests' (Marshall & Kirby, 2017; Macpherson, 2019). As a result, communities become divided into those who are native title holders and those who are not, given they are unable to meet the requirements under native title law (Marshall & Kirby, 2017).

There is continued acknowledgement that Western constructs are inadequate in the face of recognising Aboriginal and/or Torres Strait Islander law, including that 'the interpretation of traditional has been criticised as restrictive and technical' (Australian Law Reform Commission, 2015). Reliance on the existing native title framework alone is insufficient, given it is 'an ad hoc and partial response to [I]ndigenous land and resource rights in Australia' (Macpherson, 2017). The native title framework was not intended to be a comprehensive solution to Indigenous land and resource rights, as then Prime Minister Keating stated:

> [t]he Mabo decision gives little more than a sense of justice to those Aboriginal communities whose native title has been extinguished or lost without consultation, negotiation or compensation. Their dispossession has been total, their loss has been complete. (Keating, 1993)

As a result, it is clear that the current native title framework operates as a significant constraint for Traditional Owner groups in Victoria, which neglects the aspirations and livelihood opportunities of Traditional Owner groups, in light of the commercial value of water (Poh-Ling & Jackson, 2013); a fundamental 'paradigm shift' is therefore required to transcend the existing inequities in water holdings held by Traditional Owner groups so as to enable the self-determined fulfilment of the aspirations and opportunities which Traditional Owners seek (Marshall & Kirby, 2017).

An examination of the opportunities and/or constraints within the Victorian statutory context in regard to water access, use and management by Traditional Owner groups is now warranted.

12.3.2 Victorian Legislative Framework

The distribution of legislative and executive authority in Australia provides the states 'considerable legal authority' (Hobbs & Williams, 2019); this has been referred to as a 'very Australian framework of governance … [t]he cheerful term for it is "co-operative federalism" [though a] grimmer view would see it as a cockpit for interstate rivalrous self-interests' (Department of Environment and Water, 2019). *Water is Life*, at a Victorian state level, seeks to contribute to the broader National Agreement on Closing the Gap target relating to Inland Waters, reflecting negotiations underway between the Commonwealth, states and territories, and the Coalition of Peaks (National Indigenous Australians Agency, 2023). Victoria has a range of legal and policy frameworks which are utilised to acknowledge and recognise the rights of Aboriginal and/or Torres Strait Islander citizens. For example, the *Charter of Human Rights and Responsibilities Act 2006* ('the *Charter*') delineates specific obligations on public authorities, such as water corporations, to uphold the human rights of individuals when making decisions, including the cultural rights of Aboriginal Victorians. The *Charter* also recognises explicitly the distinct cultural rights of Aboriginal people, as well as their 'spiritual, material and economic relationship' to land and waters.

The TOS Act outlines a process of developing both recognition and settlement agreements between Victorian Traditional Owner groups in Victoria and the State, in which the State 'treats Aboriginal Victorians as a polity' (McMillan et al., 2020). However, as a method for enabling greater access, use and management of water for economic development of Traditional Owner groups 'it is far from satisfactory … [and does not confront the] era of aqua minima rights for Victoria's Traditional Owners' (O'Bryan, 2016, p. 593).

Despite Section 9 of the TOS Act outlining a recognition process for the rights and interests of Victorian Traditional Owner groups, which are akin to native title rights, the TOS Act excludes exclusive water rights. For Traditional Owner access, use and management of water for economic development, there is 'little to separate' both the TOS Act regime and the native title regime at a federal level; both systems focus heavily on rights to take and use water for extremely limited purposes ('rights which are not dissimilar to those [Traditional Owner group members] have as ordinary members of the public') (O'Bryan, 2016, p. 592).

Section 8A of the *Water Act 1989* (Vic) ('Water Act'), as inserted into it by the TOS Act, holds that where a Traditional Owner group has a traditional owner agreement, a person forming a part of that group has the right to take and use water on the land that is subject to the agreement. As a result, section 8A of the Water Act is rendered largely symbolic; the provision does not delineate any positive obligations on behalf of water authorities or other water stakeholders to incorporate Traditional Owner considerations. However, section 92 of the Water Act outlines that in the performing of its functions, a water corporation must consider opportunities to provide for Aboriginal cultural values and use of waterways.

Though this does overcome what Moggridge and Thompson (2021, p. 11) have referred as 'the inclusion of Indigenous values in water planning [as] rare' it continues to narrowly restrict Traditional Owner access, use and management of water away from economic development. O'Bryan (2018, p. 578) correctly argues that the operative effect of this provision of the Water Act is that it 'quarantines Traditional Owner water rights from the rest of the Act, in effect allowing them to be ignored'.

This lack of recognition and participation reflects the fact that the development of the Water Act excluded Traditional Owner groups. Ultimately, this lack of consultation led to a bias in favour of those groups with already existing water allocations, and consolidated the structural dispossession of Traditional Owner groups in Victoria, given the dispossession of land and historical allocation of water rights was linked to land tenure (O'Bryan, 2018).

These processes have had considerable consequential effects on the modern capacities of Traditional Owner groups to manage water for economic development in Victoria (Morgan et al., 2004). Such processes underscore the reasoning as to why a 'paradigm shift' is required in order to enable economic enterprise and economic certainty for Aboriginal and/or Torres Strait Islander communities in Victoria through water access, use and management (Marshall & Kirby, 2017, p. 82).

12.4 Future Opportunities: *Water Is Life* and the Victorian Treaty Process

Traditional Owner rights to water are 'coexisting with, and contesting, the state's asserted sovereignty to water' (Nikolakis & Grafton, 2021, p. 14). The Victorian *Water Is Life* policy is an example of how the State has sought to engage with Traditional Owner groups, in the context of the access to, use and management of water, within a treaty context. The policy sets out the opportunities for Traditional Owners to access and manage water for spiritual, cultural and environmental purposes.

12.4.1 Treaty

As a result of the *Advancing the Treaty Process with Aboriginal Victorians Act 2018* (Vic) the potential exists to design 'a changing environment … centred on the idea of [an] equal and ongoing relationship between nations' (McMillan et al., 2020, p. 629). In 2018 a state-wide Aboriginal Representative Body was established to represent the voice of First Peoples in the Treaty process. As part of this, the Yoorrook Justice Commission was established to inquire into historic and contemporary injustices experienced by First Peoples across all areas of social, political and economic life, and the relationship between historical injustices and ongoing systemic injustices experienced by First Nations communities in Victoria presently. In 2022, the Treaty Negotiation Framework was established which outlines who can enter Treaty negotiations, how negotiations are conducted, and what subject matter Treaty negotiations in Victoria can cover. Treaty-making in Victoria will include the negotiation of both a Statewide Treaty and local Traditional Owner Treaties. This reflects that self-determination can be exercised by all First Peoples in Victoria collectively, and also by individual Traditional Owner groups.

Clause 25 of the agreed Treaty Negotiation Framework between the Victorian Government and the First Peoples' Assembly of Victoria indicates that '[t]here are no matters that cannot or must not be agreed in the course of Treaty negotiations' (First Peoples' Assembly of Victoria & the State of Victoria, 2022). Clause 25.2 further outlines that negotiations may encompass 'land and water justice' as one of the matters that State and Traditional Owner treaty negotiators can consider. It is clear the reallocation of water holdings for Traditional Owners for commercial purposes and associated economic development can form part of the ongoing negotiation process.

12.4.2 Water Is Life

Water is Life is a unique policy document which seeks to provide a pathway to increase decision-making for Traditional Owners in the management of water and access to water. *Water is Life* 'recognises the central importance of the relationship between people and Country, embodied in the cultural water paradigm' (Department of Environment, Land and Water, 2022, p. 13). *Water is Life* provides Traditional Owners with the opportunity to submit a statement to be shared as part of the policy document, to express Traditional Owners' 'cultural and water-related values, goals, aspirations, outcomes, and any other relevant information deemed important by each Nation' (DELWP, 2022, p. 60).

From the outset, *Water is Life* recognises and acknowledges that '[c]olonisation and the establishment of the State of Victoria has had long-lasting, far-reaching and intergenerational consequences, including the dispossession of First Nations and Traditional Owners of their Country' (Department of Environment, Land and Water, 2022, p. 5). This is a marked departure for policy statements to make, at any jurisdictional level, given that Australia possesses the least formal recognition of water rights of 'any of the colonised countries' (Moggridge & Thompson, 2021, p. 11).

As will be explored, *Water is Life* provides twelve outcomes, including:

1. recognition of waterways and their surrounding lands as living, integrated, natural, and cultural entities in law, with Traditional Owners recognised as a unique 'voice' of the living entities;
2. Traditional Owners can enter formal partnership agreements with the water sector that increase the capacity and resources available to Traditional Owners to make and implement decisions about water landscape management on their Country;
3. Traditional Owners have an increased role in determining how environmental water is used for the purpose of healing Country;
4. increased involvement of Traditional Owners in public land management;
5. recognition of Traditional Owners as waterway managers for specific locations;
6. recognition of Traditional Owners as environmental water holders;
7. water is returned to Traditional Owner groups across Victoria through the issue of water entitlements for their self-determined use;
8. access to land will not be a barrier to Traditional Owners applying for or holding water entitlements;
9. Traditional Owners will be funded to pay any fees and charges associated with their water entitlements unless use is purely commercial;
10. where Traditional Owners become purely commercial users of water, they will become responsible for an increasing portion of fees and charges over time to full cost recovery;
11. Traditional Owners can hold and manage water in culturally appropriate ways; and
12. Traditional Owners have oversight of the implementation of *Water is Life*.

12.4.2.1 Opportunities

Consideration must now turn to a focus on the application of Outcome eight of *Water is Life,* which states that '[a]ccess to land will not be a barrier to Traditional Owners applying for or holding water entitlements' (DEWLP, 2022, p. 13). This is a positive step, largely in recognition of the ongoing effects of Australia's colonial history, which saw the exclusion of Aboriginal and/or Torres Strait Islander people from holding land at the time of water entitlement distribution. As a result, the separation of land and water entitlements meant that First Nations people were excluded from possessing the capacity to hold water (Mooney et al., 2020). These deliberate policy directions form the basis for why First Nations people have been deprived the capacity to 'accumulate capital and wealth that makes operating financially 'viable' properties—as defined in accordance with narrow neoclassical economics and required by settler-colonial land rights legislation or deed of grant—extremely difficult' (Hartwig et al., 2021, p. 54).

The scope of the action in which *Water is Life* seeks to attempt to remedy this issue relates to the provision of mechanisms within the existing water framework in Victoria to enable access to water for Traditional Owners 'for cultural purposes'. Parts of the water framework in Victoria that are the constraints outcome eight seeks to overcome relate to section 51 of the *Water Act*, which involves the provision of water licences attached to land. As a result, for the provision of a water licence, a Traditional Owner group would have to require ongoing land access, either through a lease or ownership (O'Donnell et al., 2022). To overcome such constraints, *Water is Life* states that the Victorian Government will seek to work with water corporations to 'streamline existing approval requirements and reduce red tape' for Traditional Owners in Victoria who are seeking to apply for a section 51 licence.

There is a strong desire to address First Nations disadvantage through the access to, use and management of water for economic development by Traditional Owner groups, through the amelioration of constraints Traditional Owner groups face (McAvoy, 2006). For example, *Water is Life* outlines in Outcome nine, that Traditional Owners will receive funding from the State to pay fees and charges which are associated with their water entitlements. This is a reasonable and proportionate policy action which can assist Traditional Owner groups in overcoming practical constraints to the access to, use and management of water. Ongoing and annual fees can be burdensome for Traditional Owner groups, with fee structures varying for use, maintenance, storage, treatment and delivery (O'Donnell et al., 2022). This aligns with existing literature which indicates that participation and transaction costs are higher for particular resource owners (Grillos, 2017; Ann Wheeler & Garrick, 2020). However, a qualification is placed on the provision of such funding by the State, in that such fees and charges will be funded 'unless [the] use [of water entitlements by Traditional Owner groups] is purely commercial' (DELWP, 2022, p. 44). This qualification requires further examination within the broader context of the constraints which *Water is Life* possesses. As will be discussed, a significant limitation of outcome nine is that, arguably, *Water is Life* possesses implied restrictions, such that where Traditional Owner groups seek to use water for commercial purposes, fees will not be ameliorated by the policy.

12.4.2.2 Constraints

There is a significant cause and effect to the function of Outcome nine; specifically, the sidelining of an enabling environment for Traditional Owner groups in Victoria to access, use and manage water for commercial purposes. This further delays the elevation of the rights of First Nations groups to water for economic development on the water reform agenda (First Peoples' Water Engagement Council, 2012).

Within this context, examining the Nation Statements of Traditional Owner groups is warranted. For example, the Nation Statement of the Bunurong Land Council Aboriginal Corporation outlines the crux of this issue:

> The economic value of water is a crucial aspect of self-determination for Bunurong people … Bunurong Elders have noted that the current market value of water and its management have disenfranchised the Aboriginal people of Australia and continue to compound the effects of colonialism … Self-determination regarding water use and allocation is crucial to rebuilding and empowering the Bunurong community through access, employment and actual control of resources. (Bunurong Land Council Aboriginal Corporation, published in DELWP, 2022, p. 89)

What is required, as Boelens (2009, pp. 316–317) describes, is the 'establish[ment of] the necessary conditions under the law (access to water and autonomy for management), in order [for Traditional Owner groups] to stay out of the way of the law'. As delineated by the above Bunurong Nation Statement, and countless other Nation Statements provided, access to water for economic purposes is considered essential. However, whilst *Water is Life* acknowledges Traditional Owner groups access to water for economic development, there is minimal evidence to suggest it has contributed to an enabling environment for Traditional Owner groups in this manner.

The lack of policy breadth to move beyond acknowledgement and enable Traditional Owner groups access to, use and management of water for economic development is not unique to *Water is Life*; similarly, the MDB *2020 Basin Plan Evaluation* outlines 'opportunities to strengthen First Nations' water access for social and cultural outcomes' without sufficient interface with the idea of enabling the economic or commercial interests of Traditional Owner groups (Murray-Darling Basin Authority, 2020).

Within the context established above, and the Nation Statements which are brought to the foreground by *Water is Life*, the question remains regarding the allocation of water and how such an approach can satisfy the 'diverse and evolving water management objectives [of Traditional Owner groups] or must [Traditional Owner groups] rely on an obscure and restrictive entitlement that privileges precolonial practices' (Jackson & Langton, 2011, p. 123).

Yet, *Water is Life*, whilst acknowledging that the allocation of water for economic development enhances economic well-being and increases both autonomy and self-determination, disavows the reallocation of existing water allocations for

Traditional Owner groups: '[t]he Victorian Government is not proposing to buy water from existing water holders for the purpose of returning water to Traditional Owners' (Department of Environment, Land, Water and Planning, 2022). What is required then, though 'highly controversial', is a policy environment which centres on concepts of both restorative and distributive justice within the context of the 'historical denial of [I]ndigenous water rights' (Jackson & Langton, 2011, p. 117).

Outcome 7 of *Water is Life* provides a commitment to developing a framework for government to fund the issuing of water entitlements for the self-determined use by Traditional Owner groups. However, there is no evidence of this framework being implemented and, as a result, no funding for Traditional Owner groups to acquire water holdings. However, at an Australian federal level, the $100 million Murray-Darling Basin Aboriginal Water Entitlements Program has recently commenced the purchase phase of the program, following extensive consultations with MDB First Nations groups. This federal process reflects the continued discourse that 'acquiring water on the market is the only pathway for substantial surface water returns to Traditional Owners' (O'Donnell et al., 2023, p. 24).

Where water is returned to Traditional Owners, per *Water is Life*, such allocations will be based on unallocated water, supplies by agreement and donations which are often limited in scope owing to small volumes of unallocated water remaining available (O'Donnell et al., 2023). Whether the quality of such unallocated water is fit for purpose has been raised, given that salinity levels may be higher in unallocated water, rendering the intended use of the water by Traditional Owner groups void (O'Donnell et al., 2022). The distribution of unallocated water also directly conflicts with the aspiration of Traditional Owner groups as delineated by the Echuca Declaration to possess water entitlements of both adequate quantity and quality given the fit for purpose concerns.

Though *Water is Life* expressly notes that its policy is not to purchase water from existing water holders for Traditional Owners for any purposes, it does speak to the Treaty between the State and Traditional Owners by outlining that this policy is 'not intended to rule out Traditional Owners seeking any future funding support for economic development purposes or to pre-empt any future Treaty negotiations'.

Yet, there remains a failure of *Water is Life* in its present iteration to confront what has been referred to as the 'most viable, immediate pathway' for Traditional Owners in Victoria to access water, that being 'via ent[ry into] the water market' (Mooney et al., 2020, p. 2). The adoption of this approach can fulfil various declarations, including the Echuca Declaration, which seeks water entitlements allowing for autonomy and which are:

> legally and beneficially owned by Indigenous nations of a sufficient and adequate quantity and quality to improve the spiritual, cultural, environmental, social and economic conditions of [Traditional Owner groups]. This is our inherent right. (Murray Lower Darling Rivers Indigenous Nations, 2010)

This reflects the commodified environment in which water holdings exist in, and therefore 'the only real power [for Traditional Owner groups in Victoria] is to own the commodity … so they can make the decision to use the water [by] keeping it in the river or by extraction' (McAvoy, 2006, p. 101). *Water is Life* can overcome the status quo policy environment by adopting the immediate pathway for Traditional Owners in Victoria to access water, via entry into the water market. Adoption of this framework could lead to an enabling environment which 'provide[s] opportunities for the development of a vital economy' to promote long-term community benefits (Marshall & Kirby, 2017, p. 161; Quentin Grafton et al., 2015).

12.5 Conclusion

This paper has sought to outline the existing opportunities and constraints in Victoria for Traditional Owner groups seeking to access, use and manage water for economic development within the context of the recent *Water is Life* policy. Opportunities which emerge from *Water is Life* include the ability for Victorian Traditional Owner groups to outline their specific 'cultural and water-related values, goals and aspirations' through the development of Nation Statements which are unique to each Traditional Owner group. Yet, despite this, Traditional Owner groups remain 'marginalised and misrepresented' relative to the access to, use and management of water for economic development in Victoria (Jobling, 2023). Constraints which emerge from *Water is Life* include the provision of holdings to Traditional Owner groups of unallocated water, of which evidence indicates that such water can be both difficult to access for Traditional Owner groups and be of an inadequate quality (O'Donnell et al., 2023).

References

Aither. (2023, February 23). *Water portfolio investigation* [PowerPoint]. Aither. https://mldrin.org/wp-content/uploads/2023/08/WEB_20230829-MLDRIN-Slide-Deck-FINAL-STC.pdf

Aither. (2022). *Water markets report: 2021–22 review and 2022–23 outlook*. Aither, Pty Ltd.

Ann Wheeler, S., & Garrick, D. E. (2020). A tale of two water markets in Australia: Lessons for understanding participation in formal water markets. *Oxford Review of Economic Policy, 36*(1), 132–153. https://doi.org/10.1093/oxrep/grz032

Australian Bureau of Statistics. (2022). *Victoria: Aboriginal and Torres Strait Islander population summary*. Retrieved April 1, 2025, from https://www.abs.gov.au/articles/victoria-aboriginal-and-torres-strait-islander-population-summary

Australian Law Reform Commission. (2015). *Connection to Country: Review of the Native Title Act 1993* (126).

Boelens, R. (2009). The politics of disciplining water rights. *Development and Change, 40*(2), 307–331. https://doi.org/10.1111/j.1467-7660.2009.01516.x

Costanza-van den Belt, M., Rao, R., Colloff, M. J., Pittock, J., & Moggridge, B. (2022). Watering of wetlands on Indigenous country in the Murray–Darling basin, Australia. *Marine and Freshwater Research, 73*(12), 1413–1425. https://doi.org/10.1071/MF22155

DELWP (Department of Environment, Land, Water and Planning). (2022). *Water is Life: Traditional Owner access to water roadmap*. Victoria State Government. https://www.water.vic.gov.au/our-programs/aboriginal-water-program/water-is-life-roadmap

Department of Environment and Water. (2019). *Murray-Darling Basin Royal Commission report*. South Australia Government.
First Peoples' Assembly of Victoria & the State of Victoria. (2022). *Treaty negotiation framework*. https://treatyauthority.au/wp-content/uploads/2024/06/Treaty-Negotiation-Framework.pdf
First Peoples' Water Engagement Council. (2012). *Options for an indigenous economic water fund (IEWF)*. Paper presentation. Indigenous Water Justice Symposium. https://scholar.law.colorado.edu/cgi/viewcontent.cgi?article=1003&context=indigenous-water-justice-symposium
Grafton, Q., Taylor, K., & Poelina, A. (2022, December 22). The lie of aqua nullius, 'nobody's water', prevails in Australia. Indigenous water reserves are not enough to deliver justice. *The Conversation*. https://theconversation.com/the-lie-of-aqua-nullius-nobodys-water-prevails-in-australia-indigenous-water-reserves-are-not-enough-to-deliver-justice-195557#:~:text=Aqua%20nullius%20denies%20the%20existence,deep%20reciprocal%20economy%
Grillos, T. (2017). Economic vs non-material incentives for participation in an in-kind payments for ecosystem services program in Bolivia. *Ecological Economics, 131*, 178–190. https://doi.org/10.1016/j.ecolecon.2016.08.010
Hartwig, L., & Jackson, S. (2020). *The status of Aboriginal water holdings in the Murray-Darling Basin: ARI report No. 2020/004*. Australian Rivers Institute, Griffith University. http://hdl.handle.net/10072/400302
Hartwig, L. D., Jackson, S., Markham, F., & Osborne, N. (2021). Water colonialism and Indigenous water justice in South-Eastern Australia. *International Journal of Water Resources Development*, 1–34. https://doi.org/10.1080/07900627.2020.1868980
Hobbs, H., & Williams, G. (2019). Treaty-making in the Australian Federation. *Melbourne University Law Review, 43*(1), 218.
Jackson, S., & Langton, M. (2011). Trends in the recognition of Indigenous water needs in Australian water reform: the limitations of 'cultural' entitlements in achieving water equity. *Journal of Water Law, 22*(2). https://www.researchgate.net/publication/287676542_Trends_in_the_recognition_of_indigenous_water_needs_in_Australian_water_reform_The_limitations_of_'cultural'_entitlements_in_achieving_water_equity
Jacobsen, R., Howell, C., & Read, S. (2020). *Australia's Indigenous land and forest estate: Separate reporting of Indigenous ownership, management and other special rights*. Australian Bureau of Agricultural and Resource Economics and Sciences.
Jasper, C. (2023). $9m pledge to advance First Nations rights in Murray-Darling raises concern over unfulfilled commitments. *ABC News*. Retrieved April 1, 2025, from https://www.abc.net.au/news/2023-04-29/first-nations-groups-urge-progress-on-water-delivery/102273040
Jobling, K. (2023). *Inspector-general of water compliance's (IGWC) regulatory policy discussion paper*. Murray Lower Darling Rivers Indigenous Nations. https://www.igwc.gov.au/sites/default/files/2023-08/mldrin-regulatory-policy-submission.pdf
Keating, P. (1993). *Parliamentary debates*. House of Representatives.
Macpherson, E. (2017). Beyond recognition: Lessons from Chile for allocating Indigenous water rights in Australia. *University of New South Wales Law Journal, 40*(3). https://doi.org/10.53637/yqwy6573
Macpherson, E. J. (2019). *Indigenous water rights in law and regulation*. Cambridge University Press.
Marshall, V., & Kirby, M. D. (2017). *Overturning aqua nullius: Securing Aboriginal water rights*. AIATSIS Press.
May, N. (2022). Indigenous fight for 'water justice' intensifies as Victoria hands back Murray-Darling entitlement, 6 March. *The Guardian*. Retrieved April 1, 2025, from https://www.theguardian.com/australia-news/2022/mar/06/indigenous-fight-for-water-justice-intensifies-as-victoria-hands-back-murray-darling-entitlement
McAvoy, T. (2006). Water—Fluid perceptions. *Transforming Cultures eJournal, 1*(2). https://doi.org/10.5130/tfc.v1i2.262
McMillan, M., Foster, D., Genovese, A., McVeigh, S., & Tehan, M. (2020). Obligations of conduct: Public law—Treaty advice. *Melbourne University Law Review, 44*(2). https://yumi-sabe.aiatsis.gov.au/sites/default/files/outputs/2024-02/McMillan-et-al-442-Advance-v2.pdf

Moggridge, B. J., & Thompson, R. M. (2021). Cultural value of water and western water management: An Australian Indigenous perspective. *Australasian Journal of Water Resources, 25*(1), 4–14. https://doi.org/10.1080/13241583.2021.1897926

Mooney, W., Hartwig, L., & O'Donnell, E. (2020). *Submission to Murray-Darling Basin water markets inquiry 2019–2021*. Australian Competition and Consumer Competition.

Morgan, M., Strelein, L., & Weir, J. (2004). *Indigenous rights to water in the Murray Darling Basin: In support of the Indigenous final report to the Living Murray Initiative*. Paper presentation. Australian Institute of Aboriginal and Torres Strait Islander Studies.

Murray-Darling Basin Authority. (2020). *The 2020 Basin Plan evaluation*. Commonwealth of Australia. https://www.mdba.gov.au/water-management/basin-plan/basin-plan-evaluations/2020-basin-plan-evaluation

National Indigenous Australian Agency. (2023). *Outcome 15—Aboriginal and Torres Strait Islander people maintain a distinctive cultural, spiritual, physical and economic relationship with their land and water*. https://www.niaa.gov.au/2023-commonwealth-closing-gap-implementation-plan/delivering-outcomes-and-targets/outcome-15

Nikolakis, W., & Grafton, R. Q. (2021). Law versus justice: The strategic Aboriginal water reserve in the Northern Territory, Australia. *International Journal of Water Resources Development*, 1–19. https://doi.org/10.1080/07900627.2021.1882406

O'Bryan, K. (2016). More aqua nullius? The *Traditional Owner Settlement Act 2010* (Vic) and the neglect of Indigenous rights to manage water resources. *Melbourne University Law Review, 40*(2), 593.

O'Bryan, K. (2018). *Indigenous rights and water resource management: Not just another stakeholder*. Routledge.

O'Donnell, E. (2021). *Water access for Aboriginal economic development discussion paper*. Murray Lower Darling Rivers Indigenous Nations, Federation of Victorian Traditional Owner Corporations, Victoria State Government, Department of Environment, Land, Water and Planning.

O'Donnell, E., Jackson, S., Jama, K., Nelson, R., Cubillo, E., & Murray Lower Darling Rivers Indigenous Nations. (2023). *Caring for country: Overcoming the twin legacies of terra nullius and aqua nullius*. Yoorrook Commission.

O'Donnell, E., Jackson, S., Langton, M., & Godden, L. (2022). Racialized water governance: The 'hydrological frontier' in the Northern Territory. *Australia. Australasian Journal of Water Resources, 26*(1), 59–71. https://doi.org/10.1080/13241583.2022.2049053

O'Neill, L., Thorburn, K., Riley, B., Maynard, G., Shirlow, E., & Hunt, J. (2021). Renewable energy development on the Indigenous estate: Free, prior and informed consent and best practice in agreement-making in Australia. *Energy Research & Social Science, 81*, 102252. https://doi.org/10.1016/j.erss.2021.102252

Poh-Ling, T., & Jackson, S. (2013). Impossible dreaming—Does Australia's water law and policy fulfil Indigenous aspirations? *Environmental and Planning Law Journal, 30*(2), 149.

Quentin Grafton, R., Horne, J., & Wheeler, S. A. (2015). On the marketisation of water: Evidence from the Murray-Darling Basin. *Australia. Water Resources Management, 30*(3), 913–926. https://doi.org/10.1007/s11269-015-1199-0

Sheehan, J., & Small, G. (2007). *Aqua Nullius* [Paper presentation]. 13th pacific-rim real estate society conference, Fremantle, Western Australia. https://www.prres.org/uploads/796/1600/Sheehan_Small_Aqua_Nullius.pdf

Strang, V. (2006). Substantial connections: Water and identity in an English cultural landscape. *Worldviews: Global Religions, Culture, and Ecology, 10*(2), 155–177. https://doi.org/10.1163/156853506777965820

Whaler, J. (2023). Aboriginal rights and interests in our waters have been overlooked. *National Indigenous Times*. Retrieved April 1, 2024, from https://nit.com.au/07-09-2023/7571/murray-darling-basin-plan-not-ideal-nswalc-aboriginal-rights-and-interests-in-our-waters-have-been-overlooked#:~:text=The%20physical%20and%20spiritual%20health,Murray%20Darling%20Basin%20

Chapter 13
First Peoples, Living Waters Cultural Exchange

Melissa Kennedy, Erin O'Donnell, Amy McCoy, Nicky Hudson, and Ben Muir

Abstract The *First Peoples, Living Waters* project is a knowledge and cultural exchange that responds to ongoing injustice caused by the denial of Indigenous water rights and jurisdiction over water in Australia. It brings Indigenous Peoples from three settler colonial states (Australia, Aotearoa New Zealand, and the US) into conversation to explore how treaty and agreement-making can lead to water law reform in settler colonial states. In the context of climate change, these conversations are especially urgent: as their waterways sicken under climate change, this directly affects Indigenous Peoples' wellbeing. Increasing water scarcity is already being used as an excuse to prevent water returns to Indigenous Peoples. This chapter shares insights from the cultural and knowledge exchange, presented as a conversation between the authors.

Keywords Indigenous · Water · Cultural exchange · Knowledge sharing · Aqua nullius · Settler colonial

M. Kennedy (✉)
School of Social Science, Monash University, Clayton, VIC, Australia
e-mail: melissa.kennedy1@monash.edu

E. O'Donnell
Melbourne Law School, University of Melbourne, Parkville, VIC, Australia
e-mail: erin.odonnell@unimelb.edu

A. McCoy
Fluent Freshwater Insights, Bend, OR, USA
e-mail: amy@fluentfreshwaterinsights.com

N. Hudson
Gunditj Mirring Traditional Owner Aboriginal Corporation, Heywood, VIC, Australia
e-mail: nicky@gunditjmirring.com

B. Muir
Dja Dja Wurrung Clans Aboriginal Corporation, Bendigo, VIC, Australia
e-mail: ben.muir@djadjawurrung.com.au

M. Kennedy, E. O'Donnell (eds.), *People, Place and Nature in Indigenous-Settler Relations*, Indigenous-Settler Relations in Australia and the World 10, https://doi.org/10.1007/978-3-032-03600-1_13

13.1 Introduction

In settler colonial states, such as Australia, Aotearoa New Zealand, and the USA, water laws largely fail to acknowledge Indigenous Peoples' sovereign rights to water (Leonard et al., 2023; Robison et al., 2018). The doctrine of discovery that dispossessed Indigenous Peoples of their lands also stole their waters (Miller et al., 2010). Aqua nullius is the assumption that water belonged to no one and no one belonged to water when the British arrived, and the consequent failure to recognise Indigenous Peoples' sovereign rights and responsibilities to water (Marshall, 2017). In Australia, this erroneous assumption is embedded within the water law statutes of each Australian state and territory, which vest rights to control water in the Crown, attempting to erase the millennia of reciprocal relationships between waterscapes and people embedded in Aboriginal laws. In all settler colonies, the failure to address aqua nullius undermines the legitimacy and sustainability of settler water laws (O'Donnell, 2023; O'Donnell et al., 2023a).

In Aotearoa New Zealand and the USA, treaties and other agreements between Indigenous Peoples and the state have provided Indigenous Peoples with (some) water rights and (some) power and authority to care for water (Curley, 2019; Ruru, 2018; Tarlock, 2010). Although no settler state has delivered full water justice to Indigenous Peoples, the USA and Aotearoa New Zealand highlight the multitude of pathways to water access and authority, as well as the transformative potential of treaty and agreement-making dialogues between the settler state and Indigenous Peoples to reshape water law and management.

In Aotearoa New Zealand, a treaty dispute settlement between Whanganui iwi and the Crown resulted in the recognition of the Whanganui River as Te Awa Tupua, a legal person. In particular, Te Awa Tupua represents an attempt to recognise and unite 'multiple ontologies of water' to improve both the legitimacy and the sustainability of water management in Aotearoa New Zealand (Morris & Ruru, 2010; Yates et al., 2017). However, the experience of Whanganui iwi also highlights the dangers of an incomplete understanding of Indigenous water ontologies, in which foundational differences can become obscured by the existing setter state water laws (Takacs, 2021).

In the USA, Tribes hold approximately 25% of water rights in the Colorado River, although their access to this water is constrained by ongoing legal disputes (Stern, 2023). Tribes with recognised water rights are playing a pivotal role in drought management in the Colorado River by selling their water rights to towns and cities, and they are slowly being enabled to play a role in water planning (Siegler, 2023). In the Klamath River basin, Tribal leadership, including Yurok and Karuk Tribes, in collaboration with environmental NGOs and settler agencies, has led to the largest dam removal in the USA, opening over 400 miles to spawning salmon for the first time in a century (Sherriff, 2024).

Despite the lack of a treaty, Indigenous Peoples in Australia have been increasingly influential in shaping settler water law and policy. In 2007, the Murray Lower Darling Rivers Indigenous Nations (MLDRIN) confederation launched the Echuca

Declaration, which reaffirmed Indigenous sovereignty over lands and waters, and defined 'cultural flows' as 'water entitlements that are legally and beneficially owned by the Indigenous Nations of a sufficient and adequate quantity and quality to improve the spiritual, cultural, environmental, social and economic conditions of those Indigenous Nations'. We note that this description by MLDRIN, and its use of terms such as water entitlements, was deliberately written in such a way as to speak to settler colonial water bureaucrats. In a rejection of aqua nullius, MLDRIN members also laid claim to these flows as the 'inherent right' of First Nations (MLDRIN, 2007). Waterways are recognised as ancestral beings with a right to life (Martuwarra RiverOfLife et al., 2021). There is growing acceptance of the idea that 'Indigenous Peoples must be fully involved in source water and watershed protection planning and operational processes including controlling Indigenous water licenses and fair allocation policies and practices' (Garma, 2008). In 2018, the Australian National Cultural Flows Research Project (a project designed and led by peak Indigenous organisations such as MLDRIN, NAILSMA and the National Native Title Council) identified three pathways to water justice: (1) increasing Aboriginal ownership of water rights; (2) increasing Aboriginal influence in water landscapes; and (3) transforming the foundations of the relationship between the settler state and Indigenous Peoples (Nelson et al., 2018).

Treaty and agreement-making has enormous transformative potential because it can finally formally acknowledge Indigenous sovereignty and legal pluralism in Australia. In Victoria, the Treaty process creates an opportunity, for the first time since colonisation, to develop a truly legitimate water law framework that delivers ecologically sustainable water management grounded in the evidence-based protocols and water laws developed by Indigenous Peoples over tens of thousands of years (Marshall, 2017; O'Donnell, 2023). The question facing Indigenous Peoples in Victoria is: what can we learn from the experience of Indigenous Peoples in other settler colonies about the role of treaty in transforming settler water laws and overturning aqua nullius?

The domestication of treaties with Indigenous Peoples in settler colonial contexts can leave Indigenous Peoples isolated and at the mercy of the settler state legal system, whose legitimacy they contest (Bhatia, 2021). One way to combat this isolation and disempowerment is by building relationships between Indigenous Peoples in other settler colonial states. In general, this knowledge and cultural exchange process is well-established across multiple venues, including the United Nations, but so far there has been only a limited focus on water.

13.1.1 *First Peoples, Living Waters: A Cultural Exchange*

The *First Peoples, Living Waters* Cultural Exchange ('the Exchange') brings together Traditional Owners from Australia, Māori from Aotearoa New Zealand and Tribes from the USA to meet and share knowledge on the care and management of water. Participants in the Exchange include representatives from:

- Gunditjmara Nation representatives from Gunditj Mirring Traditional Owner Aboriginal Corporation (Australia)
- Tati Tati First Nations (Australia)
- Taungurung Land and Waters Council (Australia)
- Eastern Maar Aboriginal Corporation (Australia)
- Wurundjeri Woi-wurrung Cultural Heritage Aboriginal Corporation (Australia)
- Wotjobaluk First Nations (Australia)
- Whanganui iwi (Aotearoa New Zealand)
- Pipa Aha Macav/Fort Mojave Tribe (USA)
- Navajo Nation (USA)
- Hopi Nation (USA)
- Havasupai Tribe (USA)

The Exchange has developed through longstanding relationships of trust, accountability, and partnership, strengthened through monthly online dialogues. The Exchange facilitates on-Country learning, which enables much deeper, place-based knowledge sharing between Indigenous Peoples. In July 2023, Traditional Owners from Victoria (Australia) and Whanganui iwi travelled to the USA, where they met with representatives from US Tribes in Arizona, learning about the Colorado River and Tribes' water rights. In November 2023, representatives from Navajo Nation, the Fort Mojave Tribe and Whanganui iwi were hosted by Traditional Owners in Victoria, meeting to discuss Indigenous water knowledge, laws, ceremony, and the possibilities and pitfalls of treaty negotiations with the setter state.

The Exchange builds on a growing international dialogue between Indigenous Peoples on the care and management of water (Leonard et al., 2023). This dialogue is both knowledge-driven and political, as it seeks the return of stolen waters and stolen authority over water. In response, the Exchange uses participant action research, which is shaped by the participants as much as the researchers. Participant action research involves working in partnership and emphasises the outcomes it achieves, and the processes it uses to get there. The primary significance of this research is its targeted, relational, and Indigenous-led nature, which prioritises the needs and experiences of the participants (Kindon, 2021; Smith, 2022).

13.2 Methodology

All authors are participants in the ongoing work of the Exchange. We are a subset of the Exchange participants, and we acknowledge that the insights we draw on in this chapter have emerged through our participation in Exchange with all the other participants. We acknowledge the gift of their wisdom, energy, and connection, which has brought joy to our work and informed our thinking, but we also acknowledge that we do not speak for them directly. In particular, we acknowledge that there are no contributing Indigenous co-authors from Tribes in the USA or iwi from Aotearoa New Zealand, and we do not speak for these participants.

This chapter draws on reflections and insights shared as part of a panel discussion at the international conference "A Profound Reorganising of Things" hosted by the Australian Centre at the University of Melbourne in November 2023. This chapter uses an edited transcript of this panel discussion to identify and explore key themes that emerged during the Exchange and the benefits and challenges the participants have faced. To do so, the chapter is structured as a conversation between the authors, so their insights are shared using their own words (see, for example, Craft et al., 2021; Rigney et al., 2021). This enables the distinct contributions and perspectives of Indigenous participants from multiple Nations and settler/invader co-authors to be clearly presented. This chapter aims to demonstrate the power and impact of participation in this Exchange on both Indigenous and settler participants, across multiple dimensions.

In this chapter, we use a range of terms for Indigenous Peoples, including First Peoples, First Nations, and Traditional Owners. We acknowledge some nuances in meaning between these terms, but we have retained the phraseology used by each participant. We recognise that not all Indigenous People use all of these terms, and we do not speak for any Indigenous People other than the authors of the paper.

13.2.1 Positionality

The authors of this chapter include both Indigenous People as well as people of settler/invader heritage. All three Indigenous co-authors have been leading water advocacy on behalf of their Country and their communities for years. Melissa Kennedy is a Tati Tati woman, the co-founder and director of Tati Tati Kaiejin, a First Nations-owned and operated environmental non-profit organisation, and an academic researcher of water justice and wellbeing. Nicky Hudson is a Gunditjmara woman, and the Aboriginal Water Officer leading the work to care for and restore waterways at Gunditj Mirring Traditional Owner Aboriginal Corporation. Ben Muir is a Wotjobaluk man and a Gatjin (Water) Officer responsible for caring for water at DJAARA (Dja Dja Wurrung Clans Aboriginal Corporation). Amy McCoy is a woman of American, white, UK-European, settler/invader heritage living on the lands of the Ute Peoples in Colorado Springs in the USA, where she works in partnership with Tribes in the Colorado River Basin on water justice reform. Erin O'Donnell is a woman of Australian, white, UK-European, settler/invader heritage living on Wurundjeri Woi Wurrung Country, who collaborates with Traditional Owners in Australia in her academic research on water justice.

We collectively acknowledge that our varied positions and lived experiences bring diverse insights and perspectives on the key insights, opportunities and challenges of participating in the Exchange. In particular, as non-Indigenous participants, O'Donnell and McCoy rely on receiving an ongoing invitation to participate in these activities, and place their skills and expertise in service of the justice agenda as defined by the Indigenous participants. Kennedy, Hudson and Muir bring their leadership and knowledge as Indigenous participants, but also rely on invitations to visit the Country of other First Nations. Collectively, we occupy spaces of belonging, host, and invited guest (see also, Butt & Local Time, 2016).

13.3 *First Peoples, Living Waters*: Insights from the Exchange

13.3.1 Cultural Connections with Water

All Indigenous participants in the Exchange undertake the work of advocacy, leadership, and caring for water as part of their professional employment, but this rarely leaves time to celebrate the strength of cultural connection with water. One of the significant outcomes of the Exchange was the opportunity for the Indigenous participants (as hosts and guests) to re-connect with culture and ceremony, to re-articulate relationships that matter with Country and with waterways, and to acknowledge the pride that comes from the connection with Country.

Kennedy: We are reclaiming our relationships with waterways. Rivers don't belong to any of us; we belong to the rivers. Our rivers, waterways, and landscapes are not possessions but kin. For Indigenous Peoples worldwide, this truth is deeply ingrained. It is the understanding of waterways not merely as resources but as ancestors and entities with a past, present, and future. This perspective is challenging for the modern, transactional approaches to environmental stewardship and invites a rethinking of humanity's relationship with nature.

Hudson: For the Gunditjmara people, water is the backbone of our cultural and environmental systems. Take the short-finned eel migration; it's a huge part of our heritage. Budj Bim's recognition as a World Heritage site highlights how important these systems are. Our ancestors built one of the world's oldest aquaculture systems, using lava flows to farm eels and fish. Without flowing water, this system cannot be used as designed. Water is like the veins of Country, as Country is alive and must be cared for as such. Water must be heard, listened to, observed, and not altered. It's more than a resource, it's a living thing that must be respected and not abused.

Muir: Water is the lifeblood of Country, like the veins through the land. It is deeply connected to the spirit, culture, and traditions of Indigenous people. This connection isn't just practical, it's sacred. That's why the call to show pride, joy, and passion for water resonates so deeply. It's about reminding everyone that water isn't just a resource; it's part of a living system that sustains everything.

The Exchange also enabled reflection on the importance of water and wellbeing. Indigenous participants framed this as a reciprocal relationship, in which healthy Country supports healthy people, and in return, people have a responsibility to care for and heal Country.

Kennedy: The degradation of waterways is not just an environmental issue; it is a personal and cultural wound that runs deep. When rivers are harmed, Indigenous communities feel this harm deeply, a connection often misunderstood by non-Indigenous societies. Discussions about disadvantage or health often adopt a Western lens, overlooking the spiritual and cultural losses tied to environmental destruction. For Indigenous Peoples, healing Country is integral to healing themselves. Recognising this connection is essential for meaningful progress.

Efforts to restore waterways are not only acts of environmental care but also acts of cultural resurgence. Viewing Country as an avenue for healing underscores the need to prioritise Indigenous-led approaches to environmental management. It moves beyond surface-level interventions to address the root causes of ecological and cultural harm, creating pathways for holistic wellbeing.

Hudson: We are the guardians of water and Country. As the Aboriginal Water Officer for the Gunditj Mirring Traditional Owner Aboriginal Corporation, this isn't just a job, it's a way of life, it's about looking after waterways and caring for Country. We care for waterways and land at the Budj Bim World Heritage site and other parts of our native title area. This work is deeply tied to culture, identity, and a sense of responsibility.

This awareness of the connection with culture, ceremony and place was not limited to the Indigenous participants. Non-Indigenous collaborators participated in conversation and, where invited, in ceremony, learning from their hosts about what connection to place involves, and how to reimagine their own relationships with place.

McCoy: Throughout our exchanges, moving through Country with culture has been a transformative personal experience. What I've heard is that for Indigenous peoples, such journeys are not merely physical but ceremonial, deeply rooted in rituals that honour the land. From the beginning to the end, these journeys are woven with acts of reverence: welcomes to Country, dancing, singing, and smoking ceremonies. This embedding of culture into movement through landscapes reinforces a profound connection to the land and its stories.

13.3.2 Shared Knowledge and Experiences

An outcome that multiple participants reflected on as surprising and life-affirming was the recognition that there is a shared language and shared understanding among Indigenous Peoples in the need to care for water and waterways, and the enormous obstacles that colonisation places in the way of undertaking this work. Many Indigenous participants had not travelled overseas before, and the sense of kinship with other Indigenous people from different continents was immediate and powerful.

Kennedy: Among First Nations people, the conversations about rivers and Country flow effortlessly, unencumbered by the need for translation. This shared understanding creates an immediate sense of comfort and kinship. While visiting

> the Colorado River Basin, standing with Navajo or Havasupai elders, hearing phrases like "my river is an ancestor of mine" or "my river is alive" are met with knowing nods by our Exchange members from the First Peoples of Australia and Aotearoa. These ideas are intrinsic and do not need explanation to us. Similarly, standing with the Māori people by the Whanganui River and hearing them say, "I am the river, and the river is me," resonates universally with other Indigenous voices. No translation is needed; the values are shared.

This ease of communication starkly contrasts with dialogues in non-Indigenous spaces, particularly when engaging with policymakers or governments. In these contexts, Indigenous perspectives often require painstaking translation into terms that align with Western frameworks, reducing sacred relationships to mere economic or ecological transactions. Despite this, the reaffirmation of shared values among Indigenous peoples across oceans is uplifting, a reminder of common threads of sustainability and respect for natural systems.

Hudson: I never really wanted to travel the world before exploring my own Country. For some people, travelling the world is a dream. But for others, the journey starts right here, at home. That's how it was for me, I wanted to know my Country before going anywhere else. But when I got the chance to join an Australian delegation heading to the United States, I knew it was an opportunity to learn, to connect, and to build relationships with other Traditional Owners and allies. The trip was all about raising awareness and standing together for water rights, a cause that spans oceans and unites us all.

Meeting the Colorado River and its people was overwhelming. What hit me hard was the heartache of realising we're all in the same fight. Whether it's about land rights, native title recognition, or water rights, the struggle is the same. We're all working to protect our waterways, the species that rely on them, and the cultural practices that tie us to the land. Water is life, and it's essential to everything, from spiritual balance to ecological health.

Muir: Visiting Lake Condah and Budj Bim is a profound experience for me, especially since I haven't been there for a while. Coming back after years and seeing the incredible work done to care for the land and water is inspiring. When the Māori people visited and sang to Budj Bim, something extraordinary happened. A flock of white cockatoos flew overhead, a moment so perfect it felt like a message from the ancestors, reminding everyone that the spirit of the land is still alive, and the connection to Country remains unbroken. Moments like these show the deep relationship between animals, people and Country. It's not just about looking after the land; it's about understanding that the land is looking after us too. The ancestors are always watching, and their presence is felt in every corner of Country.

Non-Indigenous participants observed this connection and immediate sense of kinship. For them, it opened up a perspective from which to engage with colonial ontologies.

McCoy: One striking theme is the deep observation of the unique characteristics that define the differing landscapes. From tree species to the flow of water, every element contributes to a distinct identity. This curiosity, a reverence and respect for difference, stands in stark contrast to the settler state's approach. Settler systems often use differences to critique, categorise, and create hierarchical structures, determining who benefits and who does not. Instead, the Indigenous perspective celebrates these differences, seeing them not as divisions but as opportunities for mutual curiosity and understanding.

To me, this mindset invites a re-evaluation of how societies perceive and interact with diversity and place. Rather than allowing differences to create barriers, they become a source of enrichment, fostering a sense of shared wonder, curiosity, and respect. This shift in perspective could pave the way for more inclusive and equitable relationships among people, water, and the land.

The two major outcomes of the Exchange were relationship building and the recognition that all Indigenous Peoples are in a shared fight to care for and protect water.

O'Donnell: There is a real power in relationships. Throughout our Exchange, one truth stands out: relationships are enduring. Whether formed with rivers or between people, these connections cannot be undone. They create a foundation for ongoing advocacy and collaboration. We mustn't forget that the process is the outcome. The journey of building understanding and relationships is as vital as the policies and outcomes themselves.

Hudson: Everywhere, people are working to protect water and Country. Water isn't just a resource; it's a living entity that deserves respect. That shared understanding gave me hope and reinforced how important it is for Indigenous communities to stand together. We're all on the same planet, and we all need water to survive. Recognising this brings us closer, across borders and cultures.

13.3.3 Aqua Nullius and Settler Colonial Water Constructs

Australia and the United States of America have similar histories as former British settler colonies. Techniques of violent dispossession, oppression and wholesale theft of territory that developed in the colonial invasion of the USA were then applied in Australia (Dunbar-Ortiz, 2014).

McCoy: Colonial culture expresses itself similarly in both the United States and Australia. It is marked by entitlement, a focus on economic value, and a rigid, hierarchical worldview. However, there are notable differences in how these attitudes are institutionalised. In the United States, water laws are highly structured, leaving little room for flexibility. By contrast, Australia's legal framework for water governance is less rigid, allowing for more adaptability. This distinction highlights the potential for different approaches to water management, where flexibility could open pathways for integrating Indigenous knowledge and values.

Despite these structural differences, the overarching challenge remains the same: transforming colonial systems, that prioritise exploitation and exclusion, into frameworks that honour relationships and reciprocity. By learning from Indigenous practices and perspectives, both nations have the opportunity to reimagine their approaches to water governance.

Aqua nullius remains a powerful force in both Australia and the USA. Although treaties and other agreements, including reservations, have formally created legal rights to water within the settler colonial water law frameworks of the USA (Tarlock, 2010), there is often a multi-decadal lag between the formal acknowledgement of water rights and the return of water that is available for Tribes to use (Stern, 2023). Even then, use is often more constrained than it is for other private (settler) water rights holders. In Australia, the original theft of aqua nullius has never been formally acknowledged in water law. Although native title may, under very limited circumstances, give rise to legal rights to water, this is not usually the case (O'Bryan, 2016), and the assumption that the Crown has a legitimate right to control water is generally unspoken and unchallenged in settler contexts.

O'Donnell: In settler colonial Australia, all water is Aboriginal water. This phrase that mob [Indigenous People] repeat carries deep meaning, challenging the concept of aqua nullius—the colonial assumption that water, like land, was unowned and free to be claimed. Aqua nullius dismisses the deep cultural, spiritual, and custodial ties Indigenous peoples have with waterways, an idea that is both false and unjustifiable. *Water is Life* (Department of Environment, Land, Water and Planning, 2022) stands out as the only policy document in Australia to explicitly acknowledge aqua nullius. Though it does not offer solutions, simply naming the issue is a significant first step. It signals the need to tell the truth about water's history and its rightful custodians.

McCoy: Being somebody who is schooled and steeped in the colonial culture of the Colorado River Basin, the rigidity of water laws is unquestioned. These laws are seen as immovable, like a concrete culvert that constrains the flow of water and that can only be tinkered with at the edges. This inflexibility reflects a broader colonial worldview, one that prioritises economic value, measurable outcomes, and hierarchical structures. In this framework, water laws are designed to serve specific interests, often at the expense of Indigenous Peoples and their longstanding relationships with waterways.

This exclusivity in water governance is deeply troubling to me. The realisation that Indigenous Peoples were wholly excluded from these systems has sparked my efforts to understand, dismantle, and transform the restrictive spirit of water management in the United States. Breaking down these barriers requires challenging the entrenched entitlement and economic priorities that underpin colonial water governance.

Cultural flows, the 'inherent right' of First Nations (MLDRIN, 2007), fundamentally challenge the claims of exclusive possession of the Crown regarding water. The depth of this challenge is frequently denied by settler colonial institutions, who often focus on returning water as settler water rights, without recognising that this process can create further challenges (Hartwig et al., 2021). The concept that "some" water is allocated for "cultural" purposes assumes that other water is not. These knotty problems go to the heart of the relationship between people and water, and an understanding of what water is.

O'Donnell: For many, all water is inherently cultural and connected to Country. These debates highlight the richness and complexity of Indigenous perspectives on water. They also emphasise the importance of Indigenous-led solutions. In learning from other jurisdictions, such as the Whanganui River settlement in Aotearoa (New Zealand), parallels and differences emerge. For example, the Whanganui River case involved 148 years of litigation and treaty disputes before a deed of settlement was signed. This underscores the long-term nature of these struggles, and the resilience required to achieve meaningful change.

Hudson: Getting a water license listed as cultural water might seem small, but it's a big step. It's an acknowledgment that water is more than a commodity. For us, it's a living ancestor, a vital part of our connection to Country. The fight for this recognition isn't over, but every win brings us closer to a future where water is treated with the respect it deserves. This perspective, seeing water as a living entity, challenges the colonial systems that reduce water to something to be measured, sold, or controlled. Instead, we honour water's role in sustaining life and culture. It's about relationships, not ownership, and that's a lesson worth sharing.

13.3.4 Accessing Settler State Water Rights

The Exchange provided an opportunity to share the challenges and successes in accessing water rights created by settler colonial water law. In 2022, Gunditjmara received only the second formal return of water rights to a Traditional Owner Corporation in Victoria's colonial history (Gillespie, 2022). All water returns to Traditional Owners in Victoria have been hard-fought and long-delayed (O'Donnell et al., 2023b).

Hudson: Getting water entitlements hasn't been easy. When we applied for unregulated winterfill water (water that falls in the winter, we can only access it during certain months), the process wasn't designed for us. The forms only asked if the water was for irrigation or agriculture. There was no space for cultural or ecological needs. It's frustrating, but we kept pushing and we were successful in receiving the entitlements. Now, having cultural water entitlements recognised feels like a win. It's proof that water isn't just for farming or industry, it's for Country, for our people, and for our traditions.

Muir: Water rights are a contentious issue, especially when it comes to misunderstandings between Indigenous People and non-Indigenous landowners. Farmers and other landowners sometimes assume that Traditional Owners want to take their water. But that couldn't be further from the truth. Indigenous People aren't asking to take water, they're fighting to regain rights to the water that has always been part of their traditions. It's about restoring balance, not creating conflict. Traditional Owners want to use water for the cultural practices that have been part of their lives for generations. It's not about exclusion; it's about inclusion, having their rightful place in managing and caring for the water that sustains Country.

Even when limited to water licences that have a time limit and under which water can only be used in certain ways, at certain times, and at particular locations, the allocation of water rights to Indigenous Peoples represents a transfer of power, which is why it is so often contested and controversial. This power amplifies Indigenous voices within settler colonial institutions of water management, as well as challenging the foundation of settler water constructs.

O'Donnell: Disrupting the settler state's approach to water requires a shift in mindset. When engaging with settler organisations about Indigenous perspectives on water, there is often resistance rooted in fear. Statements like "all water is Aboriginal water" can provoke concerns about loss or redistribution. These reactions mirror the violent history of colonial land and water acquisition.

One of the striking realisations when working with government bodies is how ingrained settler constructs of water management are. While collaborating with the Victorian government on the *Water is Life* policy document, a common hurdle was that there are certain "irrefutable facts" about water, a phrase I am quoting from settler water policy makers. This expression seemed to refer less to natural truths like water flowing downhill and more to the rigid, colonial frameworks governing water use. These constructs are treated as immovable, a concrete edifice against which all other perspectives are measured.

13.3.5 Walking in Two Worlds to Get Things Done

In settler colonies, Indigenous Peoples have had to become accustomed to walking in two worlds (Fish et al., 2022; Haar & Martin, 2021). The Exchange mirrored this in that it included non-Indigenous participants. Still, it attempted to 'flip the script' (see Kennedy and McCoy in this volume) so that Indigenous ways of being, communicating, and doing were the norm. Non-Indigenous participants needed to pay attention and adapt. One outcome of doing so was that it enabled discussion of the challenges faced by Indigenous People as they operate within settler colonial systems (see, e.g. Haar & Martin, 2021). The generation of water leaders participating in the Exchange is one of the first in Australia to be able to work professionally as Indigenous water leaders, building on the extraordinary efforts of earlier generations who reformed settler colonial government agencies and funding arrangements. However, this power and opportunity can also come with a cost.

Muir: My experiences, and many other Indigenous People working in government roles, the struggle often comes down to wearing two hats. On one hand, they're working for their mob, trying to represent their culture and protect Country. On the other hand, they have to operate within government systems, following rules and expectations that don't always align with cultural ways. It's a tough balancing act. Wearing the government or Catchment Management Authority [regional government agency] hat can feel restrictive. There are times when we're not allowed to speak for Country or represent our mobs the way we'd like. Instead, we have to stick to the rules of the workplace, playing a nine-to-five role that feels far removed from the cultural responsibilities we hold.

Kennedy: I see and experience the ways Indigenous leaders bear the immense responsibility of professionalising the work of cultural and environmental advocacy. We are the first generation that gets to do this work professionally, full-time as Indigenous People. While this is a profound opportunity, it comes with challenges. The demands of negotiating with States, collaborating with [settler] waterway managers, and influencing legislative change often leave little time for grassroots connections or on-the-ground collaboration. This professionalisation, while necessary, risks pulling Indigenous advocates away from the relational practices that are central to caring for Country.

Muir: Despite the challenges, many Indigenous People in these roles push forward because they know the work matters. Even when the system feels tokenistic, they focus on the bigger picture, getting results for their mob and making a difference for Country. It's not easy, and it often feels like compromising cultural values, but they see it as a necessary step to bring about real change. For those in these roles, the work is about more than just ticking boxes. It's about creating opportunities for Traditional Owners to have a voice in managing water and Country. It's about building pathways that allow cultural practices to thrive and ensuring that Indigenous perspectives are respected in decisions about land and water.

These challenges are also reflected in the academic sector (Ingram et al., 2021; Leonard et al., 2023; McGreavy et al., 2021; Moggridge et al., 2022; Moggridge & Thompson, 2019; Parsons et al., 2021; Smith, 2022). Indigenous academics have been forced to research in the settler way for too long, and this is undermining the work to connect with, care for, and heal rivers.

Kennedy: Indigenous research must centre Indigenous People, our knowledge, priorities, and ways of being. If we're talking about Indigenous research, it must be by Indigenous People, but it also has to be *for* Indigenous People and our purposes. Research that is not rooted in community priorities often fails to address the real needs of Indigenous Peoples and their Country. This misalignment stems from fundamental differences in research approaches. Western research often prioritises predefined outcomes, requiring researchers to identify goals and measures of success at the outset. In contrast, Indigenous research values relationships and process. It is about being present, listening to Country, and allowing understanding to emerge organically. For Indigenous researchers, the journey is as important as the destination, and imposing rigid structures undermines the integrity of the work.

Reorganising research to align with Indigenous methodologies is vital. This means creating space for relational approaches and valuing outcomes that may not fit neatly into Western metrics. By doing so, research can become a tool for healing and empowerment rather than a continuation of colonial structures.

13.3.6 Treaty

One of the aims of the Exchange was to share lessons about the opportunities and challenges that treaties and other agreement-making have created in the USA and Aotearoa New Zealand in addressing aqua nullius. The Exchange highlighted that there is no single example of where a treaty has overturned aqua nullius. Although there are real and significant legal rights that can flow from a treaty, these often come at the risk of subsuming Indigenous rights within the settler colonial paradigm, which is what has primarily happened with Tribes' water rights in the USA (Curley, 2019). Even where a treaty has supported a profound reimagining of the nature of rivers and waterways, as it has in the case of the Whanganui River in Aotearoa, this has required over a century of litigation by Whanganui iwi to force the Crown to recognise their treaty rights (Macpherson, 2019). Yet treaty is also the only pathway for the kind of transformative change that is needed to overturn aqua nullius.

Kennedy: Treaty-making offers a radical avenue for rethinking humanity's relationship with water. The ultimate goal is not personal rights but collective respect and stewardship. I would be happy to have zero rights to water if the entire basin starts to respect and hold genuine relationships with my rivers. What is needed is a profound shift in priorities, centring the health of the river system itself over individual or economic claims. Treaties for rivers are not just political agreements; they are acknowledgments of the river as a living entity deserving of respect and care. Such treaties could transcend colonial boundaries and bureaucratic jurisdictions, fostering collective responsibility among all. This approach calls for a return to Indigenous ways of knowing and relating to waterways, ways that honour interconnectedness, relationship, and reciprocity.

Large scale reforms of this nature are slow, and must account for a diversity of perspectives. However, the Exchange reminded us that there is no need to wait for treaty for some reforms.

O'Donnell: The release of *Water is Life* in September 2022 marked progress, but much remains to be done. For example, targeted outcome number seven in section A says we will streamline the process for water access. Some streamlining would be pretty good right about now. The slow pace of change has frustrated many, particularly given the lack of emphasis on water in the broader Treaty process underway in Victoria.

Each Indigenous community approaches Treaty differently, and individuals hold varying expectations about its potential. The Yoorrook Justice Commission's inquiry into land injustice takes a holistic view, addressing land, water, sky, and sea Country. This work could inform how water is incorporated into future treaties. While it is unclear if water will feature prominently in a statewide treaty, it may emerge in agreements between individual nations and the Victorian government.

13.3.7 Returning to Relationship: A New Vision for Caring for Water

The Exchange gave all participants an opportunity to renew our commitment to water justice and reinvigorate our thinking about the future of caring for water and Indigenous leadership. The following commentary coalesces around the themes of shared stewardship, relationships of belonging and reciprocity, the unfinished advocacy work, a glimpse of a future beyond aqua nullius, and a resurgent pride in Country and culture.

Kennedy: One of the most compelling ideas that resonates with me is the call for collective responsibility in caring for rivers. It's not up to just one mob to look after the river on their part of Country, or one government agency to make all the decisions. Our river goes for a thousand kilometres upstream and a thousand kilometres downstream of my Country. This recognition of interconnectedness underscores the need for collaboration that transcends borders and jurisdictions imposed on us by our oppressors. Achieving this vision requires a shift in how water management is approached. Rather than fragmenting responsibility, treaties for rivers, as one approach, could establish shared governance structures rooted in Indigenous knowledge systems. These treaties would honour the river's continuity, fostering partnerships that prioritise its health and longevity over short-term interests.

Indigenous voices remind us that humanity's relationship with rivers is not a matter of ownership but one of belonging. Rivers are ancestors, living entities with their own stories and rights. Restoring these relationships is not just an environmental imperative but a cultural and spiritual one. Through treaty-making, Indigenous-led research, and shared stewardship, there is an opportunity to reimagine humanity's place within the natural world. By listening to Indigenous voices, the world can move toward a future where rivers are not simply resources to be exploited but kin to be cherished and protected.

McCoy: Reclaiming water governance requires dismantling these rigid, exclusionary structures and embracing the fluid, relational nature of water itself. Indigenous perspectives offer a model for this transformation. They remind us that water is not merely a resource to be managed but a living entity with its own rights and relationships. By embedding cultural practices and mutual respect into water governance, societies can begin to address historical injustices and build more sustainable futures. The path forward is not without challenges, but the lessons from experiences such as our cultural exchanges, as well as Indigenous leadership, offer hope. By valuing difference as a source of strength and recognising the interconnectedness of all life, a new vision of water management can emerge, one rooted in respect, reciprocity, and the enduring wisdom of the land.

Hudson: The journey to secure recognition for cultural water is far from finished. Indigenous Peoples are still fighting for waterways to be treated as living entities with their own rights. This struggle is tied to the larger goal of protecting Country and preserving our cultural practices. Each step forward, whether it's through cultural exchanges, water licenses, or advocacy, lays the groundwork for a future where water and Country are respected and protected. The Gunditjmara people's work is a reminder of what's possible. Our persistence and collaboration inspire others to keep going. By sharing our stories, we are showing that water is more than just a resource. It's life itself, a sacred connection that binds us all together.

O'Donnell: The fight for water justice is far from over, but progress is being made. Each step, whether through policy, treaty, or our Cultural Exchange, contributes to a broader movement to recognise water as a living entity and a fundamental part of Country. Challenging the legitimacy of existing rights to water and water markets disrupts the settler state's assumptions and creates room for alternative approaches. However, achieving this requires persistence and creativity.

By challenging colonial constructs and embracing Indigenous knowledge, we move closer to a future where all water is understood and respected as Aboriginal water. The challenge is to envision redistributive processes that are not rooted in violence or exploitation. This requires creating spaces for understanding and dialogue, allowing people to grasp the enormity and necessity of the changes ahead. The goal is not to replicate colonial models of power but to build relationships and frameworks rooted in respect, reciprocity, and justice.

Muir: Amid all the challenges, my pride and passion for Country remain strong. Traditional Owners continue to celebrate their culture and their deep connection to water. Every act of care for Country is an act of resistance, a way of saying, "We're still here." The spirit of the land and its people is alive, and that spirit guides every step of the journey. Whether it's singing to Budj Bim, advocating for water rights, or working within government systems, the goal is the same: to honour the connection between people and Country and to ensure that the lifeblood of the land is protected for generations to come.

13.4 Conclusion

The *First Peoples, Living Waters* Cultural Exchange demonstrates the critical importance of Indigenous knowledge and leadership in addressing water justice and law reform in settler colonial states. This initiative sheds light on shared struggles, distinct cultural perspectives, and potential pathways toward meaningful change by fostering dialogue among Indigenous Peoples from Australia, Aotearoa, and the United States. This chapter explored the insights gained through these exchanges, emphasising the urgency of reimagining humanity's relationship with water in the face of climate change and systemic inequities.

Central to this project is the understanding that rivers and waterways are not objects of ownership but living entities to whom humans belong. This perspective disrupts colonial constructs that treat water as a resource to be exploited, and challenges the enduring legacy of aqua nullius. For Indigenous Peoples, waterways are ancestors, carriers of stories, and sources of identity and healing. Reclaiming relationships with rivers is a cultural imperative and a pathway toward broader environmental restoration. It underscores the interconnectedness of ecological health and Indigenous wellbeing, affirming that the degradation of waterways represents not just environmental harm but a profound cultural wound.

The conversations between the co-authors revealed the power of shared language and understanding among First Nations peoples. Unencumbered by the need for translation, these exchanges create a sense of kinship and solidarity, fostering a collective vision for water governance rooted in respect, reciprocity, and responsibility. A shared fight for water rights unites Indigenous communities across the globe, rooted in the conviction that water is not merely a resource but a spiritual and cultural lifeblood. Observing the unique characteristics of different landscapes and waterways enriches this collective struggle, reminding us of the diversity within unity. This solidarity is essential in confronting colonial parallels between countries like the United States and Australia, where rigid legal systems perpetuate inequities. However, the weight of advocacy falls heavily on Indigenous leaders, who navigate the complexities of professionalising cultural and environmental work while staying true to their communities and values. This burden underscores the need for Indigenous-led research and policymaking, which centres Indigenous priorities and ways of knowing.

Treaty-making emerges as a vital tool for reimagining relationships with water and Country. Through treaties, there is potential to dismantle the rigid, exclusionary frameworks of colonial water law and embrace fluid, relational approaches to water management. While progress has been made, as seen in the concept of cultural flows and the recognition of water as a living entity, much work remains. Securing cultural water rights and disrupting settler constructs remains long and challenging.

Amid the ongoing challenges, there is pride and resilience in the enduring connection to water and Country. Each act of care for waterways represents resistance to colonial narratives and a declaration of cultural survival. The future demands continued advocacy, collaboration, and respect for Indigenous knowledge to ensure that water justice is realised and humanity's relationship with rivers is restored to belonging, not domination. The vision of a future beyond aqua nullius is not just aspirational—it is essential for the health of the planet and its peoples.

References

Bhatia, A. (2021). Statehood, Canadian sovereignty, and the attempted domestication of indigenous legal relations. In S. Xavier, B. Jacobs, V. Waboose, J. G. Hewitt, & A. Bhatia (Eds.), *Decolonizing law: Indigenous, third world and settler perspectives*. Routledge.

Butt, D., & Local Time. (2016). Colonial hospitality: Rethinking curatorial and artistic responsibility. *Journal for Artistic Research, 10*. https://www.researchcatalogue.net/view/228399/264279/0/0

Craft, A., McGregor, D., Seymour, R., & Chiblow, S. (2021). Decolonizing Anishnaabe nibi inaakonigewin and gikendaasowin research: Reinscribing Anishnaabe approaches to law and knowledge. In S. Xavier, B. Jacobs, V. Waboose, J. G. Hewitt, & A. Bhatia (Eds.), *Decolonizing law: Indigenous, third world and settler perspectives* (1st ed., pp. 17–33). Routledge.

Curley, A. (2019). "Our Winters' rights": Challenging colonial water Laws. *Global Environmental Politics, 19*(3), 57–76.

Department of Environment, Land, Water and Planning. (2022). *Water is life: Traditional owner access to water roadmap*. State Government of Victoria. https://www.water.vic.gov.au/our-programs/aboriginal-water-program/water-is-life-roadmap

Dunbar-Ortiz, R. (2014). *An Indigenous Peoples' history of the United States*. Beacon Press.

Fish, J., Hirsch, G., & Syed, M. (2022). "Walking in two worlds": Toward an Indigenist ecological systems model for group therapy. *The Counseling Psychologist, 50*(5), 622–648. https://doi.org/10.1177/00110000221083028

Garma. (2008). *Garma International Indigenous Water Declaration*. https://www.afn.ca/uploads/files/env/garma-international.pdf

Gillespie, K. (2022). Gunditjmara Traditional Owners handed unallocated water from Fitzroy River system. *ABC News Online*. https://www.abc.net.au/news/2022-10-09/gunditjmara-given-unallocated-water-from-fitzroy-river-system/101506836

Haar, J., & Martin, W. J. (2021). He aronga takirua: Cultural double-shift of Māori scientists. *Human Relations, 75*(6), 1001–1027. https://doi.org/10.1177/00187267211003955

Hartwig, L., Jackson, S., Markham, F., & Osborne, N. (2021). Water colonialism and Indigenous water justice in South-Eastern Australia. *International Journal of Water Resources Development, 38*(1), 30–63. https://doi.org/10.1080/07900627.2020.1868980

Ingram, J. C., Castagno, A. E., Camplain, R., & Blackhorse, D. (2021). Role of professional societies on increasing Indigenous Peoples' participation and leadership in STEMM. *Frontiers in Education, 6*. https://doi.org/10.3389/feduc.2021.753488

Kindon, S. (2021). Participatory action research: Collaboration and empowerment. In I. Hay & M. Cope (Eds.), *Qualitative research methods in human geography* (5th ed., pp. 309–330). Oxford University Press.

Leonard, K., David-Chavez, D., Smiles, D., Jennings, L., Tsinnajinnie, L., Manitowabi, J., Arseneault, R., Begay, R. L., Kagaway-Viviani, A., Davis, D. D., van Uitregt, V. (Billy), Liboiron, M., Moggridge, B., Russo Carroll, S., 'Anolani Alegado, R., Tsosie, R. L., & Gomez, A. (2023). Water back: A review centering rematriation and Indigenous water research sovereignty. *Water Alternatives, 16*(2), 374–428.

Macpherson, E. J. (2019). *Indigenous water rights in law and regulation: Lessons from comparative experience*. Cambridge University Press.

Marshall, V. (2017). *Overturning aqua nullius: Securing Aboriginal water rights*. Aboriginal Studies Press.

Martuwarra RiverOfLife, Pelizzon, A., Poelina, A., Akhtar-Khavari, A., Clark, C., Laborde, S., Macpherson, E., O'Bryan, K., O'Donnell, E., & Page, J. (2021). Yoongoorrookoo: The emergence of ancestral personhood. *Griffith Law Review, 30*(3), 505–529. https://doi.org/10.1080/10383441.2021.1996882

McGreavy, B., Ranco, D., Daigle, J., Greenlaw, S., Altvater, N., Quiring, T., Michelle, N., Paul, J., Binette, M., Benson, B., Sutton, A., & Hart, D. (2021). Science in Indigenous homelands: Addressing power and justice in sustainability science from/with/in the Penobscot River. *Sustainability Science, 16*(3), 937–947. https://doi.org/10.1007/s11625-021-00904-3

Miller, R. J., Ruru, J., Behrendt, L., & Lindberg, T. (2010). *Discovering Indigenous lands: The doctrine of discovery in the English colonies.* Oxford University Press.

MLDRIN. (2007). *Echuca declaration.* Murray Lower Darling Rivers Indigenous Nations. https://www.mldrin.org.au/wp-content/uploads/2018/07/Echuca-Declaration-Final-PDF.pdf

Moggridge, B., & Thompson, R. (2019). Aboriginal voices are missing from the Murray-Darling Basin crisis. *The Conversation.* https://theconversation.com/aboriginal-voices-are-missing-from-the-murray-darling-basin-crisis-110769

Moggridge, B. J., Thompson, R. M., & Radoll, P. (2022). Indigenous research methodologies in water management: Learning from Australia and New Zealand for application on Kamilaroi country. *Wetlands Ecology and Management, 30*(4), 853–868. https://doi.org/10.1007/s11273-022-09866-4

Morris, J. D., & Ruru, J. (2010). Giving voice to rivers: Legal personality as a vehicle for recognising Indigenous peoples' relationships to water? *Australian Indigenous Law Review, 14*(2), 49–62.

Nelson, R., Godden, L., & Lindsay, B. (2018). *A multi-layer plan for cultural flows in Australia: Legal and policy design (report for the National Cultural Flows Research Project).* MLDRIN, NBAN and NAILSMA. http://culturalflows.com.au/images/documents/Law%20and%20policy.pdf

O'Bryan, K. (2016). More aqua nullius? The *Traditional Owner Settlement Act 2010* (Vic) and the neglect of Indigenous rights to manage land and water resources. *Melbourne University Law Review, 40*, 547–593.

O'Donnell, E. (2023). Water sovereignty for Indigenous Peoples: Pathways to pluralist, legitimate and sustainable water laws in settler colonial states. *PLOS Water, 2*(11), e0000144. https://doi.org/10.1371/journal.pwat.0000144

O'Donnell, E., Kennedy, M., Garrick, D., Horne, A., & Woods, R. (2023a). Cultural water and Indigenous water science. *Science, 381*(6658), 619–621. https://doi.org/10.1126/science.adi0658

O'Donnell, E., O'Bryan, K., Jackson, S., Jama, K., Nelson, R., Cubillo, E., & Murray Lower Darling Rivers Indigenous Nations (MLDRIN). (2023b). *Submission to the Yoorrook justice commission: Caring for Country: Overcoming the twin legacies of terra nullius and aqua nullius.* Melbourne Law School, Centre for Law and the Environment. https://law.unimelb.edu.au/__data/assets/pdf_file/0005/4818866/MLS-Yoorrook-LW-submission-20231117-002.pdf

Parsons, M., Fisher, K., & Crease, R. P. (2021). *Decolonising blue spaces in the Anthropocene: Freshwater management in Aotearoa New Zealand.* Palgrave Macmillan. https://doi.org/10.1007/978-3-030-61071-5

Rigney, D., Bell, D., & Vivian, A. (2021). Talking treaty: A conversation on how Indigenous nations can become treaty ready. In H. Hobbs, A. Whittaker, & L. Coombes (Eds.), *Treaty-making: 250 years later* (pp. 17–42). The Federation Press.

Robison, J., Cosens, B., Jackson, S., Leonard, K., & McCool, D. (2018). Indigenous water justice. *Lewis & Clark Law Review, 22*(3), 841–921.

Ruru, J. (2018). Listening to Papatūānuku: A call to reform water law. *Journal of the Royal Society of New Zealand, 48*(2–3), 215–224. https://doi.org/10.1080/03036758.2018.1442358

Sherriff, L. (2024). After 100 years, salmon have returned to the Klamath River—following a historic dam removal project in California. *BBC.* https://www.bbc.com/future/article/20241122-salmon-return-to-californias-klamath-river-after-dam-removal

Siegler, K. (2023). *Tribal nations were once excluded from Colorado River talks. Now they're key players*. National Public Radio (NPR). https://www.npr.org/2023/04/27/1172273665/tribal-nations-were-once-excluded-from-colorado-river-talks-now-theyre-key-playe

Smith, L. T. (2022). *Decolonising methodologies: Research and Indigenous Peoples* (3rd ed.). Bloomsbury Academic.

Stern, C. V. (2023). *Indian water rights settlements* (No. R44148). Congressional Research Service. https://crsreports.congress.gov/product/pdf/R/R44148

Takacs, D. (2021). We are the river. *University of Illinois Law Review, 2021*(2), 545–605.

Tarlock, A. D. (2010). Tribal justice and property rights: The evolution of *Winters v. United States*. *Natural Resources Journal, 50*, 471–500.

Yates, J. S., Harris, L. M., & Wilson, N. J. (2017). Multiple ontologies of water: Politics, conflict and implications for governance. *Environment and Planning D: Society and Space, 35*(5), 797–815. https://doi.org/10.1177/0263775817700395

Reflections

Maddison Miller

I am watching the waters of Birrarung dance joyfully against smooth stones. The sound of the water is music. It's a rushing, swelling, thrilling song that has been sung since the Dreaming. Cockatoos (*Giriwi* in my ancestral language) join in, screeching as they chase each other through cloudless skies. The gums sway as wind Country dances through the canopy, making leaves rustle and shimmy. The sun lays softly on my skin, providing me vitamin D and an influx of serotonin. I breathe in deeply as wind Country tickles the hairs at the back of my neck. I inhale air that carries memory of all the places far away. My bare feet rest on earth Country, grounding me to this place and an understanding that through the cyclical nature of time I occupy the same space as the ancestors before and the ancestors to come.

This place is called laughing waters, a meander of Birrarung on unceded Wurundjeri Woi Wurrung Country. This place has its own language that is sung by Country and spoken by the peoples who were created from the earth at the banks of Birrarung.[1] I feel connected to this place, and grateful to able to listen to Country's song. In this place I am unwavering, filled with a certainty and a sense of rightness, of belonging. This beautiful cup of love, filled to the brim from Wurundjeri Woi Wurrung Country is so precious. I hold it closely, knowing soon I will be walking back into the heat of the postcolonial world that will do all it can to evaporate the contents of my cup.

This book explores the many ways in which we know and relate to Country. Country has endless manifestations, Country is the water that flows through the

[1] The story of Bunjil, as narrated by Aunty Joy Murphy AO, Wurundjeri Woi Wurrung Elder, 2021. Available at: https://www.youtube.com/watch?v=kzs78j4IQBw

M. Miller
Faculty of Science, School of Agriculture, Food and Ecosystem Science, The University of Melbourne, Parkville, VIC, Australia
e-mail: maddison.miller@unimelb.edu.au

M. Kennedy, E. O'Donnell (eds.), *People, Place and Nature in Indigenous-Settler Relations*, Indigenous-Settler Relations in Australia and the World 10, https://doi.org/10.1007/978-3-032-03600-1

land, through our bodies and into the sea. Country is the wind against your cheek and the insects that fly. Country is the relationship between all things. Country is healing, Country is loving, Country is unrelenting. In these chapters we find careful considerations of how to make and unmake our conceptualizations of Country. How we explore the deeply personal and deeply communal relationships we hold with Country, unlearning and learning colonial framings to hold the capitalist extraction to account. In the coming together of many ways of knowing, many disciplinary standpoints, the authors offer insights into practices that centre connectedness, responsibility and reciprocity.

The stories laid out in this book are rich, exploring many ways in which resistance is enacted. Contextualised in contemporary political realities and our shared generational crises of climate change and biodiversity loss, these chapters show a way forward that is founded in deeply relational ways of knowing being and doing. The path to justice is one of deep care—care for Country and care for each other. These chapters shine a light on injustice and continued colonial harm that benefits almost nobody. Throughout each of these works is a deep respect and love for Country, which is woven into acts of resilience, resistance and care. The extractive frameworks that dominate the management of Country are called into question and are expertly dismantled. Emerging from knowledges deeply tied to generational observations, these offerings provoke radical action.

Birrarung continues to flow as she has since the dreaming, bringing vital waters through Country. The earth moves around the sun and suddenly the angle is just so that a thousand reflections sparkle across Birrarung's laughing waters. Country is a mirror. As I have heard from many elders, friends, and mentors when we care for Country, Country cares for us. In this moment I accept the care offered with generosity and, emboldened by the words in this book, make a promise to care in return.

www.ingramcontent.com/pod-product-compliance
Lightning Source LLC
Chambersburg PA
CBHW071953170426
43348CB00003B/20
* 9 7 8 3 0 3 2 0 3 5 9 9 8 *